# Recent Studies on Strawberries

*Edited by Nesibe Ebru Kafkas*

Published in London, United Kingdom

Recent Studies on Strawberries
http://dx.doi.org/10.5772/intechopen.98136
Edited by Nesibe Ebru Kafkas

Assistant to the Editor: İlbilge Oğuz

Contributors
Hannu Ahokas, Muhammad Sarwar, Nesibe Ebru Yaşa Kafkas, İlbilge Oğuz, Halil İbrahim Oğuz, Parween Muhammad K. Rozbiany, Shler Mahmud Taha, Alba Estela Jofre-y-Garfias, Pedro Antonio Dávalos-González, Ramón Aguilar-García, Alejandro Rodríguez-Guillén, Sandhya Gupta, Ilango Kaliappan, Sivakumar Bathula, Rushendran Rapuru, Amnon Bustan, Nir Dai, Ofer Guy, Shabtai Cohen, Sumana Neera, Refik Bozbuga, Selman Uluisik, Pinar Aridici Kara, Semiha Yuceer, Hale Gunacti, Pakize Gok Guler, Elen Ince, Hatice Nilufer Yildiz, Ozcan Tetik, Iryna Zamorska, Waseem Ahmed, Huma Qureshi Quarshi, Rafia Azmant, Nabila Chendouh-Brahmi, Salih Kafkas, Harun Karcı, Habibullah Tevfik



First published in London, United Kingdom, 2023 by IntechOpen
IntechOpen is the global imprint of INTECHOPEN LIMITED, registered in England and Wales, registration number: 11086078, 5 Princes Gate Court, London, SW7 2QJ, United Kingdom

British Library Cataloguing-in-Publication Data
A catalogue record for this book is available from the British Library

Additional hard and PDF copies can be obtained from orders@intechopen.com

Recent Studies on Strawberries
Edited by Nesibe Ebru Kafkas
p. cm.
Print ISBN 978-1-80355-198-2
Online ISBN 978-1-80355-199-9
eBook (PDF) ISBN 978-1-80355-200-2

For EU product safety concerns :
IN TECH d . o . o . , Prolaz Marije Krucifikse Kozulić 3 , 51000 Rijeka , Croatia ,
info@intechopen . com or visit our website at intechopen . com .

# Meet the editor

Nesibe Ebru Kafkas graduated from the Faculty of Agriculture, Department of Horticulture, University of Çukurova, Turkey and has worked in this department since 1996 as a researcher and lecturer. She has experience in fruit breeding and fruit metabolomics, chromatography, and in-vitro techniques. She is the author of more than 100 scientific papers (H-index:19; total citations:1600) and has been an editor of various scientific journals. She has been the Turkish coordinator for national and international EU projects such as HORIZON 2020, Cost, and Prima. She is also a member of the International Society of Horticultural Science.

# Contents

# Preface

This book presents recent developments and new production methods in strawberry cultivation. It provides a broad overview and comprehensive discussion of current challenges in production, consumption, breeding, disease and pest control, and post-harvest processing. The aim of the book is to bring together information about topics including the world distribution of strawberry plants, genetic resources, reproduction techniques, biotic and abiotic stresses and diseases caused by climate change, the challenges of control methods, and the benefits of strawberries for human health. All the chapters present detailed analyses written by experts on the subject. The book is an essential resource for virtually everyone involved in strawberry research and development, including academics, industry professionals, and farmers who produce strawberries commercially, as well as consumers.

The strawberry plant is a good source of antioxidants due to the vitamins A and C and flavonoids that it contains due to anthocyanin, which gives the fruit its bright red color, quercetin pigment and, especially, antioxidant effect. It is also known that strawberries, together with their antioxidant effect, prevent the increase of free radicals that cause various types of cancer and oxidative stress, and are a dietary fruit. As a result of these characteristics, there has been an increasing trend toward the consumption of strawberries and other berry fruits in recent years. Strawberries are among the plants suitable for intensive agricultural production, and their suitability for vertical, horizontal and soilless cultivation is attracting the attention of new investors. In terms of processing technology, they are a raw material source for many food sectors.

Special thanks are due to assistant editor İlbilge Oğuz and Author Service Manager Romina Rovan for their courteous help and dedicated efforts, without which this book would never have been published.

**Nesibe Ebru Kafkas**
Professor,
Faculty of Agriculture,
Department of Horticulture,
University of Çukurova,
Adana, Turkey

## Section 1

# Cultivation Techniques of Strawberries

Chapter 1

# Strawberry Cultivation Techniques

*İlbilge Oğuz, Halil İbrahim Oğuz and Nesibe Ebru Kafkas*

## Abstract

Among the berries, strawberries are the most commercially produced and consumed and their production and consumption are increasing in the world due to their enthusiastic aroma, taste, and biochemical properties. Strawberry is belonging to the genus *Fragaria*, from the family *Rosaceae*. It is indicated that the homeland of the strawberry is South America (Chile). It is well-known that people living in Asia, Europe, and America commonly use the wild *F. vesca*. In other regions such as Japan, North China and Manchuria, Europe-Siberia, and America there are different ecogeographic zones where alternative species are clustered. Despite its origins in the Pacific Northwest region of North America, *F. ananassa* is now grown all over the world. Strawberry is one of the most widespread berry species grown in almost every country including high altitudes of tropical regions, and subtropical and temperate areas. In this chapter, we aimed to offer new perspectives on the future of strawberry cultivation techniques by analyzing recent academic studies on strawberry production.

**Keywords:** strawberry, cultivation techniques, hydroponic culture

## 1. Introduction

Strawberry is a berry-like fruit that can be widely produced in almost all regions of the world, including tropical and subtropical regions. At the same time, its production and trade have been increasing around the world and in Turkey as it is one of the most profitable fruits due to its nutritional value and benefits for human health. In addition, the demand for marketing of fresh strawberries is higher in the world market because strawberries involve higher costs and are widely used either fresh or in prepared foods, such as fruit juices, jams, jellies, ice creams, chocolates, pies, syrups, pastries, and many beverages [1, 2]. Strawberry holds the most important place among the grape-like fruits in the world. It is preferred by consumers because it is the first fruit to ripen in the spring when no other fruit is available. For this reason, fruits can find buyers at high prices until other fruits reach the market. According to data from the Food and Agriculture Organization in 2019, world strawberry production is 8.885.028 tons. While China ranks first with 3.2 million tons of production, the USA ranks second with 1 million tons and Mexico ranks third with 861 thousand tons. 36.2% of total strawberry cultivation is provided by China in the world. Regarding the cultivation area, China leads with 126 thousand/ha, Poland is second with 50 thousand hectares, and Russia is third with 31 thousand hectares. According to world trade data for strawberries, the export amount of fresh and frozen strawberries in

2020 was 1.6 million tons. While there was a decrease in the export of frozen strawberries from Spain compared to previous years, it was estimated that the export of frozen strawberries from Mexico will increase and it will rank second with 1.7 million tons in 2020. It is predicted that the size of the world strawberry market may reach 22.450 million dollars by 2026 and that the USA alone can generate 18.370 million dollars in the world strawberry market in 2020. It is expected that world strawberry production will increase by 3.4% between 2021 and 2026. According to FAO data, an increase of 39.4% was observed between 2008 and 2018 when world strawberry production is examined. This objectifies the significant increase in the world strawberry production and market. This is an important profitability indicator in terms of strawberry production in the world [3]. In recent years, an increase of 3.1% was observed in the United States of America, which is an important country for the import of frozen strawberries or fresh strawberries, followed by Germany with an increase of 3% and Canada in the third place with an increase of 3.3%.

According to TUIK data, strawberry production in Turkey was expected to increase from 546 thousand tons to 646 thousand tons by the end of 2020 with an increase of 18.2% compared to the previous year. Turkey's foreign trade of strawberries is carried out in two ways as fresh and frozen strawberries. As most of the strawberries produced are consumed domestically, only 8.8% of the total strawberry production was exported in 2020. In addition, Turkey has become one of the most important strawberry producer countries in the world recently and exports strawberries to countries such as Russia, Serbia, Macedonia, Iraq, Bulgaria, and Romania. Although it ranks fourth in world in strawberry production, Turkey comes twelfth in terms of exports since, as mentioned before, the strawberries produced are consumed domestically. Strawberry exports in 2020 increased by 14.7% compared to the previous year and reached 47.912 tons. Strawberry exports became doubled in the last five years. Strawberry imports, on the other hand, vary from year to year but do not have a significant volume. According to the data from the Turkish Statistical Institute (thousand tons), Turkey exported 27.914 tons of strawberries in the first four months of 2021. During the first four-month period, Turkey exported about 14.543 tons to Russia with a rate of 44%, and about 4.129 tons to Iraq with a rate of 14% and the rest to other countries. Based on the above data, there was a significant increase in both strawberry production and the market in the world and in Turkey [4].

While rapid population growth and a corresponding increase in food supply motivate research about agricultural production, the aim is also to produce more with lower costs by using new production techniques for the production of strawberries. In this context, measuring plant features and determining these characteristics accurately in terms of various plant growing and breeding studies, especially in strawberry and vegetable breeding, is one of the most compelling factors for researchers and producers. One of the most important criteria in plant growing and breeding is to quickly and accurately estimate plant genetics within the plant populations to be bred and monitor plant health and development every minute. For this purpose, in a study conducted by Zheng et al. [5], remote sensing and machine learning techniques were used immensely for high-throughput phenotyping technology. Multiple sensors including high-resolution RGB (red-green-blue), multispectral, hyperspectral chlorophyll fluorescence and light detection and ranging (LiDAR) sensors were used in order to allow a range of spatial and spectral resolutions depending on the trait in question. At the same time, in this study, the computer vision and machine learning methodology for plant recognition provided great convenience in evaluating useful biological information and drawing conclusions quickly from these image data. They discovered that these tools allowed the evaluation of various morphological, structural, biophysical and biochemical features of plants. In addition, the researchers pointed out in this study that remote sensing and machine learning facilitate strawberry cultivation with features such as (1) fruit/flower detection, fruit maturity, fruit quality, internal fruit attributes, fruit shape and yield prediction; (2) leaf and canopy attributes; (3) water stress; and (4) pest and disease detection. As a result, this study showed that the use of remote sensing and machine learning technologies will provide significant convenience as they respond quickly, accurately, and effectively to future prospects for strawberries in precision agriculture. In recent years, the use of mechanization and automation systems in plant production has been increasing rapidly. This chapter aims to analyze and evaluate in detail the academic studies about soil-less agriculture, vertical farming, hydroponic method, smart greenhouses based on sensor and software technology, and harvesting and packaging methods using the same technologies, which are among the new production techniques for strawberry cultivation. Additionally, the aim was to find answers to questions about which new strategy should be used by examining the latest developments and new perspectives in world strawberry production in terms of production techniques.

## 2. New growing systems for strawberry cultivation

In a study conducted in China to use new technologies and mechanization techniques in strawberry cultivation, a small electric strawberry planting machine was developed in order to reduce the workload during planting, improve the planting technique by machine, and facilitate plant nutrition and other activities. This strawberry planter is based on placing the strawberry seedlings at the desired depth, quickly and accurately. In addition, it was reported that these non-polluting machines meet the high-capacity requirements. In that study, it was stated that the working endurance of the planting machine was 4.5 h and a success rate of 97.46% was achieved in terms of planting rate [6]. In strawberry cultivation, it is extremely important that the environment or soil to be planted has suitable conditions. Today, new technologies and methods have been used to control and accurately determine soil conditions. In a study carried out for this purpose, soil moisture content, humidity level and soil temperature were monitored using advanced systems at present. It was also indicated that, in addition to these parameters, maximum and minimum temperature and humidity values can be determined by an Arduino microcontroller, moisture sensor, temperature/humidity sensor, and GSM module. It was reported that accurate and effective data regarding the temperature and humidity of the greenhouse can be obtained with the help of a DHT 11 sensor, which was placed in the greenhouse and can communicate via SMS [7]. Considering software and information technologies in strawberry cultivation, these technologies have also begun to be used in monitoring plant diseases and pests, and suitable growing environments. For example, the use of the combination of the internet of things (IoT) with an environmental sensing and image processing device provided a very important contribution to the early and accurate detection of plant diseases and increasing yield. In order to detect and classify the disease correctly, the IoT system needs to send images of the disease to the system and provide feedback about the symptoms. In a study carried out for this purpose using a raspberry pi-based IoT device, images of disease symptoms and environmental parameters (humidity, air temperature, soil moisture, and soil pH) were examined in real-time in the MySQL database in order to detect and classify the disease accurately. After the preliminary stage, the k-mean cluster algorithm was employed in the study to segregate the afflicted part of the plant from other parts and convert it to L*, a*, and b* color space. For the classification of the diseases, multi-class SVM (Support Vector Machine) was used. Moreover, when using a gray level co-occurrence matrix, 14 different categories of color, texture, and shape properties were discovered. A total of 97.33% classification accuracy is achieved via the system. Therefore, the system used in this study will contribute significantly to the early detection of plant diseases, accurate and effective determination of environmental parameters, and yield increase [8]. In another study, Kim et al. [9] developed a common platform to collect agricultural environmental information using cloud-based technology. With the integrated system called Farm as a Service (FaaS), they monitored all data and models in agricultural enterprises. The connection and management of this system were carried out by using IoT devices. Moreover, the IoT-Hub network model was used in this study. It was stated that IoT-Hub ensures the accuracy of technology specialized for agricultural environments since this model supports efficient data transfer for each IoT device as well as communication with non-standard products, and exhibits high communication reliability even in poor communication environments. They reported that the customized FaaS system used in integrated agriculture implements different levels of specialized systems. As a result of this study, it was observed that this system played an important role in the early detection of infectious diseases regarding strawberries.

## 3. New irrigation techniques used in strawberry cultivation

Water scarcity follows population growth as a result of climate change. For this reason, using agricultural water effectively and safely has now become an unavoidable fact. This is exactly why there is a need for new irrigation methods, irrigation models, and stable water management. For example, low-cost IoT-based sensors and actuators were used to save energy and water, reduce costs, manage irrigation systems properly and effectively, reduce the effects of water-borne diseases and pests, and increase efficiency. Considering the above facts, in a study conducted by Cáceres et al. [10], proposed a control system with an economic and predictive feature that provides an advantage for irrigation periodicity. This system aimed to provide maximum efficiency, sufficient soil moisture for crops, and optimize appropriate water and energy consumption. For this purpose, the predictive controller was developed in this system to minimize damage from water-related problems and prevent energy loss by using soil moisture at different depths. Basic greenhouses, where expensive materials and smart systems are not used, are very common in China, as in many countries in the world. An intelligent planting management platform for strawberries was created based on IoT to intelligently maintain planting activities in these greenhouses. For this purpose, research was conducted in order to improve the intelligent planting system and efficient water use in IoT-based greenhouses with an intelligent planting management platform for strawberries. In order to provide computer-human interaction, a platform was developed that can accurately optimize the climate data required for the strawberry in the greenhouse, which can be controlled manually through the WeChat app on a mobile phone. In this study, the user module was added to the platform, which allows the producer to manually change the climate data in the greenhouse via IoT. This connection was based on narrowband IoT wireless transmission technology at 4G speed. Additionally, the application layer was developed with a design based on water-saving management knowledge about the strawberry. This systematic design included seven features; strawberry variety selection, planting seedlings, flower and fruit thinning, environmental control, disease and pest control, plant nutrition and fertilization, and economic irrigation. Through the deployed human-computer platform, producers reported that they were able to make adjustments to options such as regular information regarding plant cultivation, query information retrieval, cultivation management, evaluations and alternative decision-making. In addition, the outcomes of the application were summarized as follows: in comparison to the management experiences of the producers, the efficiency of water use increased by 128.55%, the production value efficiency increased by 226.31%, chemical fertilization decreased by 40%, pesticide use decreased by 61.67%, and the cost of pesticide decreased by 32.48%. Therefore, there was a significant decrease in both fertilizer and pesticide use [11].

Lozano et al. [12] conducted a study in Doñana National Park in Spain, in which they measured parameters such as evapotranspiration, crop coefficient estimation, irrigation efficiency, crop yield and water efficiency in two common strawberry cultivars using drainage lysimeters. Later, they developed an Android application that facilitated manual irrigation planning in the strawberry sector and enabled water use. A similar study was performed by measuring agro-climatic data and water requirements for evapotranspiration in strawberry plantations in the Chilean San Pedro region. In the correlation analysis, the researchers primarily used the k-means of time in the series of agro-climatic variables and the methodologically convenient evapotranspiration parameters. The periods when plants need water were classified by the researchers in order to use a water balance controller [13]. For the use of the smart strawberry irrigation system in a greenhouse in Greece, a three-step method, which records data in a network in order to verify the plant nutrition solution and plan the application, was followed. First, hardware with a ready-to-use small-scale smart irrigation prototype solution and software that was tested and evaluated on different plant species was developed, giving useful insights into larger-scale applications. Second, a reference network architecture was introduced that specifically targets smart irrigation and edge data distribution for strawberry greenhouses. Third, by adopting the reference architecture proposed in the second step, a full-scale system and a conventional strawberry irrigation system were compared in a strawberry greenhouse environment. According to the results of this study, this system gave a more accurate result when measuring the amount of soil moisture change and determining the water consumption compared to the traditional irrigation system, and also reduced the cost of irrigation [14]. The authors [15] planned a study to develop and implement an autonomous and automatic irrigation system for irrigation in a strawberry field. For this purpose, Arduino, which uses a smart irrigation system, was developed. Software was used that provides daily information to farmers about the cultivation status, solenoid valve status, soil moisture, and water tank level. The data obtained from the sensors were transferred to the microcontroller and analyzed, and the user was allowed to decide whether to water the plant or not depending on the results. Thus, according to need, the microcontroller decides whether the solenoid valves will open or not. At the same time, the water level in the water tank could be monitored through the ultrasonic sensor. It was confirmed that this system information provided real-time information about the amount of water in the tank, soil moisture in three areas, time, and date. As a result of this study, the researcher acknowledged that the functionality of the system, which was operational throughout the season, was verified by different tests.

In parallel with the developments in irrigation methods in strawberry production, plant nutrition and fertilization studies have also started to gain momentum. Due to excessive land use for production, there are significant deteriorations in soil fauna and flora. The decline in soil organic matter (SOM) due to the intensification of agricultural practices has become one of the most important threats to soil quality [16]. The decline in SOM affects a number of issues adversely. It causes a decrease in soil fertility, biodiversity, microbial activity, and aggregate stability, all of which have a negative impact on plant productivity and health. Typically, small areas with no rotation cycles restrict the agriculture of strawberries. As a result, the soil deteriorates over time as a result of constant replanting. In the areas where strawberries are produced, the decline in soil fertility as well as the emergence of soil-borne plant diseases [17], require a sustainable agronomic technique that can enhance the productivity of strawberries in replanting conditions. The use of organic material in the reclamation of land, compost or cow manure seems to be an effective strategy for enhancing soil fertility, developing soil structure, increasing microbial diversity and activities, developing the water-holding capacity of soils [18], and having a positive impact on crop yields. Furthermore, organic additions maintain or improve soil productivity even after the nutrients required by the plants have been absorbed. Composting is one of the best options available to reduce the amount of organic waste [19]. It represents the largest proportion of the total solid waste generated globally at 46% [20]. The use of compost is one of the most promising and cost-effective options for restoring the structure of degraded soils. For example, 46% of total solid waste consists of organic matter [19, 20]. The conversion of organic wastes with high biological value for soil improvement can be considered to be a sustainable soil management strategy that is cheaper than other options and is compatible with the concept of zero waste [21, 22]. In addition, composting is a much more effective application than other soil organic waste removal methods such as landfilling and incineration, and it was reported that it is more environmentally friendly than other applications in terms of restoring the soil life cycle [23]. Previous studies demonstrated that the application of compost to strawberry plants improves plant growth, yield, and fruit quality [24]. It was reported that the increase in biomass and root proliferation in the root zone of strawberries and many plants causes excessive growth in the roots and causes a decrease in yield and quality [25]. It is very interesting to understand how the photosynthetic C (carbon) and biomass of strawberry plants are distributed among the growth organs and how it affects fruit size and quality [26, 27]. Due to global climate change, there have been great changes in the soil and climate structure. For this reason, new breeding studies about drought resistance are continuing for strawberry cultivation. In this regard, iron nanoparticles and salicylic acid media were used in order to determine the optimum combination resistant to arid conditions in the tissue culture of the strawberry (*Fragaria* × *ananassa* Duch.) plant. According to the results, salicylic acid reduces the negative effect of drought in strawberry plants and positively affects plant growth in vitro conditions. In addition, a better result was obtained in strawberries using iron nanoparticles compared to the control group regarding the drought. The use of iron nanoparticles together with salicylic acid in strawberry tissue culture studies will make strawberry plants more resistant to drought stress before they are transferred to the field [28]. In addition to the tissue culture studies related to drought stress resistance mentioned above, the use of dazzling technologies in irrigation systems along with plant nutrition is increasing day by day. It is important for plant development to plan plant nutrition programs and evaluate the results together with irrigation applications for strawberries. Therefore, sensor networks were used recently to measure soil moisture, electrical conductivity EC (measures the total salts in solution), and climate data in both soil-less cultivation and traditional cultivation. These systems provide fast and reliable information to manufacturers at the right time. This provides the opportunity to intervene immediately in any negative situation in plant breeding, providing significant advantages in plant development and productivity. Wireless sensor networks can offer cost-effective solutions against adverse situations such as water shortage and climate change that strawberry producers may face. In a study conducted in the USA, controlled irrigation systems were installed on two commercial strawberry farms in Maryland using wireless sensor networks. Then, the efficiency of irrigation practices in these two commercial farms and the effects of irrigation practices on plant growth and fruit quality were compared. Sensor-controlled irrigation was based on measuring the volumetric moisture values taken from the sensors placed in the root zone of the plant on the predetermined soil sets. Using a software program (Sensorweb™; Mayim LLC, Pittsburgh, PA), control nodes were enabled to apply irrigation for a specified period of time when the average sensor reading drops below the set point. Real-time root zone humidity, temperature, and electrical conductivity values were recorded at 15-minute intervals by monitoring/control nodes, and information was transmitted to both producers and researchers thanks to the internet using Sensorweb™. Irrigation volumes, plant growth, fruit yield, and fruit quality parameters were evaluated for two commercial production systems on farms. The utility of the wireless sensor networks system for spring frost warning was also tested by researchers by comparing the accuracy of canopy-based temperatures and on-farm weather station data with satellite (e.g., Skybit™) data [29].

## 4. Strawberry production methods

### 4.1 Hydroponics

The use of the hydroponics method in herbal production dates back many years. During the Babylonian period (605–562 BC), plants on the terraces were irrigated from the Euphrates River using pumps. It is understood from historical records that the Aztecs living in Tenochtitlan in the 40s BC produced plants on man-made islands by a method called chinampas, in which plants were directly in contact with water [30]. Until the present, chinampas have been producing 40.000 t/yr. of vegetables and flowers annually, consequently, the FAO recognized chinampas as a globally important agricultural historical heritage [31, 32]. As is known, hydroponic plant production systems are one of the fastest growing sectors in horticulture. The use of hydroponic crop production methods has increased by 20% worldwide between 2016 and 2019. In addition, the production value in dollar terms increased from 6.9 to 8.3 × 109 dollars in the same interval, and it is estimated that these values will increase up to 45% in 2025 [33]. In addition, the annual growth rate is estimated to increase by 6.8% between 2019 and 2024 in areas such as the United States, Canada, Germany, United Kingdom, China, Brazil, the Middle East, and South Africa [34, 35]. Strawberry is more popular than any other fruit in the world. Traditional strawberry farming, on the other hand, has major difficulties with productivity and plant loss due to soil-borne plant diseases. Besides, the chemicals needed to treat soil are quite toxic. That is why growers have opted for hydroponic strawberry cultivation. With this approach, the issues such as pest and disease control, high productivity in the field and good quality of the fruit are considerably achieved. In regions where the climatic conditions are ideal, the soil is not a concern thanks to hydroponic farming. Soil-borne pathogens and the damaging impacts of pests are the main reason why hydroponics in greenhouse strawberry cultivation are demanded nowadays. Hence, owing to hydroponics, growers can increase yields and enhance product quality by using correct fertilization and water management [36–39]. For this method, producers have generally used the varieties such as Rubigen, Sabrina, Festival, and Albion.

Cocopeat is a growing medium used frequently by gardeners and especially for hydroponics. Cocopeat is made free from sand and out of coconut husk which is why it is appropriate for use in agriculture, particularly hydroponics. Thanks to its high water-holding capacity, %100 organic feature and pH of 5.7–6.5, Cocopeat is one of the best products to be used in agriculture. For the greenhouse cultivation of hydroponic strawberries, diagonal planting with 13 seedlings placed at 15 cm spacing is used for generally around 12 thousand seedlings per decare. Planting begins in October, with the first harvest occurring in December. This process continues until the middle of June. At the end of this process, approximately 10 tons of product can be obtained if suitable garden management methods are used and a correct plant nutrition program is implemented. Producers have become more aware of their production processes. Despite the ongoing rise in input costs, growers continue to crop since they still make a profit. Bumblebees are used to enhance pollination and fertilization, especially in greenhouse production. By avoiding the use of pesticides, bumblebees improve product quality while also contributing to natural production. Since the initial expense of growing strawberries in a greenhouse is so expensive, easiness, and productivity are critical. Strawberry production is generally limited to a six-month season, however, by using hydroponics in cultivation, this may be prolonged up to 12 months. Thanks to this benefit, fresh strawberries may be brought to market for a period of 12 months. When strawberries are properly cared for, production and efficiency in hydroponics may be four times higher than in traditional agriculture. Regardless of the significant initial investment, hydroponics has been used in the Mediterranean region by gardeners for the cultivation of strawberries in recent years.

This method is preferred since customers are prepared to spend the most and there is a lot of demand [40]. In addition, in greenhouse hydroponic strawberry cultivation, it was determined that frigo seedlings are advantageous in terms of yield and tubed seedlings are advantageous in terms of earliness [41]. Likewise, frigo seedlings are advantageous in terms of productivity and tubed seedlings are advantageous in terms of early maturing. Briefly, the productivity has been enhanced and fresh strawberries for the market have been available for over a year, thanks to the adoption of hydroponic technology in the strawberry greenhouse culture. Moreover, with this approach, plants have been protected from soil-borne pathogens and pests while the nutrition they provide has been boosted [4].

Hydroponic systems can technically be classified into two groups. The first is open systems that provide nutrients directly and once to plant roots, and the other is closed systems that provide a continuous and cyclical supply of nutrients to plants. In addition, the nutrient solution given once in open systems comes into contact with the plant roots continuously or occasionally. The media substrates and nutrient solutions used in this system are used only once; that is, they are not reused. Some advantages of open systems are that the plant nutrition solution application is simpler and the risk of infection is less for the plant [42]. In closed systems, plant nutrition solutions are applied to the plant roots, which are given alternately to the plants and collected in containers, as a liquid substrate or as a liquid solution. Substrates used can be organic (such as coconut fiber, rice husk, sawdust, and charcoal) or inorganic (pumice, sand, gravel, and ground brick). In this plant production method, the use of water and nutrients is generally the best, and the disadvantage is the need for electricity [35, 43]. However, in recent years, attempts were made to minimize these disadvantages by using renewable energy. Furthermore, additional applications such as sufficient oxygen uptake of plant roots, suitable temperature environment, adequate nutrient supply, and increasing the activity of beneficial microorganisms for the plant are rapidly applied. Thus, smart farming methods are used in soilless culture in order to understand and follow the communication between plant roots and shoots in strawberry plants in a deeper and more detailed way, and as a result, new technology-based trends were developed to improve plant roots [44]. In another research, strawberry production was carried out using a hydroponic system in tunnel greenhouses to protect against the harmful effects of rain. In this study, a significant increase in yield was achieved by protecting plants from the harmful effects of rain and reducing disease and pest pressure [45, 46]. Hydroponics is a fairly new method that can produce products in and out of season under fully controlled conditions without soil. As is well known, plant feeding and fertilization processes are carried out entirely through the irrigation system in this production model.

In agricultural production, applications such as hydroponic production, vertical agriculture, or soilless agriculture are also increasingly popular for strawberry production [47, 48]. Production methods of vertical farming and hydroponics can be applied in smaller areas and use 95% less water and nutrients than traditional strawberry production methods [49]. At the same time, hydroponic production methods are more advantageous than other methods, since production can be closer to consumption centers in arid and semi-arid conditions regardless of soil quality [50]. Besides, this production model has many advantages such as more efficient and correct use of water management, production throughout the year, higher yields and minimizing the use of pesticides compared to soil culture [51]. It is extremely important to apply for an accurate and effective plant nutrition program in strawberry production with the hydroponics system. In this system, remote-controlled automation systems have been used in strawberry and tomato production in recent years [52, 53]. In this method, the properties of irrigation water and the accuracy of these properties are extremely important. In remote sensing systems, the turnkey solution collects information about the growth of plants in soilless strawberry cultivation and makes predictions accordingly. Previously, a compact sensor with an oscillator circuit was used to monitor the irrigation status and concentration of fertilization of the plants [54]. Moreover, there was a noticeable increase in the use of light emitting diode (LEDs) technology in strawberry production in recent years. Some researchers report that LED lights can be used alone or in combination with other light systems to increase plant behavior, yield, and fruit quality. In a study conducted for this purpose, three different light systems (LED blue, LED red, and fluorescence neon tubes as control) were used to evaluate the effect on plant growth and fruit quality in soilless strawberry production. According to the results, blue LED light with a wavelength of 400–500 nm promotes biomass accumulation, especially at the root and crown level. In addition, fruit set (65 g plant$^{-1}$) in plants treated with blue light was 25% higher than plants in the control group (45 g plant$^{-1}$) and with red light (35 g plant$^{-1}$). There was no change in the main quality traits of the fruit, but it was determined that the color and anthocyanin amounts were low as a result of both applications. As a result of this study, it was reported that the use of blue light increases fruit yield by keeping fruit quality stable [55].

Since the plant nutrient solution is used repeatedly in closed hydroponic systems, root exudates, which have an intraspecific allelopathic effect, accumulate in the strawberry roots over time and inhibit plant growth by causing an autotoxic effect on the plants. In a study, electro-degradation (ED) was applied to the culture solution in order to degrade these root exudates in strawberries and to increase fruit yield and quality. During this application, four types of nutrient solutions were applied. These include renewed, non-renewed, and non-renewed with direct current electrode gradation (DC-ED) and finally non-renewed with alternative current electro-degradation (AC-ED). While 25% standard Enshi nutrient solution was added to the culture solutions which were renewed every three weeks, DC- and AC-ED were applied to the non-regenerated solutions. It was reported that the fruit yield obtained with the renewed solution (225.9 g plant$^{-1}$) and the yield obtained from the non-renewed and AC-ED solution provided statistically similar results. Fruit yield was decreased to approximately half (114.0 g plant$^{-1}$) in the non-renewed solution compared to the renewed solution, but plants treated with the non-renewed solution with DC-ED produced an intermediate yield between the non-renewed and renewed solution. The growth performances of the plants treated with the renewed nutrient solution were higher than the plants treated with the non-renewed solution with DC-ED. Briefly, it was indicated that nonrenewed and AC-ED nutrient solutions may have a positive effect on fruit development, yield, and quality in strawberries [56]. In addition to the hydroponic method mentioned above, new methods such as vertical farming are also used in strawberry production.

### 4.2 Vertical farming (VF) in strawberry production

In recent years, the area of arable land has been declining gradually due to the increase in the human population, urbanization, pollution, and soil erosion. While the world population living in urban areas is 60%, it is estimated that this rate may increase to 68% in the 2050s with the increase in immigration in the 2030s [57, 58]. Vertical farming will be an important factor in solving problems occurring due to these challenges. Regarding advanced farming techniques such as vertical farming, controlled-environment agriculture is performed so that high yields are obtained by using fewer resources in a restricted area [54, 59, 60]. Unfortunately, industry-based agricultural practices disrupt the natural structure of the soil and increase the erosion rate (10–40 times). Moreover, according to some studies, it is estimated that these agricultural production methods can reduce clean water resources by about 70%. However, in VF applications, high-efficiency production can be achieved using much less water and space. It was stated, for example, that a Japanese agricultural tool named Mirai provides information on 25.000 m$^2$ of the indoor agricultural farm to producers and academics. This agricultural vehicle provides 40% energy and significant water savings [61]. As a leader of VF, the aviation farm has increased the agricultural product yield of New York by 390 times and 95% water savings [62]. Carbon dioxide is a very vital factor in agricultural production. As a result, a new toolkit was built with wireless communication that can be handled by mobile phones to properly determine carbon dioxide estimation in vertical farming. Moreover, these devices provide automatic observation for all developmental stages of plants [63, 64]. In recent years, hydroponic vertical farming has become the most advanced, environmentally-friendly agricultural production technique that does not harm biodiversity. Ways to achieve this are to focus on deserts, which make up one-third of the world. Based on this idea, Chinese and Norwegian experts are currently working on the application of these production methods in the deserts of Dubai, Qatar, Jordan, and China. The most important argument for achieving this goal is the use of technologies such as IoT

[65, 66]. In recent years, due to negative factors such as climate change in agricultural lands, pollution of soils due to the intensity of agricultural practices, use of agricultural lands as settlement areas, and deterioration of soil structure, plant production companies, or their investors have pioneered the use of horizontal farming techniques, which is a different and alternative method for strawberry production.

### 4.3 Using horizontal systems in strawberry production

Horizontal systems can be on the ground or have the potential to be stacked on top of each other. Since fruit harvesting and other agricultural tasks are easier at breast- or neck height, horizontal systems are preferred. Containers, pots, bags, and gutters can be used as a medium for plant growth. Since the plants need regular irrigation and fertilization, the preferred environments for growing plants should have sufficient depth and high water-holding capacity. In this method, in contrast to field production, the ability to store water and nutrients is limited. It was reported by experts that the use of galvanized metal, which cause high concentrations of zinc accumulation, should be avoided as a plant-growing medium. In this system, as the plants grow, they make use of wire and other support systems so the plants can stand upright to provide convenience during harvest [67]. In addition to the use of hydroponics, vertical farming, and horizontal systems in strawberry production, there has been an increase recently in the use of quite new techniques such as robot technology, artificial intelligence, and machine learning in the harvesting process.

### 4.4 New harvesting techniques used in strawberry production

As is known, harvesting is one of the most important criteria in determining product quality and productivity in agricultural production. It is reported that 3.1 billion USD product loss is expected every year in the USA due to the lack of qualified manpower [68]. According to data from the United States Department of Agriculture, 14% of agricultural input costs are spent on manpower. At the same time, the labor cost in industrial agriculture can reach up to 39% [69]. Deep studies are still needed about the production of more specific sensors for the effective use of automation systems in fruit harvesting. For example, strawberries are berry-like fruits that can be consumed in any season of the year. However, the use of manpower in strawberry harvesting and packaging processes is one of the most effective factors in increasing fruit prices [70]. Since the strawberries grown in greenhouse conditions are harvested with robotic harvesting technology, the cost is reduced. A robot named Agrobot was developed for this harvesting process and this robot can harvest and pack strawberries from rows of plants [71]. For example, Agrobot's SW 6010, which is a semi-automatic robot model, was reported to provide significant convenience in strawberry harvesting. Another example, Tektu T-100, is another strawberry harvesting robot model that can be charged with electricity and is environmentally friendly [70]. In recent years, due to the decrease in qualified personnel and manpower in agricultural production, the necessity of minimizing losses during the harvest, and reasons such as time and cost savings force farmers to use agricultural robotic technologies. Recently, some researchers reported that fruit-picking robots are being developed by private companies rather than academic researchers. Moreover, the problems in seasonal fruit picking jobs necessitate the use of automation systems in fruit harvesting. In the last 5 years, there was a significant increase in the number of companies producing fruit-picking robots. Advanced vision systems, image processing techniques and artificial intelligence are used in the harvesting of berries, pome fruits, apples, and stone fruits. For example, the mobile robot can pick up strawberry fruits growing on strawberry pads several feet above the ground and can sort them by size or weight and place them in fruit baskets as they move. RGB (Red-Green-Blue) cameras with three-dimensional (3D) features are used to determine the position and ripening times of the fruits. The robot gently harvests with the help of an arm that imitates the human arm extending from the bottom up, has padded soft-grip plastic claws and can rotate 90° to pluck the fruit from the stem. It can harvest soft fruits at a rate of 11.500 berries (between 180 and 360 kg) in a 16-hour day, well beyond the 50 kg typically collected by a human [72]. In light of this information, it is expected that there will be an increase in the use of robotic technologies in fruit harvesting.

In this context, many studies were carried out around the world. A deep learning algorithm, Rotate-YOLO (R-YOLO) was developed to perform real-time location and harvesting of strawberries regarding a strawberry robot technology that performs the strawberry harvest. In addition, with the help of the bounding box, it accurately detects the plucking point with an angle to follow the direction of the strawberry and harvests it gently. It was reported that the robot, customized for the harvest of ridge-planted strawberries, with fiber sensors on its end-effector to control its speed, avoids real-time distance measurements. As a result, the researchers stated that the robot using Rotate-YOLO (R-YOLO) was successful in correctly identifying strawberries at a speed of 0.056 per second, with a 640 × 480 resolution RGB camera and a rate of 94.43% [73]. Furthermore, the monitoring system used in the strawberry harvest monitors the ripening time of the strawberry, reduces possible injuries during the harvest and detects diseases and pests at the right time. Therefore, the strategic advantages of agricultural production are implemented in different strawberry production applications by using innovative technologies. Experts are working on a system that can monitor strawberry cultivation in real ecologies, as well as access more accurate information about the harvest time of the strawberry plant and make the right decision. The system recommended above has a design that analyzes and stores the climatic data for the strawberry and images recorded through the IoT-Edge-AI-Cloud concept [74, 75]. The IoT-Edge device, Arduino and Raspberry Pi will be sufficient to install such systems at affordable costs. Even if the strawberry producer expands their production area, the system operated through AI-Cloud can easily be expanded as well according to the purpose and demand. The system can effectively evaluate the climate data in strawberry cultivation by reasoning with an artificial memory of the maturity stages of the strawberry plant. All data obtained in strawberry production (such as harvest time, disease detection, and production data) are analyzed by evaluating the data transferred to the integrated interface. In a study, ecological data and images of strawberries from hydroponic strawberry production were obtained using the IoT-Edge module and transferred to a nano-sized private AI-Cloud-based analysis station module and visualized to determine when to harvest. The monitoring and analysis results were envisioned with an integrated interface supply for major data such as fluctuating yields, harvest periods, and pest diagnosis. The suggested system is based on the idea of AI-Cloud. This concept helps server container to be scaled up quickly and simply as it grows. The suggested system was put to the test in a home where Seolhyang strawberries were grown using hydroponics. Over the course of four months, 1.316.848 actual environmental data points pertaining to 13 data kinds were monitored. Using 1575 strawberry photos from the Smart Berry Farm and a Google Images search, the harvest time was predicted with a high accuracy rate of 98.267% [76].

## 5. Conclusion

In conclusion, the use of new methods and techniques in strawberry cultivation has become a necessity. The reason for this is the importance of strawberry in human nutrition, based on its good taste and aroma, pleasant smell, and also the increased demand and supply in world strawberry production and the market since strawberry is a dietary fruit. Due to the mismanagement of cultivated soils and water resources used in strawberry cultivation around the world, the decrease in these production areas, the pollution of soil due to excessive use, and the rapid increase in labor and input costs, researchers and producers have been searching for new alternative production models in strawberry production. For this reason, new production methods have begun to be developed for smart agriculture systems that are close to large consumption centers, and use less water and soilless farming methods. These new production methods are expected to achieve maximum efficiency by reducing manpower costs and minimizing the damage caused by human error, especially in harvesting and other processes. Consequently, this chapter shed light on studies performed about smart strawberry production models based on new technologies and will be a reference for information technology and artificial intelligence-based studies in strawberry production from planting to harvesting and packaging.

## Author details

İlbilge Oğuz[1]*, Halil İbrahim Oğuz[2] and Nesibe Ebru Kafkas[1]

1 Faculty of Agriculture, Department of Horticulture, University of Çukurova, Adana, Turkey

2 Faculty of Agriculture, Department of Horticulture, Adıyaman University, Adıyaman, Turkey

*Address all correspondence to: ilbilge94@gmail.com

## References

[1] Kafkas E. Strawberry growing in Turkey: Current status and future prospects. In: VIII International Strawberry Symposium 1156. Leuven, Belgium: ISHS Acta Horticulturae; 2016. pp. 903-908. DOI: 10.17660/ActaHortic.2017.1156.133

[2] Husaini AM, Zaki FA. Strawberries: A general account. Husaini, AM, Neri, D editiors. Strawberry: Growth, Development and Diseases. CABI, Wallingford, United Kingdom. 2016. pp. 1-9

[3] FAO. 2021. Available from: https://www.fao.org/faostat/en/#home [Accessed on 4 May, 2021]

[4] Oğuz Hİ, Oğuz İ. Greenhouse Cultivation of Fruits in the Mediterranean Region of Turkey. Adiyaman Province, Turkey: Iksad Publications; 2021. pp. 93-128 ISBN: 978-625-7562-09-6

[5] Zheng C, Abd-Elrahman A, Whitaker V. Remote sensing and machine learning in crop phenotyping and management, with an emphasis on applications in strawberry farming. Remote Sensing. 2021;**13**(3):531

[6] Wu D, Hu J. Design and test of miniature electric directional transplanter for strawberry. International Core Journal of Engineering. 2021;7(3): 468-476

[7] Elenzano JJ. Smart farming for lowland strawberry (*Fragaria x ananassa*) production. Turkish Journal of Computer and Mathematics Education (TURCOMAT). 2021;**12**(3):1797-1804

[8] Pavel MI, Kamruzzaman SM, Hasan SS, Sabuj SR. An IoT based plant health monitoring system implementing image processing. In: Proceedings of the 2019 IEEE 4th International Conference on Computer and Communication Systems (ICCCS). Piscataway, NJ, USA: Institute of Electrical and Electronics Engineers (IEEE); 2019. pp. 299-303

[9] Kim S, Lee M, Shin C. IoT-based strawberry disease prediction system for smart farming. Sensors. 2018;**18**(11):4051. DOI: 10.3390/s18114051

[10] Cáceres G, Millán P, Pereira M, Lozano D. Smart farm irrigation: Model predictive control for economic optimal irrigation in agriculture. Agronomy. 2021;**2021**(11):1810. DOI: 10.3390/agronomy11091810

[11] Li Q, Cao B, Wang X, Wu JJ, Wang YK. Systematic water-saving Management for Strawberry in basic greenhouses based on the internet of things. Applied Engineering in Agriculture. 2021;**37**(1):205-217. DOI: 10.13031/aea.14095

[12] Lozano D, Ruiz N, Gavilán P. Consumptive water use and irrigation performance of strawberries. Agricultural Water Management. 2016;**169**:44-51. DOI: 10.1016/j.agwat.2016.02.011

[13] Carrasco R, Soto I, Seguel F, Osorio Valenzuela L, Lagos C, Flores C. Water balance analysis in plantations of strawberries, in the commune of sanpedro. In: 2017 CHILEAN Conference on Electrical, Electronics Engineering, Information and Communication Technologies (CHILECON). Vol. 2017. Pucon, Chile: IEEE; 2017. pp. 1-5. DOI: 10.1109/CHILECON.2017.8229715

[14] Angelopoulos CM, Filios G, Nikoletseas S, Raptis TP. Keeping data at the edge of smart irrigation networks:

A case study in strawberry greenhouses. Computer Networks. 2020;**167**:107039

[15] Aurel GI. Arduino platform for optimized irrigation in strawberries fields. In: 9th International Conference on Modern Power Systems (MPS). Cluj-Napoca, Romania: IEEE; 2021. pp. 1-5. DOI: 10.1109/MPS52805.2021.9492687

[16] Batlle-Bayer L, Batjes NH, Bindraban PS. Changes in organic carbon stocks upon land use conversion in the Brazilian Cerrado: A review. Agriculture, Ecosystems and Environment. 2010; **137**(1-2):47-58. DOI: 10.1016/j.agee.2010.02.003

[17] Neri D, Sugiyama N, Inujima A. Effects of organic residues on strawberry root growth. International Journal of Fruit Science. 2005;**5**(1):127-137. DOI: 10.1300/J492v05n01_12

[18] Canali S, DiBartolomeo E, Trinchera A, Nisini L, Tittarelli F, Intrigliolo F, et al. Effect of different management strategies on soil quality of citrus orchards in southern Italy. Soil Use and Management. 2009;**25**:34-42. DOI: 10.1111/j.1475-2743.2008.00191.x

[19] Hargreaves JC, Adl MS, Warman PR. A review of the use of composted municipal solid waste in agriculture. Agriculture, Ecosystems and Environment. 2008;**123**(1-3):1-14. DOI: 10.1016/j.agee.2007.07.004

[20] Hoornwcg D, Bhada-Tata P. What a Waste: A Global Review of Solid Waste Management. Washington, D.C., United States: World Bank; 2012. Available from: http://hdl.handle.net/10986/17388

[21] Lim SL, Lee LH, Wu TY. Sustainability of using composting and vermicomposting technologies for organic solid waste biotransformation: Recent overview, greenhouse gases emissions and economic analysis. Journal of Cleaner Production. 2016;**111**:262-278. DOI: 10.1016/j.jclepro.2015.08.083

[22] Curran T, Williams ID. A zero waste vision for industrial networks in Europe. Journal of Hazardous Materials. 2012;**207**:3-7. DOI: 10.1016/j.jhazmat.2011.07.122

[23] Saer A, Lansing S, Davitt NH, Graves RE. Life cycle assessment of a food waste composting system: Environmental impact hotspots. Journal of Cleaner Production. 2013;**52**:234-244. DOI: 10.1016/j.jclepro.2013.03.022

[24] Wang SY, Lin SS. Composts as soil supplement enhanced plant growth and fruit quality of strawberry. Journal of Plant Nutrition. 2002;**25**(10):2243-2259. DOI: 10.1081/PLN-120014073

[25] Baldi E, Toselli M, Eissenstat DM, Marangoni B. Organic fertilization leads to increased peach root production and lifespan. Tree Physiology. 2010;**30**(11): 1373-1382. DOI: 10.1093/treephys/tpq078

[26] Poorter H, Niklas KJ, Reich PB, Oleksyn J, Poot P, Mommer L. Biomass allocation to leaves, stems and roots: Meta-analyses of interspecific variation and environmental control. The New Phytologist. 2012;**193**(1):30-50. DOI: 10.1111/j.1469-8137.2011.03952.x

[27] Martre P, Bertin N, Salon C, Génard M. Modelling the size and composition of fruit, grain and seed by process-based simulation models. The New Phytologist. 2011;**191**(3):601-618. DOI: 10.1111/j.1469-8137.2011.03747.x

[28] Akbar Mozafari A, Havas F, Ghaderi N. Application of iron nanoparticles and salicylic acid in in vitro culture of strawberries (*Fragaria* × *ananassa* Duch.) to cope with drought stress. Plant Cell, Tissue, and Organ Culture. 2018;**132**(3):511-523

[29] Lea-Cox JD, Belayneh BE. Using wireless sensor networks in commercial strawberry production. In: 2015 ASABE/ IA Irrigation Symposium: Emerging Technologies for Sustainable Irrigation-A Tribute to the Career of Terry Howell, Sr. Conference Proceedings. St. Joseph, Michigan, United States: American Society of Agricultural and Biological Engineers; 2015. pp. 1-11

[30] González Carmona E, Torres Valladares CI. La sustentabilidad agrícola de las chinampas en el Valle de México: caso Xochimilco. Revista Mexicana de Agronegocios. 2014;**34**:699-709. Available from: http://www.redalyc.org/pdf/141/14131514005.pdf. (In Spanish)

[31] Arano CR. Hidroponía: Algunas páginas de su historia. Horticultura internacional. 2007;**58**:24-33 (In Spanish)

[32] FAO. Sistema de Informacion sobre el Uso del Agua en la Agricultura de la FAO. Rome, Italy Available from: http://www.fao.org/nr/water/aquastat/wateruse/indexesp.stm: FAO; 2016 Accessed 16 May, 2019

[33] Shahbandeh M. Hydroponics: Forecasted Market Value Worldwide 2016-2025. Hamburg Available from: https://www.statista.com/statistics/879946/global-hydroponics-market-value/: Statista; 2018 Accessed 19 August, 2019

[34] Intelligence M. Hydroponics Market|Growth, Trends, and Forecasts (2019-2024). Dublin, Ireland Available from: https://www.mordorintelligence.com/industry-reports/hydroponics-market: Research and Markets; 2019 Accessed 19 August, 2019

[35] Cifuentes-Torres L, Mendoza-Espinosa LG, Correa-Reyes G, Daesslé LW. Hydroponics with wastewater: A review of trends and opportunities. Water and Environment Journal. 2021;**35**(1):166-180. DOI: 10.1111/wej.12617

[36] De Cal A, Martinez-Treceno A, Saltoa T, Lopez Aranda JM, Melgarejo P. Effect of chemical fumigation on soil fungal communities in Spanish strawberry nurseries. Applied Soil Ecology. 2005;**28**:47-56. DOI: 10.1016/j.apsoil.2004.06.005

[37] Palencia P, Gine Bordonaba J, Martinez F, Terry LA. Investigating the effect of different soilless substrates on strawberry productivity and fruit composition. Scientia Horticulturae. 2016;**203**:12-19. DOI: 10.1016/j.scienta.2016.03.005

[38] Demirsoy L, Serçe S. Strawberry culture in Turkey. III Balkan Symposium on Fruit Growing 1139 ISHS, Belgrade, Serbia. 2015. pp. 479-486 10.17660/ActaHortic.2016.1139.82

[39] Martinez F, Oliveira JA, Calvete EO, Palencia P. Influence of growth medium on yield, quality indexes and SPAD values in strawberry plants. Scientia Horticulturae. 2017;**217**:17-27. DOI: 10.1016/j.scienta.2017.01.024

[40] Demirsoy L, Mısır D, ve Nafiye ADAK. Topraksız Tarımda Çilek Yetiştiriciliği. Anadolu Ege Tarımsal Araştırma Enstitüsü Dergisi. 2017;**27**(1):71-80. Available from https://dergipark.org.tr/tr/pub/anadolu/issue/31211/336149. (In Turkish)

[41] Adak N, ve Pekmezci M. Topraksız Kültürle Çilek Yetiştiriciliğinde Fide Tipleri ile Yetiştirme Ortamlarının Erkencilik ve Verim Üzerine Etkileri. Akdeniz Üniversitesi Ziraat Fakültesi Dergisi. 2011;**24**(2):67-74. Available from https://dergipark.org.tr/tr/pub/akdenizfderg/issue/1567/19415. (In Turkish)

[42] Jones B. Hydroponics a Practical Guide for the Soilless Grower. 2nd

ed. Florida: CRC Press; 2005. p. 55. DOI: 10.1017/CBO9781107415324.004

[43] Lee S, Lee J. Beneficial bacteria and fungi in hydroponic systems: Types and characteristics of hydroponic food production methods. Scientia Horticulturae. 2015;**195**:206-215. DOI: 10.1016/j.scienta.2015.09.011

[44] Balliu A, Zheng Y, Sallaku G, Fernández JA, Gruda NS, Tuzel Y. Environmental and cultivation factors affect the morphology, architecture and performance of root systems in soilless grown plants. Horticulturae. 2021;**7**(8):243. DOI: 10.3390/horticulturae7080243

[45] Rohloff J, Nestby R, Folkestad JA, Iversen TH. Influence of rain cover cultivation on taste and aroma quality of strawberries (Fragaria ananassa Duch.). Journal of Food Agriculture and Environment. 2004;**2**:74-82

[46] Claire D, Watters N, Gendron L, Boily C, Pépin S, Caron J. High productivity of soilless strawberry cultivation under rain shelters. Scientia Horticulturae. 2018;**232**:127-138. DOI: 10.1016/j.scienta.2017.12.056

[47] Jensen M. What Is Hydroponics? Tucson, Arizona: Controlled Environment Agriculture Center; 2013. Available from: http://ag.arizona.edu/ceac/what-hydroponics

[48] Khan N, Ray RL, Sargani GR, Ihtisham M, Khayyam M, Ismail S. Current Progress and future prospects of agriculture technology: Gateway to sustainable agriculture. Sustainability. 2021;**13**(9):4883. DOI: 10.3390/su13094883

[49] Nishimura T, Okuyama Y, Matsushita A, Ikeda H, Satoh A. A compact hardware design of a sensor module for hydroponics. In: Proceedings of the 2017 IEEE 6th Global Conference on Consumer Electronics (GCCE), Nagoya, Japan, 24-27 October 2017. Piscataway, New Jersey: IEEE; 2017. pp. 1-4

[50] Bellows AC, Brown K, Smit J. Health Benefits of Urban Agriculture. Venice, CA, USA: Community Food; 2003

[51] Resh HM, Howard M. Hydroponic Food Production: A Definitive Guidebook for the Advanced Home Gardener and the Commercial Hydroponic Grower. Boca Raton, Florida, United States: CRC Press; 2012

[52] Tripodi P, Massa D, Venezia A, Cardi T. Sensing technologies for precision phenotyping in vegetable crops: Current status and future challenges. Agronomy. 2018;**8**:57. DOI: 10.3390/agronomy8040057

[53] Sarraf M, Kataria S, Taimourya H, Santos LO, Menegatti RD, Jain M, et al. Magnetic field (MF) applications in plants: An overview. Plants. 2020;**9**:1139. DOI: 10.3390/plants9091139

[54] Zhou J, Reynolds D, Websdale D, Le Cornu T, Gonzalez-Navarro O, Lister C, et al. CropQuant: An automated and scalable field phenotyping platform for crop monitoring and trait measurements to facilitate breeding and digital agriculture, BioRxiv. 2017:161547. DOI: 10.1101/161547

[55] Nadalini S, Zucchi P, Andreotti C. Effects of blue and red LED lights on soilless cultivated strawberry growth performances and fruit quality. European Journal of Horticultural Science. 2017;**82**(1):12-20. DOI: 10.17660/eJHS.2017/82.1.2

[56] Talukder MR, Asaduzzaman M, Tanaka H, Asao T. Electro-degradation

of culture solution improves growth, yield and quality of strawberry plants grown in closed hydroponics. Scientia Horticulturae. 2019;**243**:243-251. DOI: 10.1016/j.scienta.2018.08.024

[57] United Nations Department of Economic and Social Affairs. 68% of the World Population Projected to Live in Urban Areas by 2050, Says UN. New York, NY, USA: United Nations Department of Economic and Social Affairs; 2018

[58] Khan N, Siddiqui BN, Khan N, Khan F, Ullah N, Ihtisham M, et al. Analyzing mobile phone usage in agricultural modernization and rural development. International Journal of Agricultural Extension. 2020;**8**:139-147. DOI: 10.33687/ijae.008.02.3255

[59] Touliatos D, Dodd IC, McAinsh M. Vertical farming increases lettuce yield per unit area compared to conventional horizontal hydroponics. Food and Energy Security. 2016;**5**:184-191. DOI: 10.1002/fes3.83

[60] Bin Ismail MIH, Thamrin NM. IoT implementation for indoor vertical farming watering system. In: Proceedings of the 2017 International Conference on Electrical, Electronics and System Engineering (ICEESE), Kanazawa, Japan, 9-10 November 2017. Piscataway, New Jersey: IEEE; 2017. pp. 89-94

[61] Siregar B, Efendi S, Pranoto H, Ginting R, Andayani U, Fahmi F. Remote monitoring system for hydroponic planting media. In: Proceedings of the 2017 International Conference on ICT for Smart Society (ICISS), Tangerang, Indonesia, 18-19 September 2017. Piscataway, New Jersey: IEEE; 2017. pp. 1-6

[62] Liang J-Y, Chien Y-H. Effects of feeding frequency and photoperiod on water quality and crop production in a tilapia–water spinach raft aquaponics system. International Biodeterioration and Biodegradation. 2013;**2013**(85):693-700. DOI: 10.1016/j.ibiod.2013.03.029

[63] Ibayashi H, Kaneda Y, Imahara J, Oishi N, Kuroda M, Mineno H. A reliable wireless control system for tomato hydroponics. Sensors. 2016;**16**:644. DOI: 10.3390/s16050644

[64] Theopoulos A, Boursianis A, Koukounaras A, Samaras T. Prototype wireless sensor network for real-time measurements in hydroponics cultivation. In: Proceedings of the 2018 7th International Conference on Modern Circuits and Systems Technologies (MOCAST), Thessaloniki, Greece, 7-9 May 2018. London: IEEE; 2018. pp. 1-4

[65] Lyra DA, Ismail S, Butt KURB, Brown J. Evaluating the growth performance of eleven' Salicornia Bigelovii' populations under full strength seawater irrigation using multivariate analyses. Australian Journal of Crop Science. 2016;**10**:1429. DOI: 10.3316/informit.501299902956210

[66] Elsheshtawy Y. Cities of sand and fog: Abu Dhabi's global ambitions. In: The Evolving Arab City: Tradition, Modernity and Urban Development. Oxfordshire, UK: Routledge; 2008. pp. 258-304

[67] Lantz W, Swartz H, Demchak K, Frick S. Season-long Strawberry Production with Everbearers for Northeastern Producers. College Park, Maryland: University of Maryland; 2010. p. 401

[68] Bronars SG. A Canishing Breed: How the Decline in US Farm Laborers Over the Last Decade Has Hurt the US Economy and Slowed Production on American Farms. New York. Available from:

https://research.newamericaneconomy.org/wpcontent/uploads/2015/08/PNAE_FarmLabor_August-3-3.pdf: New American Economy; 2015 Accessed on 10 July, 2015

[69] Wiggins MF. Farm labor and the struggle for justice in the eastern United States. In: Latinx Farmworkers in the Eastern United States. Vol. 2020. Berlin/Heidelberg, Germany: Springer; 2020. pp. 227-251. DOI: 10.1007/978-3-030-36643-8

[70] Defterli SG. Review of robotic technology for strawberry production. Applied Engineering in Agriculture. 2016;**32**:301-318. DOI: 10.13031/aea.32.11318

[71] Bolda M. Robotic Strawberry Harvester on the Move. California. Available from: https://ucanr.edu/blogs/blogcore/postdetail.cfm?postnum=8162: UC ANR; 2012 Accessed on 17 August, 2012

[72] Bogue R. Fruit picking robots: Has their time come? Industrial Robot: The International Journal of Robotics Research and Application. 2020;**47**(2):141-145

[73] Yu Y, Zhang K, Liu H, Yang L, Zhang D. Real-time visual localization of the picking points for a ridge-planting strawberry harvesting robot. IEEE Access. 2020;**2020**(8):116556-116568. DOI: 10.1109/ACCESS.2020.3003034

[74] Han JS, Hong YJ, Kim JW. Refining microservices placement employing workload profiling over multiple Kubernetes clusters. IEEE Access. 2020a;**8**:192543-192556. DOI: 10.1109/ICDCS47774.2020.00173

[75] Han JS, Park S, Kim JW. Dynamic OverCloud: Realizing microservices-based IoT-cloud service composition over multiple cloud. Electronics. 2020b;**2020**(9):969. DOI: 10.3390/electronics9060969

[76] Park S, Kim JW. Design and implementation of a hydroponic strawberry monitoring and harvesting timing information supporting system based on Nano AI-cloud and IoT-edge. Electronics. 2021;**10**(12):1400. DOI: 10.3390/electronics10121400

Chapter 2

# Winter Strawberries Production in Greenhouse Soilless Culture under an Arid Climate – Cultivars, Phenology, Physiology, and Consequent Practices

*Ofer Guy, Nir Dai, Shabtai Cohen and Amnon Bustan*

## Abstract

Fifteen years of research and development aimed at the production of high-quality early-winter strawberries in the Negev desert, are concluded. This goal required synchronization of seasonal yields with the peaks of the markets' demand, November–February. For this purpose, selected infra short-day (ISD) cultivars were used. Rooted plug plants with enhanced capacity of simultaneous fast vegetative growth and fruiting were produced. Production was carried out in greenhouses and various protocols aimed at yield enhancement were tested. Currently, drip-fertigated, eye-level hanging soilless system, with plant density at about 200,000 plants $ha^{-1}$, combined with biological pest management program, are employed. This system brought about a two-fold increase in yield compared to soil culture (80–100 t $ha^{-1}$), significantly reduced risks of soil- and air-borne diseases, reduced the use of fungicides, and eased extensive labor demands. However, beyond substantial refinements of practices, a significant proportion, 15–25% of the seasonal high-quality yield, was harvested during the earlier two months with no reduction in the total yield or fruit quality. In this chapter, the state of art in early-winter strawberry production is portrayed, including principles, approaches, and methods that have been used and improved during the project. Additionally, ideas for further possible enhancement are discussed.

**Keywords:** early winter strawberries, *Fragaria ananassa*, greenhouse, hung soilless systems, infra short day (ISD), Negev desert

## 1. Introduction

Strawberry (*Fragaria* x *ananassa*) is among the most appreciated fruit species due to its extraordinary appearance, shiny red color, sweet-sour balanced taste, and complex rich aroma. Most of the cultivars belong to the 'June-bearing' group, the floral induction of which is sensitive to short days (SD) and low temperature. In regions of a temperate climate, fruit production takes place during spring and early summer.

 

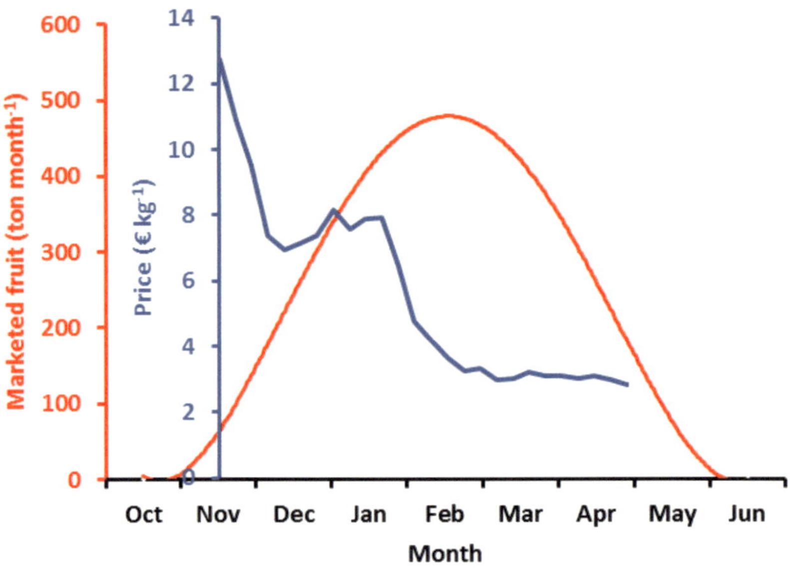


**Figure 1.**
*A typical scenario of the dynamic strawberry marketing season in Israel, referring to produce quantities (red) vs. prices at the farm gate (blue) during mid-November to late April (values represent averages of 10 years of data, provided by the Plant Production & Marketing Board of Israel).*

However, substantial demands for 'off-season' strawberries have triggered significant efforts to deliver the fruit as early as possible. Mediterranean climate regions harbor great potential for early season strawberry production; while the photoperiod noticeably changes with the seasons, the winter can be mild enough to allow plant growth. Californian breeding programs of short-day cultivars, and the development of plastic tunnels practices enabled the delivery of strawberry fruit to the markets already in early March.

Nevertheless, the exceptionally high produce prices at the very beginning of the marketing season, as clearly demonstrated in **Figure 1**, have promoted further research and development efforts aiming at strawberry marketing earlier than before. In Israel, during the recent 15 years, high-quality strawberries are available for export and local markets already in mid-November. The initial produce prices are extremely high, about 12 € $kg^{-1}$; they drop with the increasing production to a high level of 7–8 € $kg^{-1}$ during December–January, and drop again to about 3 € $kg^{-1}$ for the rest of the season, when the production peaks or the quality degrades (**Figure 1**). These achievements are primarily founded on continuous breeding programs that took place since 1980s at ARO (and later on, also by private breeders), and on consistent efforts carried out at regional R&D centers (Darom, Ramat Negev, and Central & Northern Arava) aiming to develop innovative technologies and practices that support, altogether, significantly higher yields, higher and safe fruit quality, and easier labor work. Most of the results have been published in Hebrew in annual reports of each R&D center. In this chapter, we gather the most up-to-date information, describe and discuss the principles and practices that have led to the successful early strawberry production in Israel, and illustrate lines for possible future progress of this culture.

## 2. Varieties

The modern strawberry, *Fragaria* x *ananassa*, was first found in the 1760s in Versailles gardens, France, via a cross of *Fragaria virginiana* from eastern North America and *Fragaria chiloensis*, which was brought from Chile [1]. Typically to species originated in temperate climate regions, temperature and photoperiod are the most important environmental factors that regulate the transition from vegetative to floral growth in strawberries [2]. Under temperate climate conditions, strawberry plants usually enter a vegetative phase and develop runners during the warmer and longer summer days. With the decreasing temperature and day length during autumn, flower initiation and dormancy are simultaneously induced in the young developing crowns [3, 4]. These young crowns, called daughter plants, are rooted and planted towards spring, develop a foliar canopy and realize their reproductive potential during early summer (June-yielding types). Nevertheless, the modern strawberry harbors substantially broad genetic diversity, including a wide sensitivity ranges to photoperiod or temperature [4–7]. Moreover, individual plants may possess significant plasticity in response to combinations of temperature and photoperiod.

To obtain significant early fruiting, two prerequisites must be fulfilled: early and extensive reproductive induction; and, avoided bud dormancy. The first has been met by breeding within the SD strawberry types. This way, young crowns that are detached from the mother plant on the end of August, when the declining photoperiod is still long (13 h), will continuously exhibit floral induction throughout the young plant establishment from August to October. Bud dormancy in many strawberry cultivars is facultative or considered a quantitative trait; hence, avoiding the conditional dormancy could be gained by breeding, by which genotypes that yield continuously for 7 months from November to May are selected.

Four decades ago, a few Californian varieties, mostly 'Fresno', 'Tioga', 'Douglas' and 'Tuft', the harvest of which began in late January, dominated the Israeli market. The breakthrough in obtaining earlier fruiting was accomplished through the establishment of a local (ARO, Israel) breeding program led by Izsak and Izhar [8], in which day neutral (DN) types were hybridized with SD strawberry types, resulted in the release of the first Israeli cultivars 'Nurit' and 'Rachel' in 1977, and during the 1990s—the commercial cultivars 'Tamar', 'Hadas', 'Yael' and 'Malach'. All these are very early cropping cultivars patented as infra short-day (ISD) strawberry types [9]. Unlike the Californian short-day types, the Israeli varieties are less sensitive to temperature and can initiate their reproductive phase under the relatively warm and long-day conditions prevailing in Israel during August and September.

Over the last two decades, the ARO breeding program has been renewed in an attempt to enhance the already high commercial value of the previously developed local varieties [10]. The ARO's in-house germplasm collection has been restored and its genetic base extended with new cultivars from different sources. As a part of this program, approximately 60 specific crosses have been carried out every year, resulting in 6000–10,000 seedlings that have been screened each year under various commercial conditions. While early fruiting is indeed compulsory for strawberry breeding programs in Israel, many other traits are carefully overlooked in order to develop cultivars bearing premium quality fruits suited for local and European markets. These traits include: plants with open canopies and long fruit pedicels for easier harvest; attractive fruit appearance with a bright red color, medium to large size, and a conical or heart shape; firm texture with extended shelf life; and, desired sweet/sour balance with excellent aromatic flavor. On top of these, tolerance to Powdery mildew (*Sphaerotheca macularis*) and anthracnose fruit rot (*Colletotrichum acutatum*), two major diseases affecting plant vigor and postharvest quality gains an increasing attention as a strategic means to overcome the challenge. Since there are no simple selectable markers for the resistance against these diseases, at least one resistant parent is always combined in the basic crosses, and clones with good field tolerance to those diseases are selected.

The attempts made by the ARO strawberry breeding programs, as well as simultaneous efforts made by the private sector, have yielded a growing number of qualified cultivars, all of which bearing early fruiting and desired quality trait (**Figure 2**). Among

**Figure 2.**
*Representative cultivars released by the Israeli ARO strawberry breeding program for commercial use in the recent decade.*

cultivars released at about 2005, 'Tamir' was selected as a vigorous, early-ripening, and highly productive cultivar, with orange-red, large, pretty, and very sweet fruit. 'Tamir' also does very well in hanging, soil-less growing systems. 'Barak' has a medium growth habit and long fruit stalks, which make it easier to harvest. It also exhibits good resistance to powdery mildew and mites. Its fruits are uniform in size, medium, and have an attractive appearance, with a strong red color, excellent flavor and aroma, and extended shelf-life. 'Yasmin' is one of the earliest Israeli cultivars, the harvest of which can begin by mid-November. It has a moderate-sized canopy and continuous waves of flowering and fruiting. The fruits are large to medium-sized, bright red, juicy and sweet. Yasmin's sibling-cultivar Shani has a large, erect growth habit. It produces relatively high annual yields of firm, large, juicy fruit. The cultivars Gili, Matan and Rocky were released in 2015; these three are characterized by a medium-size canopy bearing firm, large, red, and sweet fruits with an extended shelf-life. In addition, Rocky displays good tolerance to powdery mildew and high and stable yields along the season.

The most recently released (2020) cultivars of the ARO breeding program are Tammuz and Lavi (**Figure 2**); both cultivars exhibit vigorous growth habits, long fruit pedicels, early fruiting (late-November to early-December), high yield, continuous and stable production course, and above all, highly attractive fruit appearance, sweetness (Brix 8.5-10) and rich flavor, firm texture, and long shelf life. In addition, private breeders released competitive commercial strawberry cultivars based on the same genetic origin, among which 'Aya1', 'Daniel', 'Peles', 'Shaked', and 'Rotemi' (Yosef's Farm, Israel), and 'Yuval', 'Orly', 'Noa' and '6050' (Fertiseeds Ltd., Israel) have been evaluated and are available for growers.

Along with the continuing challenge to enhance early yielding cultivars, several points require special attention. Most of the breeding programs, including at ARO, are carried out under the traditional low-tunnels cultivation rather than on the advanced hanging soilless culture technology (which will be discussed below). Certain cultivar traits may significantly differ fitting to either set of conditions. For instance, while a vigorous plant canopy is undesired in the traditional cultivation, as it decreases the access for pollinating insects, promotes foliar diseases and makes harvest less efficient, this trait is considered an advantage in the hung soilless system, the fruits and vegetative canopy of which are separated, as it leads to higher yields and to better fruit quality. Concurrently, early strawberry plants grown under greenhouse conditions are more exposed to powdery mildew infection and red spider mites, and hence, more resistant cultivars are preferred [11]. In addition, considerable year-to-year variations in the weather conditions, particularly during the critical autumn season, and the impatient strawberry industry, both prevent the desired adequate long-term evaluation of new cultivars. On the one hand, the rapid rotation among the leading cultivars makes it difficult to point to uniquely promising and stable ones. On the other hand, simultaneously growing several sound cultivars in a season appears a good strategy to deduct failures in the performance of one cultivar or another and accomplish an acceptable operation over a whole season.

## 3. Propagation

An appropriate preparation of plug plants is essential to obtain significantly early yields. The leading principles include the establishment of a massive and functioning root system, adequate foliage, and an early reproductive readiness. The high investments in the growing system (as described later) and the consequent expectation for high outputs require high accuracy and uniformity levels in preparing the plug plants.

### 3.1 Mother plants

The mother plant material is derived from apical meristems that are first grown in tissue culture; the emerging seedlings are then tested for virus infection while grown in an insect-free core greenhouse. In a second stage, all the 'core plants' undergo a tight supervision by the breeders to guarantee its genetic identity and cleanliness, in a "true to type" process, before further propagated in a foundation greenhouse to produce clean mother plants. After 2–3 months of cold treatment, the bare-rooted mother plants are rooted in 7–7–8 cm pots and then distributed among farmers for the last propagation step. For propagation, plugged mother plants are transferred in late April to a hung coir channel system at a planting density of a single mother plant per 1 m row length, at height of 2 m above ground, and at 1.5 m distance between rows, thus providing maximum light interception and aeration throughout the propagation process. Coconut fibers packed in 1 m long growbags are the preferred growth medium. The mother plants are planted on top of the growbag and after a short rooting period, they start producing runners. Crowning joints of the runners are planted, while still attached to the mother plant, in empty holes between mother plants along the growbags. After rooting, these buds produce secondary runners that fall as curtains on both sides of the hung growbags. On early September, 100–150 daughter plants per meter-row, depending on the cultivar, are selected and retrieved from these secondary runners. During the propagation period, water is supplied at 4–6 mm day$^{-1}$ using 1.6 l h$^{-1}$ emitters, five emitters per 1 m dripline length. The irrigation water is of a high quality (desalinated water), fortified with a liquid composite NPK fertilizer Mor 4:2.5:6 (ICL, Israel) plus calcium and magnesium at 1 and 0.5%, respectively, and at a fertilizer concentration of 60–120 ppm N. Fertigation is operated four times a day, which can be raised up to 6 pulses a day under heat wave events.

### 3.2 Daughter plants

To provide stable temperature and moisture conditions for the root primordia, selected stolons carrying daughter plants are immersed in water and cleaned from auxiliary runners and sick leaves. Cuttings are then planted on moist Zohar-6 mixture (Tuff Agricultural Cooperative Society Ltd., Merom Golan, Israel) containing fertilizers. Rooting takes place in plastic trays of 77 conical cells, 2.5"-deep (50 ml). The trays are placed on tables under net tunnels that reduce ambient radiation by 50%. During the rooting process, which lasts about 21 days, both the irrigation frequency and the shading regime are reduced gradually. Low capacity (7.5 l h$^{-1}$) (CoolNet Foggers$^{TM}$ Netafim, Israel) provide the moisture required for a successful rooting process. The initial fogging intensity is gradually reduced from 48 pulses per day (2 min per pulse) during the first 3 days, to 6 pulses per day at the end of the rooting phase. This procedure is essential for adequate seedling hardening. Nutrient supply, for supporting a well-established root system and canopy, is carried out daily from the second week of the rooting phase applying a liquid composite NPK (6:6:6) Shefer fertilizer (ICL, Israel) at 60 ppm N.

### 3.3 Transplantation

After about 21 days, typically toward the end of September, the root system fully occupies the conical cells, the canopy comprises a few healthy leaves, and the plug plants are ready for transplantation into the final soilless growth system. However, in case of forecasted heat wave events, it is recommended to keep the seedlings under cool shaded nursery conditions for a longer time and delay the transplantation. The target growbags are carefully prepared: the condensed coir, which serves as the growth medium, is soaked in water to regain maximal volume, washed from excess salts, and recharged with fertilizer. Upon planting, plug plants with roots and growth medium are transferred from the trays and plugged into the new growbag sockets, positioned in a 45 degree angle to maximize canopy spread. This procedure of plug plants preparation is principal for successful transplantation as it minimizes stresses associated with former practices involved with plant uprooting and re-rooting, thus saving the precious time required for recuperation; using the plug plant method, the seedling continues growing with no interruption, with significantly better chances to enhance early production.

### 3.4 Floral induction

The concrete ability of well-established strawberry plants to produce considerably early fruit yield depends on the intensity of the floral induction of its crowns' primordia at the earliest stage of development possible. Most of the strawberry cultivars used in Israel for early fruiting are of an ISD origin, which potentially provides floral induction once leaves are exposed to shortening photoperiods of 13 h and below. Nevertheless, temperature is a dominant and not less important determinant of floral induction. High temperatures usually support vegetative vigor in the expense of initial reproductive processes, and *vice versa*. Low temperature, and particularly low night temperature, seem to have a crucial significance promoting the onset of flowering and, furthermore, on the subsequent floral induction of the auxiliary crowns, thus enabling a continuous bloom and fruit production during winter [12].

The response of strawberry cultivars to chilling conditions was extensively studied and significant progress has been made in elucidating the genetic factors and molecular mechanisms underlying its complex flowering responses (for a review see, [13]. However, most of the experiments in Israel that examined the effects of stable low temperature treatment were implemented in cold rooms and failed to produce consistent results. Presumably, the physiological impacts of the circadian rhythm of light and temperature, including seasonal effects on this regime, may be more important than the absolute temperatures alone [14–16]. Light must be adequately supplied during the chilling treatment to avoid carbon deficiency and etiolation. The circadian temperature rhythm is required to support sufficient carbon assimilation during the day under relatively high temperature, as well as its translocation and allocation during the lower night temperature. In Israel, where the autumn nights are cool while the days are still warm enough, both floral induction and carbon supply are favored. Consequently, arid locations at relatively high altitudes are advantageous for the production and establishment of daughter plants, since the cooler nights enhance floral induction and earlier fruit production.

Nevertheless, among the ISD cultivars, the relationships between floral induction and the temperature regime during the seedling establishment phase is yet obscure, due to two major reasons: (1) the substantially fluctuating temperature regime during September–October in Israel; and (2) the rapid rotation of commercial strawberry cultivars. Unequivocally, gaining a precise insight into a cultivar's response to temperature would require 2–3 years of investigation under phytotron conditions that provide various predetermined temperature and photoperiod combinations. In the absence of this kind of research, we conducted a meta-analysis of the early marketable

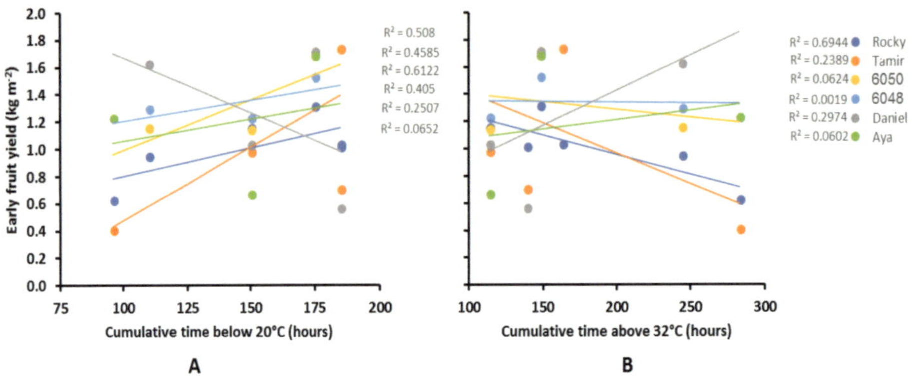


**Figure 3.**
*Meta-analyses of the relationships between early yield (during November–December) and the ambient cumulative time under cool (hours below 20°C; A) or warm (hours above 32°C; B) during seedlings establishment in August–September of the corresponding years (2013–2021, Ramat Negev).*

yield (November–December) of six commercial cultivars *vs*. the cumulative time under warm (hours above 32°C) or under cool (hours below 20°C) during plug plants establishment (August–September) in the corresponding years (**Figure 3**).

We assumed that greater cumulative cool hours would promote higher early yields, and that the accumulation of warm hours would delay fruit production. Although the expected response to cool temperature could be noticed, it substantially differed between cultivars (**Figure 3A**); only two cultivars (Tamir and 6050) exhibited a significant response, three cultivars responded positively but mildly, whereas one cultivar (Daniel) displayed a clear opposite response. The expected negative response of the early yields to cumulative high temperatures was visible only for Tamir and Rocky, but cultivars Aya1, 6048, and 6050 were unaffected, while 'Daniel', again, exhibited a clear positive response (**Figure 3B**). This rough analysis indicates an extensive variability in the sensitivity or responsiveness of the floral induction to temperature among the Israeli ISD cultivars. Further research is required in order to provide better matching of cultivars to specific climatic conditions.

## 4. Soilless culture technologies

### 4.1 Plant density

Out-of-season strawberry cultivation requires considerable investments in shelter from typical winter weather patterns and events, such as rain, hail and limiting low temperature. To compensate with these expenses, the crop productivity must be significantly boosted. A major means to achieve this goal is to substantially increase plant density.

In the traditional soil-cultured strawberry, plant density ranges from eight to nine plants $m^{-2}$. Within this range, plant density is determined by cultivar characteristics; however, the boundaries of this range are defined by aboveground inter-plant space considerations, such as light interception and mutual shading, and by the space necessary for technical operation (spray, harvest, etc.). In fact, in soil-grown strawberry, like in other cover crops, the canopy volume tends to be rather bi-dimensional, with the upper foliage layer receiving most of the light while shading the lower layers.

This canopy structure escalates the risks of plant diseases due to poor ventilation and proximity to the soil.

The transition of strawberry cultivation from soil to soilless cultures have significantly mitigated these constraints. Shifting strawberry cultivation to the three-dimensional space provides significantly higher plant density, enhanced light interception, and better aeration to larger canopy parts for longer time periods. As a result, crop productivity is expected to substantially rise, and the probability and severity of diseases - to decline.

Different approaches and technologies of soilless culture have been developed and examined in strawberries for decades, since the 1970s [17]. In this chapter, we describe the current perception and the recent design of strawberry soilless culture in southern Israel. It should be stated, however, that the technology is consistently 'under construction', as further enhancement is steadily sought.

### 4.2 Eye-level hung crop

The traditional cultivation of soil-grown strawberry is a labor-intensive crop, as fruit harvest requires daily-long hours of bending. Thus, lifting the strawberry culture from the ground closer to the workers' hand-reach brings significant ergonomic advantages easing harvest and other farming activities, which may appeal seasonal workers. One way to do so is placing the growth media containers on fixed shelves along the row, a method which requires an additional costly construction. An alternative solution, broadly accepted in Israel, is the eye-level hung system, in which the growth media containers are hung along the row using the existing greenhouse construction plus some accessory cables. This way, the system height is more versatile and can be adjusted once in a while to the average workers' height. Additionally, the ground below remains free or can be used for a secondary shade crop. Another advantage is the relatively flexible distance between rows; while a distance of 65 cm was found optimal for light interception, it can be temporarily and locally modified to facilitate ad-hoc workers' convenience. The hanging system should be designed to generate the uninterrupted slope (2.5%) required to guarantee a satiated water drainage. Therefore, the system height is about 1.9 m at the beginning of a 20 m long row, and ends at about 1.4 m aboveground.

### 4.3 System design

Designing the soilless culture system must obey several principal prerequisites, as follows: adequate rhizosphere space; minimal weight; rigid structure; efficient drainage; and, optimal plant positioning. In addition, the growth medium properties must be light (low self-weight when dry), chemically neutral and stable, well-aerated with adequate water retention and rapidly drained. These principles ensure optimum canopy and root development and function, provide crop uniformity, and prevent technical failures, all of which are essential for high and stable crop performance throughout the growing season.

The soilless system described here is comprised of two major elements: a rigid plastic support gutter (carrier), and a growbag—coir (compact coconut fibers) wrapped by a plastic sleeve. The gutter profile is designed in a wide U shape (11 × 5 cm), with a special mid-bottom drainage duct (**Figure 4A**). The upper edges of the vertical walls form a narrow wing; holes drilled every 1 m serve for hanging the gutter on cables attached to the greenhouse structure (**Figure 4B**). The gutters are supplied in 5 or 6 m long units that are attached (using special adaptors) in a slope

**Figure 4.**
*Descriptive pictures of the hung soilless strawberry culture. A: Gutter system profile; B: growbags and drip line.*

(~2.5%) to allow the continuous flow of the drainage water. The growbag packages, 1 m long each, are perforated in the bottom side to allow drainage, while two holes in the upper side, one at each longitude edge, allow the insertion of the drip irrigation pipeline through the package and above the coir throughout the row (**Figure 4A**). On delivery, the coir medium is dry, compressed, and a bit hydrophobic. For priming, the growbags are soaked in water until fully swelled, mount well above the gutter's walls, and the growth medium completely fills the growbag. As a result, the growbag profile gets a trapezoid shape. Eleven or 13 planting holes per meter are located in a zigzag manner on both sloping sides of the growbag (**Figure 4B**).

This plant positioning in the space generates a density of 170,000–200,000, compared to 60,000–80,000 plants $ha^{-1}$ in field-grown strawberries. Furthermore, the fruit quality is enhanced due to the improved environmental conditions surrounding the plant canopy and fruit, including better light interception, reduced humidity, and a significant reduction in saprophytic diseases such as botrytis and aspergillus.

### 4.4 Growth media

The suitability of various growth media for hung soilless strawberry culture was thoroughly evaluated during the late 1990s [17–19]. Among these were peat from Germany, perlite, coir, and coir mixed with polystyrene foam flakes (30%). As no significant differences in crop performance, yield, and fruit quality were observed between the growth media, the coir medium was preferred due to its lower cost and convenient packaging. In addition, during the recent years, coir quality has been greatly improved, containing more homogeneous fine fibers and less salts. It should be noted that while other growth media can be more easily reused for several growing seasons, the careful disinfection procedure required is costly and labor demanding. Undoubtedly, the recycling of the growth media, including the growbags or not, remains an important challenge from both environmental and economic aspects.

## 5. Climate

### 5.1 Solar radiation

In Israel, only 500 km from north to south, the cumulative solar irradiation reaching the atmosphere during a given day do not vary among locations, but it significantly fluctuates during the annual cycle. However, due to differences in overcast, the ground-reaching solar radiation substantially increases from north to south, as well as with the distance from the Mediterranean Sea. These differences are demonstrated in **Figure 5** by means of monthly cumulative solar radiation records of 19 years (1999–2017) from three research centers, located at: 1. the northwestern coastal Negev (Besor); 2. Negev Highlands (Ramat Negev); and, 3. Rift Valley, Arava (Paran). During the critical winter months (November–February), the amounts of ground-reaching solar radiation are 7–19% greater at Ramat Negev, and 15–40% greater at Paran, compared to the northern coastal region of Besor center (**Figure 5**). Apparently, these differences provide some advantage to southern regions. Yet, the temperature regime might restrict this advantage.

### 5.2 Temperature regimes

The parameter of mean daily temperature might be misleading, as it hides a lot of important information about the actual temperature regimes. For example, during most of the season (from July to April), the mean daily temperature at Ramat Negev and Besor is quite similar, with slightly lower winter values at Ramat Negev. From November to late February, Paran values also merge with those of the other two regions (**Figure 6**). However, the extremum temperature means reveal significant differences between the three regions. Unequivocally, mean RN minimum temperatures are the lowest throughout the season, while Besor is the warmest during winter,

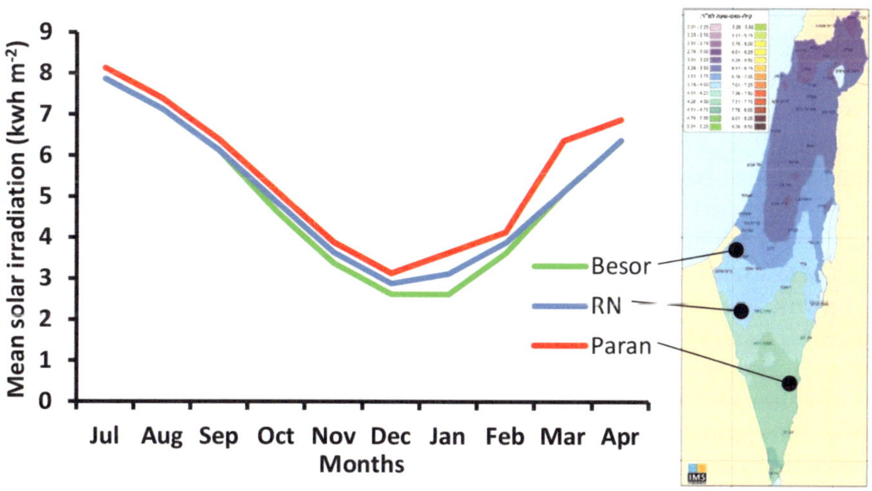


**Figure 5.**
*Mean daily ground-reaching solar irradiation during the strawberry season at Besor, Ramat Negev (RN), and Paran research sites that represent three different strawberry-growing regions in southern Israel. The map indicates differences in ground-reaching irradiance in February. Data and map were extracted from solar radiation Atlas [20], expressing means of 19 years (1999–2017).*

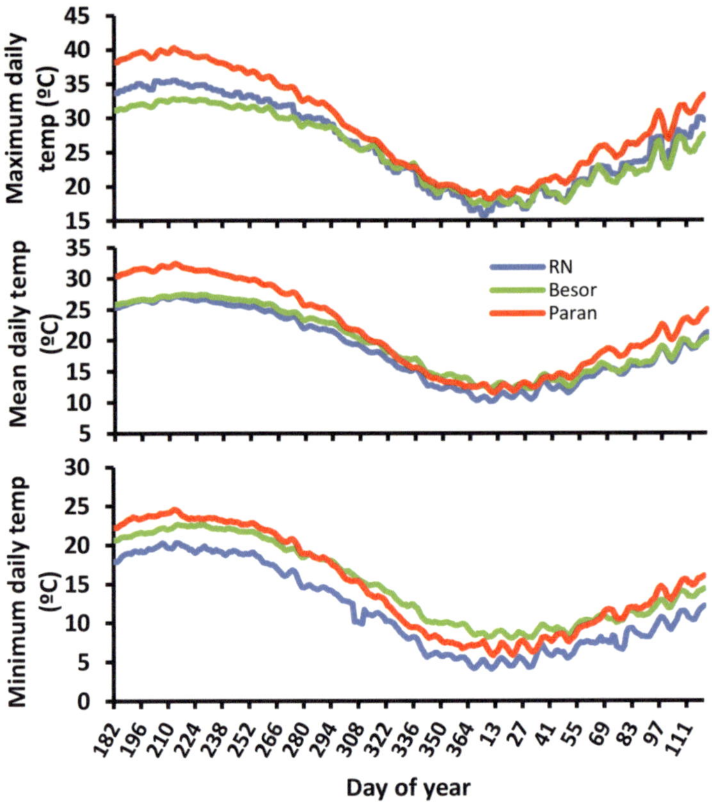


**Figure 6.**
*Temperature regime during the strawberry season, from launching propagation on 1 July (DOY 182) until the end of the harvest period on 30 April (DOY 120) at the northwest Negev (Besor), Ramat Negev (RN), and the Arava Valley (Paran). Data present means of maximum, mean, and minimum daily temperature over years 2010–2021, using official meteorological stations.*

and Paran is the warmest during fall and spring (**Figure 6**). Paran displays the highest mean maximum temperatures throughout the season, but from mid-November to late January, they are quite similar to those in the other two regions. During summer and early spring, Ramat Negev is warmer than Besor, but during December–January, daily maxima at Ramat Negev are the lowest (**Figure 6**).

The differences in solar radiation and temperature regimes introduce opportunities, as well as certain limitations for strawberry cultivation in each region. Relatively low temperature during summer and early fall enhance floral induction in most of the cultivars. Therefore, propagation and plug plants preparation are advantageous at the cooler Ramat Negev. Planting can take place first at Ramat Negev, as early as at early September, with the advantage of earlier harvest. However, low night temperatures often restrict plant development and fruiting during late December and January. The optimum planting time at the northern coastal Negev (Besor) would be late September, with the advantage of more stable mid-season (January–February) cropping due to the milder winter temperature. In the Arava Valley (Paran), the strawberry season is pretty much confined to late December to February; with optimal temperature and favorable solar radiation both plant and fruit development are enhanced, and fruit quality (Brix) is excellent. Nevertheless, the rapidly increasing temperature does not permit season extension there, whereas most of the yield of the two other regions is produced during March and April. The climate differences between the three regions may open opportunities for cooperation, exploiting the relative advantages of each region while covering possible gaps in production, altogether generating a complete and complementary strawberry marketing season from November until April. Thus, propagation and production of highly productive plug plants can take place at Ramat Negev for the three regions, whereas the production season can be allocated to Besor and Ramat Negev for the early- and late-season, and Besor and Arava Valley for the mid-season marketing.

## 6. Fertigation

Principally, soilless crop cultivation is highly susceptible to fluctuations in water availability due to the restricted rhizosphere volume and the risk of rapid salinization processes. In contrast, this technology easily permits collection and reuse of the drainage water. Irrigation water quantity varies from 3 to 6 mm day$^{-1}$, depending on the current daily weather and evapo-transpiration conditions. The electric conductivity (EC) of the drainage water is carefully monitored, as it is a good indicator for the maintenance of an optimum drainage coefficient of about 50%; a drainage EC increment above a certain predetermined threshold above the irrigation water EC indicates the occurrence of salinization and a need to raise the daily irrigation and the drainage rates, and *vice versa*. To avoid transient water shortage during the day, the daily amount is supplied through three to four pulses. Drip lines, with 1.6 l h$^{-1}$ emitters every 20 cm allocate the irrigation water along the growbags, and the coir medium provides the hydraulic conductivity required for uniform water distribution among plants.

Strawberries are highly sensitive to salinity. Yield depletion occurs already at EC of 1 dS m$^{-1}$, with a yield decrease rate of 33% per every further unit of EC increase [21, 22]. Chloride ions are the major salinity component causing damage [23], suggesting not to exceed irrigation water EC of 1.2 dS m$^{-1}$ and 1% Cl of the leaf dry matter. Therefore, when growing strawberries in warm arid climates, the highest water quality, as well as the lowest EC possible are recommended. Following a few failures with various mixtures of local brackish and fresh water, we exclusively use desalinated water at 0.5 dS m$^{-1}$. However, in recycling hydroponic systems using greater water rates, where the risk of salinization is reduced higher EC levels may be considered [24, 25].

The alkalinity of the desalinated water used for irrigation is relatively high (pH 7–8), above the optimum pH for strawberry. Alkaline growth media restrict the absorption of essential micronutrients (e.g., Fe, Mn, Zn, and Cu), resulting in vegetative and reproductive growth retardance and poor yield and quality [26]. Earlier attempts to resolve this problem included the use of high- $NH_4^+$ fertilizers; however, the combination of those with occasional high temperature was followed by gradual plant deterioration, which was related to a cascade of rapid sugar depletion and anoxia during $NH_4^+$ metabolism in the roots [27]. This was overcome using a micronutrient-fortified composite liquid fertilizer with sulfuric acid to adjust for the desired pH. In the recent years, we apply commercial liquid composite fertilizer comprised of 4:2.5:6+6 (N:P:K) with a N– $NO_3^-$ :N– $NH_4^+$ ratio of 9:1, fortified with 600, 300, 150, 22, and 16 mg kg$^{-1}$ of Fe, Mn, Zn, Cu, and Mo, respectively, plus Ca and Mg at 2% and 1%, respectively. Fertilizer concentration varies from 60 to 150 ppm of N, adjusted to plant size and crop stage. The fertilizer is applied through the irrigation system (fertigation) constantly in every irrigation pulse and provides a stable pH at the desired range of 6–7.

In order to save water and fertilizer and reduce environmental consequences of the drainage water, considerable attempts have been made to recycle the irrigation water. Nevertheless, obtaining this goal requires a careful Nano-filtration of the water to avoid the spread of diseases, as well as continuous monitoring and adjustment of water quality and nutrient concentration and composition. The technological challenge and the subsequent costs did not justify the benefits, so far. Therefore, the drainage water are monitored, collected, and utilized to irrigate other crops, nearby, thus increasing greater use efficiency.

## 7. Pollination

Flowers of commercial strawberry cultivars are hermaphrodite and self-fertile. However, self-pollination through gravity and wind is often partial since the pollen might drop on many, but not necessarily all of the pistils [28, 29]. The fertilized ovules, through auxin release, promote receptacle development and formation of fleshy tissue [30]. However, non-fertilized ovules fail to develop, resulting in fruit deformations that raise the proportion of non-marketable produce [31]. In addition, there is a relationship between the number of fertilized ovules (achenes) and berry weight [30, 32]. Insects, mostly honeybees (*Apis mellifera* L.) serve as complementary pollinators of strawberry [33–35]. Nonetheless, beehives in the recent years have been facing the severe worldwide problem of the colony collapse disorder [36], which has significantly decreased the beehives availability for pollination, and consequently brought about a dramatic rise of the rental prices [37]. Additionally, honey bees foraging is temperature-limited [38, 39] and hence, pollination performance sharply declines under high day temperature during fall, as well as under low winter temperatures. Indeed, greenhouse strawberry crops often suffer from inadequate flower fertilization, so alternative insect pollinators that replace or act together with honeybees seemed reasonable.

Bumble bees (*Bombus terrestris* L.) have several attributes that are beneficial for pollination [40, 41]. Therefore, bumble bees have been introduced and have successfully replaced honey bees as pollinators of greenhouse vegetable [41] and orchard crops [38, 42–44]. Today, *B. terrestris*' colonies are a "ready-made shelf product", easily

**Figure 7.**
*A bumble foraging on a strawberry flower. (Photograph by O. Guy.)*

suitable for marketing and transport to any given greenhouse or habitat [41]. The efficiency of bumble bees as a greenhouse-strawberry pollinator was found comparable to that of honey bees [45, 46]. In greenhouse-strawberries in Israel, *B. terrestris* displays considerable advantages compared to honeybees, particularly in respect of foraging under harsh weather conditions (**Figure** 7).

## 8. Plant protection

Alongside the expanding export to Europe during late 1990s, the approach to plant protection issues has drastically shifted from massive use in pesticides to biologically based integrated pest management. For this purpose, specific beneficial insects such as wasps and predator mites against strawberry pests make chemical use against pests almost unnecessary.

The mostly common pest in greenhouses-grown strawberry is the red spider mite (*Tetranychus urticae*) that feeds from chloroplasts in the leaves, leaving typical yellowish signs. At high infestation levels, leaves and whole plants are severely damaged and fruit production declines. Although red spider mite is the main pest throughout the growing season, its population substantially declines during cold periods. Therefore, the risk of red spider mite infestation of winter strawberries occurs mainly during crop establishment in October-November or during March-April, and is subject to weather fluctuations. To date, red spider mites are effectively treated using the predator mite *Phytoseiulus persimilis* (Bio-Bee, Israel). These predatory mites are applied at the beginning of the season, when it is still warm, and a balance is reached between the pest and the predator populations. As long as this balance is preserved, no pesticides are needed. However, once disrupted (e.g., in incidents of hot and dry weather), the balance must be re-established using a pesticide spray.

More recently, infestation by Castor thrips (*Scirtothrips dorsalis*) has been increasingly observed. This pest causes blackening of main leaf veins and petioles and to growth retardation. Such infestation may be acute mainly during autumn, when high temperatures and humidity stimulate pest population and, if not treated timely and appropriately using Abamectin or Spinosad, the damage to the young seedlings might be quite severe.

Strawberries are susceptible to a wide range of plant diseases. The hung growbags culture, which provides enhanced aeration and light interception for the canopy, reduces risks of canopy diseases related to high air humidity or long-lasting dew. Strawberry gray mold (*Botrytis*), a well-known disease under cold and moist conditions, hardly occurs on fruit hanging in the air. In a similar way, anthracnose (mainly *C. acutatum*), which often spread under warm and humid conditions when grown on plastic-covered soils, rarely strikes the hung strawberry culture. In contrast, powdery mildew (*S. macularis*), which harms greenhouse crops primarily in the fall, requires careful attention. Practically, susceptible cultivars, such as 'Tamir', are gradually withdrawn from the lists of cultivars recommended to farmers. Avoiding contaminated young plug plants is of primary importance. Under conditions of warm and dry days followed by cool and humid nights, powdery mildew can be efficiently prevented using sulfur vaporizers during the night. In case of disease outbreak, mild chemical sprays that are even permissible for use according to the organic standard are quite effective.

Soil-borne diseases are extremely rare in the hung growbag technology, particularly when propagation is carried out according the methods described above. Among soil diseases, *Macrophomina phaseolina* has recently become a threat endangering increasing areas of various crop species, including soil-grown strawberries [47, 48]. This pathogen might be introduced to a hung strawberry greenhouse only through contaminated plug plants, which must be carefully avoided. So far, no chemicals were found adequately effective against *Macrophomina* and hence, in case it does occur, an early removal of the infected growbags should easily solve the problem.

Nevertheless, the main and long-term strategy is breeding for disease-resistant cultivars. Alongside the approach of minimum use of chemicals and no plant growth regulators, this crop protection strategy supports high yields of high quality off-season strawberries.

## 9. Source-sink relations

The term source-sink relations points to the balance between supplying plant organs and receiving ones in respect to any kind of an essential resource, such as water, nutrients, carbohydrates, or other secondary metabolites. Strawberry plants are simple and small, lacking organs or tissues specialized in reserve accumulation, such as tuber, trunk or others. Therefore, source-sink relations in the strawberry plant are fragile; when demands exceed supply, a competition between sink organs would immediately occur, with crucial consequences on further plant development, including fruit yield and quality [49–51].

The case of out-of-season, early yielding strawberries, is particularly interesting in respect of the carbohydrate balance. Naturally, under their original temperate climate, young strawberry plants develop a considerable vegetative biomass during the early spring, which would adequately support the carbohydrate requirements of the emerging reproductive phase. Nevertheless, when strawberry seedlings are manipulated to early fruit bearing, the carbohydrate balance might be extremely brittle. While possibly adequate to successfully support initial reproductive development (flower bud differentiation, initiation, and even fruit growth and development), the foliar biomass existing at that stage might be too small to reinforce further plant development. The typically declining temperature and the descending light availability during autumn both limit carbon assimilation, and hence, further restrict plant growth. In extreme cases, crop development and production might be restrained for a long period. Under less extreme situations, the fruit yield might considerably fluctuate over time with the alternating vegetative and reproductive flushes, with negative consequences on the marketing.

The strawberry growing season of 2020–2021 at Ramat Negev provides an example of the interactions between the temperature regime and the current fruit yield, fruit size, and TSS during the season (**Figure 8**). Compared to the temperature means of the recent decade (**Figure 6**), November 2020 was considerably cold, particularly at night (**Figure 8A**). However, the rest of the winter (December–January) was consistently warmer than the average of the former decade, with average maximum and minimum temperatures above 20°C and 7°C, respectively (**Figures 6** and **8A**). In contrast, February and March 2021 were quite normal. The relatively cold November repressed fruit growth and ripening, giving rise to lower early yielding in the five cultivars examined. Significant differences between cultivars occurred during December, with 'Aya' and 'Daniel' displaying two peaks while the other three cultivars exhibited moderate to low yields. In January, 'Rocky' and '415' emerged to yield, while 'Aya', 'Daniel' and '6050' produced low fruit yield. From February and later on, all five cultivars showed increasing yield levels that followed the rising but fluctuating temperature (**Figure 8A** and **B**).

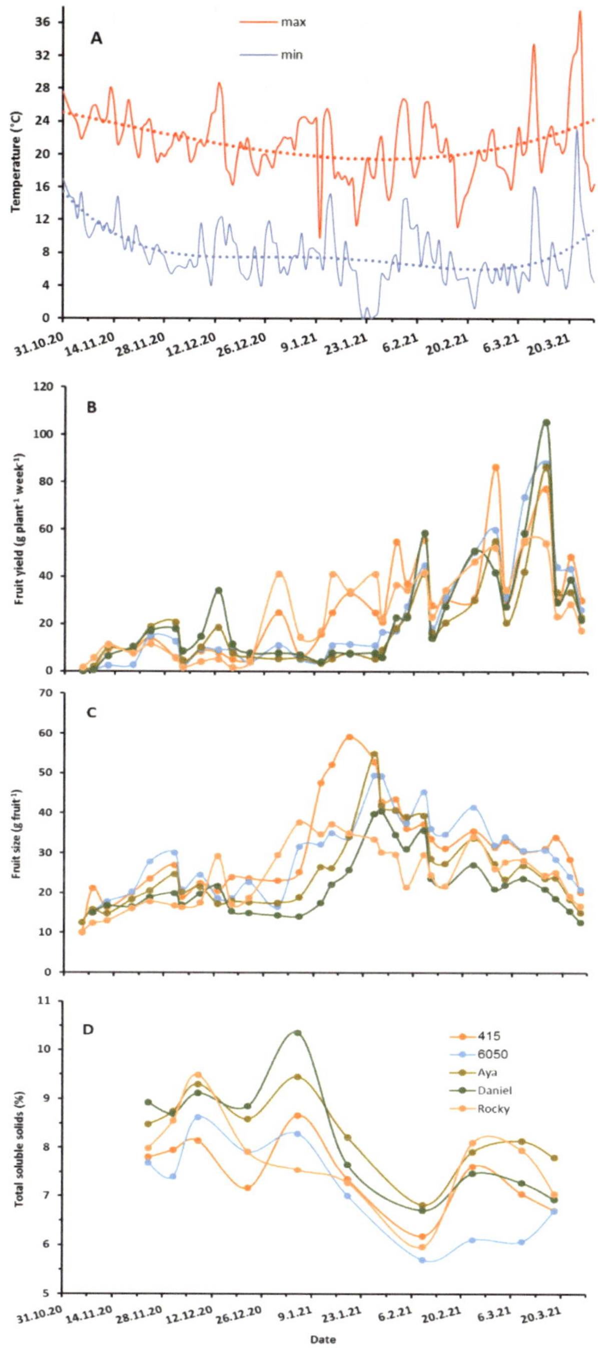


**Figure 8.**
*Yield parameters of five early-yielding strawberry cultivars during the 2020–2021 season at Ramat Negev, Israel. Daily maximum and minimum temperatures from 31 October 2020 until 27 March 2021 (A); weekly fruit yield (B); mean fruit size (C); the average concentration of total soluble solids in the fruit (D).*

Fruit size (**Figure 8C**) was relatively small until late December, ranging from 15 to 30 g per fruit. It steadily increased during January, and then gradually declined until the end of the season. Beyond being a varietal trait, fruit size negatively corresponds with the number of fruit per plant. Thus, fruit size is suppressed at the beginning of the fruiting season by the large number of fruit relative to the current plant canopy size. During January, fruit size peaked with the depletion in the number of fruit, however, it decreased again when plants' productivity was restored from February and later on.

Fruit TSS (**Figure 8d**) negatively reflected the changes in fruit yield (**Figure 8b**) during the season, but also expressed the current plant capacity. Thus, TSS was quite high until late December, as long as the fruit yield was low; it surged at the beginning of January as a result of a relatively long period of low yield *vs*. increased plant canopy and capacity, but then sharply declined in response to the rising fruit size and, later on, fruit yield. At the end of the season, TSS was partially restored with the warming weather and increasing plant capacity. Unfortunately, the frequency of TSS measurements did not allow a direct analysis of its relationships with other yield parameters (**Figure 8**).

There are several possible practices that might assist avoiding imbalanced source-sink relations. The most critical one is planting well-established seedlings that already have significantly developed root system, and a considerable number of healthy leaves. Suitable cultivars should display a good balance between extensive vigor and adequate tendency to reproductive development; excessive vigor or ample reproductive potential should be avoided. Heating [52] and additional artificial light [53] can be considered when necessary and economically feasible. In addition, a temporary application of fortified nitrogen nutrition, especially in the beginning of the season before night temperatures drop, was found to support the vegetative growth, thus increasing the leaves/fruit ratio (data not shown).

Recent research efforts have opened deep insight into mechanisms governing sink-source relations and carbohydrate translocation in strawberry [54, 55]. Nonetheless, the effects of leaf canopy manipulations may vary considerably between cultivars, production systems, and with varying time and duration of application [51]. Consequently, recurrent efforts will be needed to fit an appropriate solution to each combination of cultivar, climate, and growing system.

## 10. Concluding remarks

Substantial research and development efforts have been made in southern Israel during the recent two decades aiming at early winter production of strawberries. An inter-disciplinary augmentation of cultivar selection, refined propagation methods, highly-intensified cultivation technologies, enhanced pollination, and sheltered growing conditions has brought about to a consistently high proportion (15–25%) of very early production (November–December) of high-quality produce, followed by a considerable yield proportion during January and February, without any yield reduction during the main production season of March and April (**Figure 9**). However, in order to further increase the proportion of the early winter yield, and to mitigate undesired yield fluctuations during the winter, more research is required, focusing on the following challenges:

1. Thorough examination of candidate cultivars for early winter fruiting, including particular study of their floral induction response to temperature.

2. Enhancement of floral induction during plug plants establishment using controlled temperature at the growth medium and in the vicinity of the crowning zone, according to cultivars' chilling requirements.

**Figure 9.**
*Heavy yields in the mid of the fruiting season Hung strawberry culture. Photograph by Nir Dai.*

## Funding

This research was funded by KKL-JNF and the Israeli Ministry of Agriculture.

## Author details

Ofer Guy[1*], Nir Dai[2], Shabtai Cohen[1,3,4] and Amnon Bustan[1]

1 Ramat Negev Desert Agro-Research Center, Ramat Negev Enterprises Ltd., Halutza, Israel

2 Plant Sciences, Agricultural Research Organization, Volcani Center, Israel

3 South R&D, Habsor Farm, Israel

4 Central and Northern Arava R&D, Israel

*Address all correspondence to: ofer-guy@mop-rng.org.il

## References

[1] Dubey A. Origin, taxonomy and distribution. In: RY RMS, Dubey AK, Pandey V, editors. Strawberries – Production, Postharvest, Management and Protection. Boca Raton, FL, USA: CRC Press, Taylor & Francis Group; 2019. pp. 37-48

[2] Yamdagni RSA. Climatic requirements. In: RY RMS, Dubey AK, Pandey V, editors. Strawberries–Production, Postharvest, Management and Protection. Boca Raton, FL, USA: CRC Press, Taylor & Francis Group; 2019. pp. 161-167

[3] Konsin M, Voipio I, Palonen P. Influence of photoperiod and duration of short-day treatment on vegetative growth and flowering of strawberry (Fragaria 3 ananassa Duch.). The Journal of Horticultural Science and Biotechnology. 2001;**76**(1):77-82

[4] Sønsteby A, Heide O. Dormancy relations and flowering of the strawberry cultivars Korona and Elsanta as influenced by photoperiod and temperature. Scientia Horticulturae. 2006;**110**:57-67

[5] Rivero R, Remberg SF, Heide OM, Sønsteby A. Environmental regulation of dormancy, flowering and runnering in two genetically distant everbearing strawberry cultivars. Scientia Horticulturae. 2021;**290**:110515

[6] Serce S, Hancock J. The temperature and photoperiod regulation of flowering and runnering in the strawberries, *Fragaria chiloensis*, *F. virginiana*, and *F.* x *ananassa*. Scientia Horticulturae. 2005;**103**:167-177

[7] Stewart PJ, Folta KM. A review of photoperiodic flowering research in strawberry (*Fragaria* spp.). Critical Reviews in Plant Sciences. 2010;**29**(1):1-13

[8] Izsak E, Izhar S. The Israeli Strawberry Industry. Acta Hortic. 1989;**265**:711-714

[9] Izhar S, Izsak E. Inventor. Infra short-day strawberry types. US Patent patent 5444179. 1995

[10] Dai N, Tanami Z, Slotzky S, Lifshitz C, Koch-Dean M, Freeman S, et al. 'Tamir' and 'Herut', Two Novel Israeli Strawberry Cultivars Bred to Produce Early High Quality Fruits for Local and European Markets. Leuven, Belgium: International Society for Horticultural Science (ISHS); 2009

[11] Lifshitz C, David N, Shalit N, Slotzky S, Tanami Z, Elad Y, et al. Inheritance of powdery mildew resistance in strawberry lines from the Israeli germplasm collection. P. 70-74. In: Takeda, F., D.T. Handley, and E.B. Poling (ed.). *Proc. 2007 N. American Strawberry Symposium*. North American Strawberry Growers Association, Kemptville, ON Canada

[12] Hancock JF. Strawberries. In: Erez A, editor. Temperate Fruit Crops in Warm Climates. Dordrecht: Kluwer Acad. Publ; 2000. pp. 445-455

[13] Heide OM, Stavang JA, Sønsteby A. Physiology and genetics of flowering in cultivated and wild strawberries – A review. The Journal of Horticultural Science and Biotechnology. 2013;**88**(1):1-18

[14] Guttridge CG. *Fragaria* x *Ananassa*. In: Halevy A, editor. CRC Handbook of Flowering. III. ed. Boca Raton, FL, USA: CRC Press; 1985. pp. 16-33

[15] Heide OM. Photoperiod and temperature interactions in growth and

flowering of strawberry. Physiologia Plantarum. 1977;**40**(1):21-26

[16] Verheul MJ, Sønsteby A, Grimstad SO. Interactions of photoperiod, temperature, duration of short-day treatment and plant age on flowering of *Fragaria* x *ananassa* Duch. cv. Korona. Scientia Horticulturae. 2006;**107**(2):164-170

[17] Lantz W, Swartz H, Demchak K, Frick S. Alternative Cultural Systems for Growing Everbearers: An Overview. In: Mertz T, editor. Season-Long Strawberry Production with Everbearers for Northeastern Producers. TKM Marketing Inc; 2010. pp. 42-56

[18] Cantliffe DJ, Castellanos JZ, Paranjpe AV. Yield and quality of greenhouse-grown strawberries as affected by nitrogen level in coco coir and pine bark media. Proceedings of the Florida State Horticultural Society. 2007;**120**:157-161

[19] Cohen S, Ezekiel H, Shmuel D, Matan E, Dinar M, Aaron Y, et al. Development of agricultural practices for greenhouse-grown strawberries. Annual report. In Hebrew Available from: https://mopdarom.org.il/ 2000

[20] Muskatel H, Khalfon N. Solar Radiation Atlas of Israel. Israel Meteorological Service. 2020. (https://ims.gov.il/en/node/122) in Hebrew

[21] Ehlig C, Bernstein L. Salt tolerance of strawberries. Proceedings of the American Society for Horticultural Science. 1958;**72**:198-206

[22] Maas EV, Hoffman GJ. Crop salt tolerance current assessment. Journal of Irrigation and Drainage Engineering. 1977;**ASCE 103**(1221):115-134

[23] Barroso MM, Alvarez CE. Toxicity symptoms and tolerance of strawberry to salinity in the irrigation water. Scientia Horticulturae. 1997;**71**(3-4):177-188

[24] Bar-Tal A, Keinan M, Aloni B, Karni L, Oserovitz Y, Gantz S, et al. Relationships Between Blossom End Rot and Water Availability and Ca Fertilization in Bell Pepper Fruit Production. Leuven, Belgium: International Society for Horticultural Science (ISHS); 2001

[25] Bar-Yosef B, Markovich T, Levkovich I. Pepper Response to Leachate Recycling in a Greenhouse in Israel. Leuven, Belgium: International Society for Horticultural Science (ISHS); 2001

[26] Malekzadeh Shamsabad MR, Roosta HR, Esmaeilizadeh M. Responses of seven strawberry cultivars to alkalinity stress under soilless culture system. Journal of Plant Nutrition. 2021;**44**(2):166-180

[27] Ganmore-Neumann R, Kafkafi U. The effect of root temperature and nitrate/ammonium ratio on strawberry plants. II. Nitrogen uptake, mineral ions, and carboxylate concentrations. Agronomy Journal. 1985;**77**(6):835-840

[28] McGregor SE. Insect pollination of cultivated crop plants. 1976 Available from: http://www.beeculture.com/content/pollinationhandbook/strawberry-1.html

[29] Pion S, De Oliveira D, Paradis RO. Agents pollinisateurs et productivité du fraisier 'Redcoat'. Phytoprotection. 1980;**61**:72-78

[30] Nitsch JP. Growth and morphogenesis of the strawberry as related to auxin. American Journal of Botany. 1950;**37**(3):211-215

[31] Vincent C, De Oliveira, D, Bélanger, A. The management of insect pollinators and pests in Québec strawberry plantations. In: Bostonian NJ, Wilson LT,

Dennehy TJ. Monitoring and integrated management of arthropod pests of small fruit crops. pp. 177-192, Intercept; 1990

[32] Albano S, De Oliveira D, Salvado E, Quaresma M, Borges P, Mexia A. Avaliação da importância dos insectos polinizadores na melhoria da produção na cultura do morango. Actas da Associação Portuguesa de Horticultura. 2005:121-129

[33] Free JB. Insect Pollination of Crops. London: Academic Press; 1993

[34] Goodman RD, Oldroyd BP. Honeybee pollination of strawberries (*Fragaria* x *ananassa* Duchesne). Australian Journal of Experimental Agriculture. 1988;**28**(3):435-438

[35] Nye WP, Anderson JL. Insect Pollinators Frequenting Strawberry Blossoms and the Effect of Honey Bees on Yeild and Fruit Quality. Journal of the American Society for Horticultural Science. 1976;**99**(1):40-44

[36] vanEngelsdorp D, Traynor KS, Andree M, Lichtenberg EM, Chen Y, Saegerman C, et al. Colony collapse disorder (CCD) and bee age impact honey bee pathophysiology. PLoS One. 2017;**12**(7):e0179535

[37] Delaplane KS, Mayer DF. Crop pollination by bees CABI Publishing. Wallingford, UK. 2000

[38] Sapir G, Baras Z, Azmon G, Goldway M, Shafir S, Allouche A, et al. Synergistic effects between bumblebees and honey bees in apple orchards increase cross pollination, seed number and fruit size. Scientia Horticulturae. 2017;**219**:107-117

[39] Vicens N, Bosch J. Weather-dependent pollinator activity in an apple orchard, with special reference to *Osmia cornuta* and *Apis mellifera* (Hymenoptera: Megachilidae and Apidae). Environmental Entomology. 2000;**29**(3):413-420

[40] Bostanian NJ, Wilkaniec Z, Radajewska B. Solitary bee *Osmia rufa* L. (Apoidea, Megachilidae) as pollinator of strawberry cultivated in an unheated plastic tunnel. Acta Horticulturae. 1997;**439**:489-493

[41] Dafni A, Kevan P, Gross C, Goka K. *Bombus terrestris*, pollinator, invasive and pest: An assessment of problems associated with its widespread introductions for commercial purposes. Applied Entomology and Zoology. 2010;**45**:101-113

[42] Sapir G, Goldway M, Stern R. Supplementing bumblebees to 'Mauritius' lychee improves yield. Scientia Horticulturae. 2019;**251**:162-166

[43] Stern RA, Rozen A, Eshed R, Zviran T, Sisai I, Sherman A, et al. Bumblebees (Bombus terrestris) Improve 'Hass' Avocado (Persea americana) Pollination. Plants. 2021 Jul;**10**(7):1372

[44] Zisovich AH, Goldway M, Schneider D, Steinberg S, Stern E, Stern RA. Adding bumblebees (*Bombus terrestris* L., Hymenoptera: Apidae) to pear orchards increases seed number per fruit, fruit set, fruit size and yield. The Journal of Horticultural Science and Biotechnology. 2012;**87**(4):353-359

[45] Abrol DP, Gorka AK, Ansari MJ, Al-Ghamdi A, Al-Kahtani S. Impact of insect pollinators on yield and fruit quality of strawberry. Saudi Journal of Biological Sciences. 2019;**26**(3):524-530

[46] Zaitoun ST, AL-Ghzawi AA, Shannag HK, Rahman A, AL-Tawahaa RM. Comparative study on the pollination of strawberry by bumble

bees and honey bees under plastic house conditions in Jordan valley. Journal of Food, Agriculture and Environment. 2006;**4**(2):237-240

[47] Cohen R, Elkabetz M, Paris HS, Gur A, Dai N, Rabinovitz O, Freeman S. Occurrence of *Macrophomina phaseolina* in Israel: Challenges for Disease Management and Crop Germplasm Enhancement. Plant Disease. 2022 Jan 31;**106**(1):15-25

[48] Pickel B, Dai N, Maymon M, Elazar M, Tanami Z, Frenkel O, et al. Development of a reliable screening technique for determining tolerance to *Macrophomina phaseolina* in strawberry. European Journal of Plant Pathology. 2020;**157**(4):707-718

[49] Casierra-Posada F, Torres I, Blanke M. Fruit quality and yield in partially defoliated strawberry plants in the tropical highlands. Gesunde Pflanzen. 2013;**65**:107-112

[50] Hansen P. Source-sink Relations in Fruits IV. Fruit Number and Fruit Growth in Strawberries. Leuven, Belgium: International Society for Horticultural Science (ISHS); 1989

[51] Sønsteby A, Woznicki TL, Heide OM. Effects of runner removal and partial defoliation on the growth and yield performance of 'Favori' everbearing strawberry plants. Horticulturae. 2021;7(8)

[52] Kadir S, Sidhu G, Al-Khatib K. Strawberry (*Fragaria × ananassa* Duch.) Growth and productivity as affected by temperature. HortScience. 2006;**41**:1423-1430

[53] Yoneda A, Yasutake D, Hidaka K, Muztahidin NI, Miyoshi Y, Kitano M, et al. Effects of supplemental lighting during the period of rapid fruit development on the growth, yield, and energy use efficiency in strawberry plant production. International Agrophysics. 2020;**34**(2):233-239

[54] Hidaka K, Miyoshi Y, Ishii S, Suzui N, Yin YG, Kurita K, et al. Dynamic analysis of photosynthate translocation into strawberry fruits using non-invasive $^{11}C$-labeling supported with conventional destructive measurements using $^{13}C$-labeling. Frontiers in Plant Science. 2019 Jan 9;**9**:1946

[55] Liu H-T, Ji Y, Liu Y, Tian S-H, Gao Q-H, Zou X-H, et al. The sugar transporter system of strawberry: Genome-wide identification and expression correlation with fruit soluble sugar-related traits in a *Fragaria × ananassa* germplasm collection. Horticulture Research. 2020;7(1):132

Chapter 3

# Response of Strawberry (*Fragaria X ananassa* Duch.) Flowering and Yield to Photoperiod

*Parween Muhammad K. Rozbiany and Shler Mahmud Taha*


## Abstract

The study was carried out during the growing seasons 2016-2018, collage of Agricultural Engineering Science—Salahaddin University—Erbil. For studying the effect of photoperiod on flowering and yield of two cultivars of strawberry, Festival and Albion were covered with black clothes for (0, 2 and 4 hours). The experiment was laid out in randomized complete block design with three replications for each treatment. Ten plants per experimental unit were arranged randomly in 54 plots. The data were analyzed using (SAS) program. As a result, when propagating the strawberry, the number of runners produced by the adult plants is an important consideration. Parameters significantly increased at photoperiod for 4 h included: for Festival in the first season, number of flowers, plant-1, viability pollen grain% and fruit set% in the first and the second seasons for Festival, (fruit dry weight and dry weight%) for Albion increased in the first season, (fruit fresh weight, fruit size and fruit length) for Albion in second season, (number of fruits. plant-1 and fruit diameter) for Festival in first season, marketable fruits% for Albion in second season, Yield. plant-1 (g) and yield, hectare-1 (Kg) significantly increased at photoperiod 4 h for Festival (100.591) in second season and (96.633) in first season.


**Keywords:** photoperiod, *Fragaria X ananassa* Duch., flowering, yield, Albion, festival

## 1. Introduction

### 1.1 Effect of photoperiod on the yield of strawberry plant

Young plants grown at 10 and 12 h photoperiod during 21 days at day temperatures between 12 and 18°C achieved complete flower emergence, while no flowers emerged at a photoperiod of 16 h. More detailed knowledge is required on the reaction of strawberry plants to photoperiods between 12 and 16 h. Greenhouse production provides the opportunity to optimize both day and night temperatures for flower production. Among the subset, the day and night time temperature had the greatest effect. Fruits, resulting from the primary flowers, especially those to the coolest day time temperature, were always the largest and required the most days from on thesis to harvest [1]. However, an interaction of photoperiod and temperature in the flower initiation of this plant has been demonstrated in many studies [2]. Different threshold photoperiods and temperatures have been reported for different strawberry cultivars [3, 4].

Le Mière et al. [2] studied the influence of photoperiod and temperature on inflorescence and flower initiation through the autumn in the strawberry cv. Elsanta. The percentage of flower initiation or final flower number in the primary, secondary, or tertiary inflorescences is not impacted by photoperiod. The final flower number in the primary inflorescence is slightly affected by temperature. For flowering bud initiation, the lowest number required of photoinductive cycles varied between 7 and 24 [5]. Darnell [6] reviewed and found the significant effect of interaction between temperature and photoperiod on the production of strawberry plant. Sønsteby and Nes [3] showed maximum flowering in "Korona" and "Elsanta" at 15°C and 24 days with 8 h photoperiod. However, for successful greenhouse production, more detailed knowledge is required about the effects of photoperiod and temperature and their interactions on flower and inflorescence emergence [7].

In June bearing strawberry cultivars, flowering is induced by short photoperiod, which also reduces vegetative growth. Plants were subjected to different photoperiods (12, 13.5, or 15 h) to be successful in 12 and 13.5 h photoperiod and number of flowers and yield were increased by lengthening the treatment [4]. The critical photoperiod for flower induction in "Korona" is in the range of 12 and 15 h. The confusion in the literature about the floral groups of the grown strawberry is actually caused by the strong interaction of the photoperiod and temperature; at one temperature checked, a cultivar may behave day-neutral, but it works out that it requires a photoperiod at any other temperature. The critical duration of the day depends heavily on temperature; in all photoperiods, flowering is prevented at high temperatures above 24°C, short days promote flowering at moderate temperatures between 14 and 20°C, while cooler temperatures induce flowering 14°C independently of photoperiod [7].

Nishiyama and Kanahama [8] examined the photoperiod and temperature impact on flower bud initiation (FBI) of ever-bearing and day-neutral strawberries (*Fragaria x ananassa* Duch. cv. "Hecker" and cv. "Summer berry"). Consequently, flower bud initiation (FBI) was totally limited by the 16th week. After that, under 24 or 8 h photoperiods at 20/15°C or 30/25°C, these plants were grown. The plants recommenced flower bud initiation (FBI) in these environments, excluding under the 8 h photoperiod at 30/25°C. The results demonstrated that the flowering response of these cultivars is quantitative under low temperature and qualitative under high temperature. Verheul et al. [6] investigated on strawberry cv. Korona. No flowers emerged in plants exposed to photoperiods of 16, 20, or 24 h or to an SD treatment for 14 days. An SD treatment (10 or 12 h photoperiod) of 28 days resulted in highest numbers of inflorescences and flowers per plant, while an SD treatment of 21 days resulted in the highest numbers of flowers per inflorescence.

The effects of photoperiod (12, 13, 14, 15, or 16 h) day (6, 9, or 12°C) and the interactions between flower and inflorescence emergence were investigated by exposing 4-week-old runner plants of strawberry cultivars Korona and Elsanta during a period of 3 weeks. A daily photoperiod of 12 or 13 h resulted in the highest number of plants with emerged flowers. A photoperiod of 14 h or more strongly reduced this number, while no flowers emergence at a photoperiod of 16 h [9]. Strawberry cv. Hecker and cv. Summer berry plants were grown at 30/25°C under an 8 h photoperiod from June to September (for 16 weeks). As a result, flower bud initiation was completely inhibited by the 16th weeks. Then, these plants were grown under 8 or 24 h photoperiod at 30/25°C, the plants resumed FBI. These results indicate that the flowering response of these cultivars is qualitative at high temperature and quantitative at low temperature [10].

Sønsteby and Heide [11] studied the perpetual-flowering F1-hybrid "Elan" that is propagated by seed. The results demonstrated a marked quantitative LD response across range of temperature from 9 to 27°C. Seedlings were response to LD stimulus at an early stage germination. The critical day length from early flowering strongly enhanced by short-day condition in combination with high temperature. In controlled environment, [12] investigated the environmental control of flowering in the ever-bearing (perpetual-flowering) diploid strawberry *Fragaria vesca* ssp. *semperflorens* cultivars "Rügen" and "Baron Solemacher." At temperatures ranging between 9°C and 27°C, seed-propagated plants were exposed to 24 h LD and 10 h SD environments. There was a quantitative LD response of flowering. The formation of runner was occasionally monitored in short day at high temperature. The study found an evident interaction with temperature in both cases. Rising temperature causes an increase in the photoperiodic responses, in both cases.

Environmental control of flowering and runnering in three Constantin *Fragaria chilonesis* population with geographic origins in Alaska, Oregon, and Chile have been studied. All populations were short-day plants at intermediate temperature (15 and 21°C), while at low temperature 9°C, the Alaska and Chile populations were essentially day-neutral. At a day temperature of 18°C, flowering increased with increasing photoperiod (10, 16 or 20 h) had no effect on flower development at 18°C. [13] stated that flowering is substantially advanced and the number of leaves produced before flowering is reduced by night interruption (3 h light in the middle of a 14 h daily dark period). For SD strawberry cultivars in particular, high temperature (greater than 26°C) would suppress the effect of short photoperiod and retard flowering or reduce flower initiation ratio [14]. Strawberry LD plants can be further classified into four flowering scenarios, strong-day-neutral: cultivars flower at the same rate in a photoperiod from 12 to 24 h; intermediate day-neutral: cultivars have 100% flower under 12 h day length; weak day-neutral: cultivars have significant reduction in flower initiation when photoperiod is shorter that 12 h (Hamano *et al.,* 2015); and some of the DN cultivars can show facultative LD response under lower temperatures (appr. 17°C) [15]. Sønsteby and Heide [16] demonstrated the capability of fractional induction. Berry yield varied in parallel with flowering in the field and was always higher in plants raised under SD conditions. After autumn planting, all studied cultivars flowered most abundantly in plants raised in SD and intermediate temperatures. Flowering was earliest in "Nobel" and "Rumba."

### 1.2 Effect of photoperiod on the flowering of strawberry plant

Gast and Pollard [17] exhibited that leaf growth, mean temperature and inflorescence, and flower number increased in SD strawberry by applying row covers over plants from autumn to the beginning of bloom in spring. Fortuna showed a higher sensitivity of this genotype to light and the highest values of the TSS/TA ratio in exposed and nonexposed fruits [18]. The minimum temperature/maximum day length for flower bud initiation likely varies among cultivars [4]. Conversely, stolon or runner emergence generally occurs under days longer than 10 hand temperatures above 20°C. The decline in temperature and photoperiod with the decline in vegetative growth of some strawberry cultivars observed in the field was correlated for estimating this impact biometrically. Furthermore, no direct morphological observations can expose the effect of the growth potential change on vegetative growth since it happens throughout the decline of vegetative growth [18].

Kader [19] displays that yield is increased in the autumn by renovation immediately after harvest in short-day strawberry (postharvest defoliation) without a vernlazation period. On the other hand, the yield is reduced in the following year when the postharvest defoliation is declined in SD strawberry and discovered that highest yield is obtained when renovation occurred 14–28 days after last harvest in short-day strawberry. The development of flower bud is optimal at higher temperatures (19–27°C) [2]. The end dormancy strength differs with cultivar. Tehranifar and Battey [20] expressed that excessive chilling prevents and/or delays flower bud initiation. Therefore, yield is reduced and harvest is delayed in SD strawberry by excessive chilling. Inhibitory long-day process in the leaves regulates the flower bud initiation in short-day strawberry [21], as proven by the positive impact of postharvest defoliation in short-day cultivars on flower bud initiation [22] and the manipulation of phytochrome with FR and R light [23]. By the application of gibberellins, flowering inhibition can be mimicked. Le Mière et al. [2] indicated that there was no positive correlation between temperature and rate of progress to fruiting in "Elsanta." However, while the size of crown correlated positively to yield, it was not related to fruiting time. Nonetheless, the yield is declined through a decrease in canopy size by warmer temperatures. Mori [24] discovered that the numbers of achenes per fruit (for all flower positions) and temperature through the ovule/pistil flowering bud initiation period were inversely related. When temperatures were 16/11°C (day/night during FBD), the maximum number of achenes per fruit was found.

Nishiyama and Kanahama [8, 21] explained that several cultivars in DN strawberry react as qualitative long-day plants at high temperatures (>27°C), quantitative long-day plants at lower temperatures (10 to 25°C), and DN at temperatures below 10°C. However, high temperatures decrease the flowering of DN types. Mochizuki et al. [25] investigated a forcing system for production of strawberries in winter in Japan. The study exposed nursery plants, which are grown in pots or plug trays, to slight nitrogen deficiency for promoting FBI. A petiole $NO_3$-N sap test was utilized for testing plant nitrogen status. For preconditioning nursery plants and promoting FBI, the plants were subjected to low temperature and SD treatments. For preventing heavy dormancy and promoting continued FBI, plants are kept at temperatures above 5°C. Strawberries are grown under long-day and mild temperatures for growth and continued flowering and fruit production. Throughout the early part of an SD night-chilling treatment, the application of fertilizer delayed FBI.

Sufficient plant nitrogen status is vital for increasing growth and FBI and FBD in strawberry grown on field. Nevertheless, there was no impact of spraying urea in short-day strawberry, in the autumn, through the FBI period, on yield the following summer in field grown plants. Nonetheless, the efficiency of utilized fertilizer enhanced and plant crowns production increased when granular fertilizer nitrogen was used at renovation in perennial, SD strawberry systems [26]. Serçe and Hancock [27] actively studied the genetic basis for remountainy. There is no basis for continuing to classify these plants as "day neutral" because of the variable response of flower bud initiation in remontane strawberry to temperature. Experiments on producing fruit during the winter on the short-day cultivar "Korona" in Norway have been effective utilizing SD treatments for inducing FBI [28] However, productivity was highly reliant on the quality plant; preconditioning plants to short-day escalated the formation of branch crown and subsequent yield. The temperature threshold is differed by cultivar. In addition, several cultivars in Nordic area have no flowering bud initiation under LD at temperatures as low as 9°C. Little flowering bud initiation happens below

10°C and above 25°C, the ideal temperature for flowering bud initiation is between 15 and 18°C under SD condition [9, 28].

Sønsteby and Heide [10] declared that these plants respond more like long-day plants at all temperatures excluding those below 10°C. The FBI is either directly affected by soil temperature or indirectly through its impact on vegetative growth. In comparison to plants that were not defoliated, early renovation resulted in yield escalation by up to 41%. Flower bud initiation is declined or postponed by the crowns treatment of SD strawberry with red light (600–702 nm). The application of fertilizer in the spring when further FBI and FBD happen has not been successful at rising the harvest while decent nutrition throughout the autumn period of FBI is required in SD strawberry [29, 30]. The manipulation of the growing and fruiting season of strawberry has been effective by the utilization of row covers and tunnels. If the tunnel or row cover offers an appropriate temperature and day length that is suitable for fruiting and flower bud initiation, remnant strawberries can have their fruiting season advanced or extended. For instance, once temperatures get too high under tunnels in the United Kingdom and in the continental United States of America, flowering in day-neutral strawberries stops [31]. Flowering was advanced by 1 week and flower number was raised to double, once the fertilization of the plants started 7 days after the beginning of the SD period although the number of crowns was not affected by the treatment. Even though high nitrogen status throughout the SD inductive period surges the FBI, in 201 strawberries, FBI is inhibited when plants are at a high nitrogen status directly prior to the SD flower inductive period [32]. The amount and period of flowering were impacted by time of the application of nitrogen fertilizer when groups of short-day strawberry with a low fertility system were fertilized with additional nitrogen for a period of time relative to 28 days' SD floral induction period [13]. Antioxidant accumulation is a very important issue since these compounds have been extensively associated to antioxidant capacity and, therefore, to the healthy properties of strawberry fruits [33].

The effect of photoperiod (10, 12, 16, 20, or 24 h), day temperature (12, 15, 18, 24, or 39°C), the number of short days (14, 21, or 28) days plant age (4, 8, or 12 weeks) and their interactions on flower and inflorescence emergence were investigated in strawberry cv. Korona. No flower emerged in plant exposed to photoperiod of 16 h, 20, or 24 h or to short-day treatment for 14 days. All plants exposed to short days at daily photoperiod of 10 or 12 h for 21 days or longer emerged flower at treatments between 12 and 18°C [13]. Lately, it revealed that short-day cultivar "Honeoye" was insensitive to photoperiod at 14–20°C [34]. For stimulating long-day reactions in a naturally SD, day extension or night interruption can be utilized. The expression of reminting or repeat flowering is powerfully influenced by temperature and cultivar variation in rate of FBD. The DN cultivar "Tribute" was insensitive to photoperiod at 14–23°C, but required LD at higher temperatures (Bradford *et al.*, 2010), similar to what has been observed by others [11].

Sønsteby and Heide [35] studied the impact of temperature and photoperiod on flowering, growth, and fruit yield in red raspberry cultivars (*Rubus idaeus* L.) "Autumn Treasure," "Erika," and "Polka." The plants were grown in a controlled environment in various day-length conditions and temperature for 42 days. By raising temperature over the 15–25°C, "Erika" and "Polka" displayed an improved flowering, fruit maturation and yield, whereas photoperiod had no substantial impact on flowering and fruit yield in "Autumn Treasure," flowering advanced and increased fruit yield resulted in the LD conditions 20 h at all temperatures in "Erika" and "Polka." Via decreasing the number of nodes created prior to flowering in "Autumn Treasure," a realization improved the transition to flowering with no impact fruit yield. In general, the higher the light exposure, the higher the antioxidant content and capacity [36]. Also, it has been reported that light intensity upregulates flavonoid biosynthesis in strawberry [32, 37] leading to the accumulation of anthocyanin, flavones, and total phenolic [38]. To regulate flowering in strawberry, temperature and photoperiod are the most vital environmental factors and their impacts have been comprehensively investigated [10]. Cervantes et al. [39] declared that at temperatures higher than 27°C with a critical photoperiod of 15 hours, "Elan" is a qualitative long-day plant. It is also suggested that all recurrent flowering (RF) cultivars are qualitative long-day plants at high temperatures of 27°C, quantitative long-day plants at intermediate temperatures (between 10 and 27°C), and DN at temperatures below 10°C. For affecting fruiting season and time of flowering, several berry crops are treated by changing growth and environment in commercial production systems. Also, [40] established that light incidence affects strawberry fruit quality (flavor and antioxidant content) and that strawberry fruit response to light conditions (as measured by the plasticity index) is genotype-dependent. Any cultural methods that allow fruit to be exposed to light may result in enhanced fruit quality in strawberry cultivars such as Fortuna [41–45].

## 2. Result and discussion

Some blooming characteristics for Festival cv. rose dramatically at photoperiod 4 h, such as blossom number. Similar to (Hidaka et al., 2014), plant-1 (32.200) in the second season, viability pollen grain percentage (80.156%) in the first season, and fruit set percentage (86.120%) in the second season (**Figure 1**).

Some fruit parameters increased significantly at photoperiod 4 h, such as fresh weight of strawberry fruit (13.782 g) in the second season and dry weight of strawberry fruit (3.467 g) in the first season; dry weight percentage recorded the highest value (29.515%) for Albion cv. in the first season, while Festival cv. recorded the

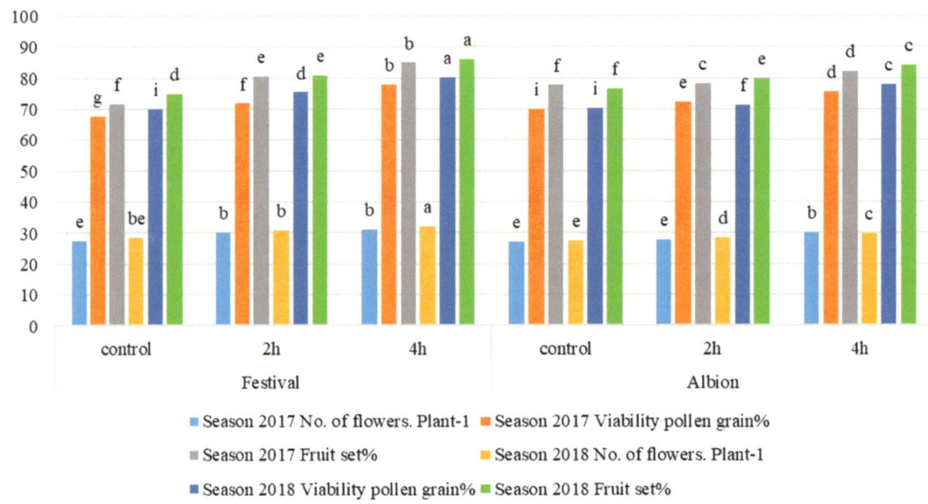


**Figure 1.**
*Effect of photoperiod on some flowering parameters of two cultivars in two seasons 2017–2018.*

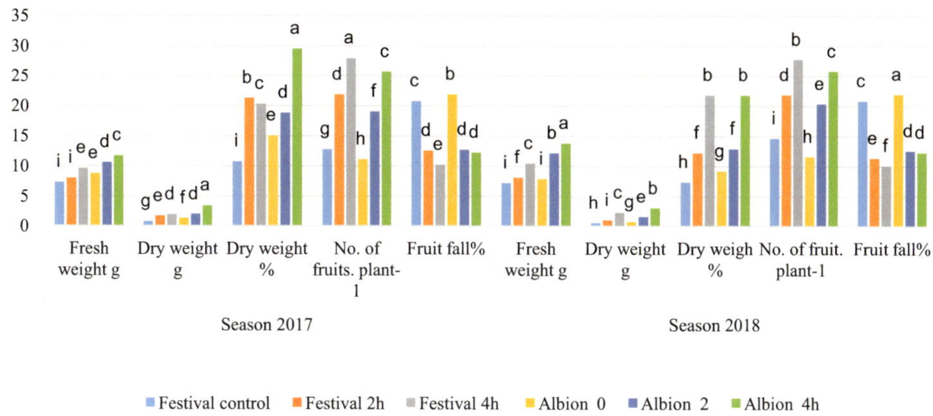


**Figure 2.**
*Effect of photoperiod on some fruit parameters of two cultivars in two seasons 2017–2018.*

lowest value (7.357%), Number of fruits. Plant-1 significantly increased (27.943) for Festival cv. at photoperiod 4 h in the first season and the lowest value recorded (11.111) for Albion cv. at control treatment in the first season. Finally, fruit fall percentage recorded the highest value (21.942%) for Albion cv. at control treatment in the second season and the lowest value (10.108%) for Festival cv. at photoperiod 4 h in (**Figure 2**).

Strawberry fruit diameter increased significantly to 3.984 cm for Festival cv. at photoperiod 4 h in the first season of planting, while fruit length increased dramatically to 4.688 cm in the second season for Albion cv. at photoperiod 4 h, and fruit size recorded the highest value (4.307 $cm^3$) in the second season for Albion cv. at photoperiod 4 h and the lowest value recorded (1.321 $cm^3$) for Festival cv. at (**Figure 3**).

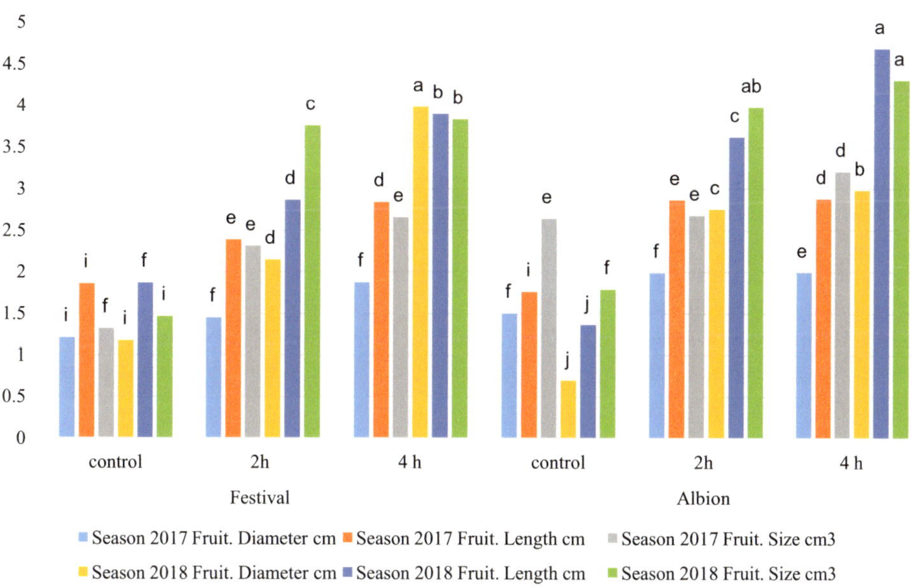


**Figure 3.**
*Effect of photoperiod on some fruit parameters of two cultivars in two seasons 2017–2018.*

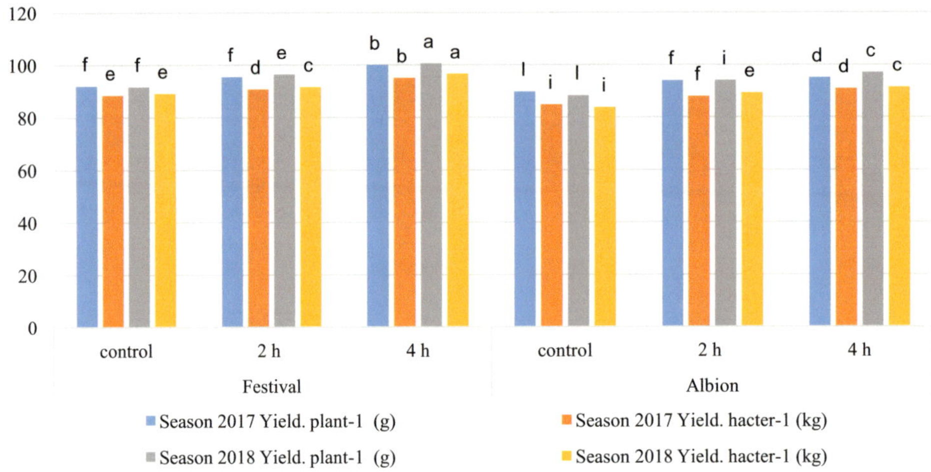


**Figure 4.**
*Effect of photoperiod on yield of two cultivars in two seasons 2017–2018.*

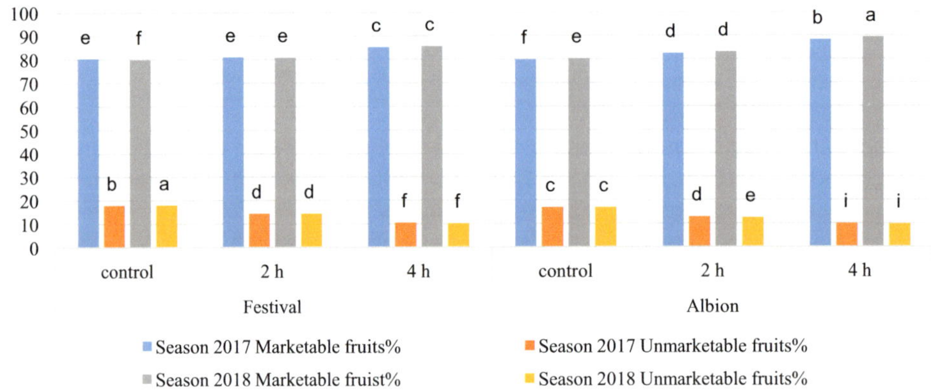


**Figure 5.**
*Effect of photoperiod on marketable fruits% and unmarketable fruits% of two cultivars in two seasons 2017–2018.*

Yield per plant-1 (g) and yield per hectare-1 (Kg) rose considerably at photoperiod 4 h for Festival cv. (100.591) in the second season and (96.633) in the first season, respectively (**Figure 4**).

Strawberry marketable fruits percentage increased significantly (89.344%) for Albion cv. in the second season at photoperiod 4 hours, while unmarketable fruits recorded the highest value (17.932%) for Festival cv. at control treatment in the first season and the lowest value (9.769%) for Albion cv. in the second season at photoperiod 4 hours (**Figure 5**).

## 3. Conclusions

The results confirm our previous finding [38] that flowering and fruiting in the strawberry plant are promoted and advanced by SD during early stages of plant growth and development. Photoperiod is one of the primary factors eliciting hormonal changes that stimulate flowering and fruit set. Covered plants for 4 h in day had a significant effect on most vegetative growth, flowering, and yield parameters. Generally, photoperiod via shorting day had significant effect on the studied parameters of the two (Festival and Albion) cultivars of strawberry plant. Festival cultivar responded to photoperiod more than Albion cultivar in fruit and yield parameter and that index to tolerance and good response of Festival cultivar.

## Acknowledgements

All praise and thanks are due to the Lord of universe Allah for giving me inspiration, patience, time, strength and knowledge to complete this study. I am deeply indebted to Prof. Dr. Shler Mahmood Taha for her excellent advice, support, motivation, encouragement and patience that extended to me in the success of this study. I would also like to thank IntechOpen for their kind assistant. Extracted search from A Dissertation "Effect of Photoperiod and Foliar of Calcium on Growth, Flowering and Yield Quality Storage of Two Strawberry Cultivars (Fragaria X ananassa -Duch)" By Parween Muhammad Kareem Rozbiany and Prof. Dr. Shler Mahmud Taha. Erbil-Iraq. May 2020.

## Author details

Parween Muhammad K. Rozbiany* and Shler Mahmud Taha
College of Agricultural Engineering Science Salahaddin University, Erbil, Iraq

*Address all correspondence to: parwwen.kareem@su.edu.krd

## References

[1] Kumakura H, Shishido Y. Effects of temperatures and leaf position on the translocation and distribution of 14C-photosynthatcs in fruiting strawberry cv. Morioka-16. Journal of Japan Society of Horticultural Science. 1994;**62**:833-838

[2] Le Mière P, Hadley P, Darby J, Battey NH. The effect of temperature and photoperiod on the rate of flower initiation and the onset of dormancy in the strawberry (*Fragaria x ananassa* Duch.). Journal of Horticultural Science. 1996;**71**:361-371. (2)

[3] Sønsteby A, NES A. Short days and temperature effects on growth and flowering in strawberry (*Fragaria*× *ananassa* Duch.). The Journal of Horticultural Science and Biotechnology. 1998;**73**:730-736

[4] Konsin M, Voipio I, Palonen P. Influence of photo-period and duration of short-day treatment on vegetative growth and flowering of strawberry (*Fragaria* × *ananassa* Duch.). The Journal of Horticultural Science and Biotechnology. 2001;**76**:77-82

[5] Hancock JF. Strawberries. New York: CAB International; 1999

[6] Darnell RL. The Strawberry: A Book for Growers. Gainesville, FL: Dr. Norman F. Childers Publications; 2003. pp. 3-10

[7] Verheul MJ, Sonsteby A, Grimstad SO. Interactions of photoperiod, temperature, duration of short-day treatment and plant age on flowering on *Fragaria* × *ananassa* Duch. cv. Korona. Scientia Horticulture. 2006;**107**:164-170

[8] Konsin M, Voipio I, Palonen P. Influence of photo-period and duration of short-day treatment on vegetative growth and flowering of strawberry (*Fragaria* × *ananassa* Duch.). The Journal of Horticultural Science and Biotechnology. 2001;**76**:77-82

[9] Manakasem Y, Goodwin PB. Responses of day neutral and June bearing strawberries to temperature and day length. The Journal of Horticultural Science and Biotechnology. 2001;**76**:629-635

[10] Nishiyama M, Kanahama K. Effects of temperature and photoperiod on flower bud initiation of day-neutral and ever bearing strawberries. Acta Horticulturae. 2002;**567**:253-255

[11] Verheul MJ, Sønsteby A, Grimstad SO. Influence of day and night temperature on flowering of *Fragaria x ananassa* Duch. cvs. Korona and Elsanta. Science Horticulture. 2007;**112**:200-206

[12] Sønsteby A, Heide OM. Long-day control of flowering in ever bearing strawberries. The Journal of Horticultural Science and Biotechnology. 2007a;**82**:875-884

[13] Sønsteby A, Heide OM. Quantitative long-day flowering response in the perpetual-flowering F1 strawberry cultivar Elan. The Journal of Horticultural Science and Biotechnology. 2007b;**82**:266-274

[14] Sønsteby A, Heide OM. Temperature responses, flowering and fruit yield of the June-bearing strawberry cultivars Florence, Frida and Korona. Scientia Horticulturae. 2008;**119**:49-54

[15] Sønsteby A, Opstad N, Myrheim U, Heide OM. Interaction of short day and timing of nitrogen fertilization

on growth and flowering of 'Korona' strawberry (*Fragaria× ananassa* Duch.). Scientia Horticulturae. 2009;**123**:204-209

[16] Sønsteby A, Solhaug KA, Heide OM. Functional growth analysis of 'Sonata' strawberry plants grown under controlled temperature and day length conditions. Scientia Horticulturae. 2016;**211**:26-33

[17] Garcia KP. Physiology of flowering and diurnal net photosynthetic response in American strawberry cultivars under controlled environment. Journal of Agricultural and Food Chemistry. 2016;**43**:1489-1492

[18] Sønsteby A, Heide OM. Flowering performance and yield of established and recent strawberry cultivars (*Fragaria × ananassa*) as affected by raising temperature and photoperiod. The Journal of Horticultural Science and Biotechnology. 2017;**25**:367-375. (18)

[19] Gast KLB, Pollard JE. Row covers enhance reproductive and vegetative yield components in strawberries. HortScience. 1991;**26**:1467-1469

[20] Robert F, Risser G, Petel G. Photoperiod and temperature effect on growth of strawberry plant (*Fragaria × ananassa* Duch.): Development of a morphological test to assess the dormancy induction. Scientia Horticulturae. 1999;**82**:217-226

[21] Kader AA. Quality and its maintenance in relation to the postharvest physiology of strawberry. In: Quality and Its Maintenance in Relation to the Post-harvest Physiology of Strawberry. Portland: Timber Press; 1991;**21**: 145-152

[22] Tehranifar A, Battey NH. Comparison of the effects of GA3 and chilling on vegetative vigor and fruit set in strawberry. Acta Horticulturae. 1997;**409**:629-630

[23] Zhang L, Strik BC, Martin LW. Date of renovation affects fall and summer fruiting of June-bearing strawberry. Advanced Strawberry Research. 1992; **11**:47-50

[24] Guttridge CG. *Fragaria ananassa*. In: Halevy AH, editor. Handbook of Flowering. Boca Raton , FL: CRC Press; 1985;**3**:16-33

[25] Vince-Prue D, Guttridge CG. Floral initiation in strawberry: Spectral evidence for the regulation of flowering by long-day inhibition. Planta. 1973;**110**:165-172

[26] Le Mière P, Hadley P, Darby J, Battey NH. The effect of temperature and photoperiod on the rate of flower initiation and the onset of dormancy in the strawberry (*Fragaria x ananassa* Duch.). Journal of Horticultural Science. 1996;**71**:361-371

[27] Mori T. Effect of temperature during flower bud formation on Achene number and fresh weight of strawberries. Journal of the Japanese Society for Horticultural Science. 1998;**9**:396-399

[28] Nishiyama, M. & Kanahama, K. (2000). Effects of temperature and photoperiod on flower bud initiation of day-neutral and ever bearing strawberries. IV International Strawberry Symposium 567.

[29] Mochizuki T, Yoshida Y, Yanagi T, Okimura M, Yamasaki A, Takahashi H. Forcing culture of strawberry in Japan – production technology and cultivars. Acta Horticulturae. 2009;**842**:107-109

[30] Strik B, Buller G, Righetti T. Influence of rate, timing and method of nitrogen fertilizer application on uptake

and use of fertilizer nitrogen, growth, and yield of June-bearing strawberry. Journal of American Society of Horticultural Science. 2004;**129**:165-174

[31] Serçe S, Hancock JF. Inheritance of day-neutrality in octoploid species of Fragaria. Journal of American Society of Horticultural Science. 2005;**130**:580-584

[32] Sønsteby A, Heide OM. Environmental control of flowering and runnering in three contrasting populations of *Fragaria chiloensis*. The Journal of Horticultural Science and Biotechnology. 2009;**84**:267-272

[33] Sønsteby A, Heide OM. Dormancy relations and flowering of the strawberry cultivars Korona and Elsanta as influenced by photoperiod and temperature. Scientia Horticulturae. 2006;**110**:57-67

[34] Verheul MJ, Sønsteby A, Grimstad SO. Influence of day and night temperature on flowering of *Fragaria x ananassa* Duch. cvs. Korona and Elsanta. Science Horticulture. 2007;**112**:200-206

[35] Lieten P. The effect of nutrition prior to and during flower differentiation on phyllody and plant performance of short day strawberry Elsanta. Acta Horticulturae. 2002;**567**:345-348

[36] Opstad N, Sønsteby A. Flowering and fruit development in in a field experiment with two fertilizer strategies. Acta Agriculture Scandinavian Section B. 2008;**58**:297-304

[37] Dale A, Luby JJ, Hancock JF. Breeding day neutral strawberries for Northern North America. Acta Horticulturae. 2002;**567**:133-136

[38] Atmani D, Chaher N, Atmani D, Berbouc M, Debbache N, Boudaoud H. Flavonoids in human health: From to biological activity. Current Nutrition & Food Science. 2009;**5**:225-237

[39] Bradford E, HAncock, J.F. & Warner, R.M. Interactions of temperature and photoperiod determine expression of repeat flowering in strawberry. Journal of the American Society for Horticultural Science. 2010;**135**:102-107

[40] Sønsteby A, Heide OM. Effects of photoperiod and temperature on growth, flowering and fruit yield in annual-fruiting red raspberry cultivars (*R. idaeus* L.). European Journal of Horticultural Science. 2012;**77**:97-108

[41] Li P, Ma F, Cheng L. Primary and secondary metabolism in the sun exposed peel and the shaded peel of apple fruit. Physiologia Plantarum. 2013;**148**:9-24

[42] Kadomura-Ishikawa Y, Miyawaki K, Takahashi A, Masuda T, Noji S. Light and abscisic acid independently regulated FaMYB10 in *Fragaria x ananassa* fruit. Planta. 2015a;**241**:953-965

[43] Feng F, Li M, Ma F, Cheng L. Phenylpropanoid metabolites and expression of key genes involved in anthocyanin biosynthesis in the shaded peel of apple fruit in response to sun exposure. Plant Physiology and Biochemistry. 2013;**69**:54-61

[44] Heide OM, Stavang JA, Sønsteby A. Physiology and genetics of flowering in cultivated and wild strawberries–a review. The Journal of Horticultural Science and Biotechnology. 2013;**88**:1-18

[45] Cervantes L, Ariza MT, Gómez-Mora JA, Miranda L, Medina JJ, Soria C, et al. Light exposure affects fruit quality in different strawberry cultivars under field conditions. Scientia Horticulturae. 2019;**252**:291-297

Section 2

# *In Vitro* Propagation of Strawberries

Chapter 4

# *In Vitro* Propagation and Conservation of *Fragaria* Species

*Sandhya Gupta*


## Abstract

The genus *Fragaria* L. belongs to the family Rosaceae. The basic chromosome number is 7 (x = 7). Around 20 species of diploid, tetraploid, hexaploid and octoploid are found in the genus. The species of *Fragaria* are commonly known as strawberries. The genus is found in the temperate regions of the northern hemisphere as well as South America. The most extensively distributed species in the genus is *F. vesca* L. and the commonly cultivated strawberry is *F.* x *ananassa* Duch. While strawberries are native to temperate climates, some varieties can be grown in subtropical climates. *Fragaria* is a vegetatively propagated crop. The easiest and most direct method for conservation of the strawberry plants is in a field gene bank. Their germplasm remains at risk of loss due to biotic and abiotic factors including climate change. Besides, this approach does not result in the distribution of healthy, virus-free plants. *In vitro* techniques are in place to propagate and conserve *Fragaria* germplasm. *In vitro* storage may be done in cold conditions, or liquid nitrogen as meristem or shoot tip. In this review chapter, tissue culture propagation technique, various aspects and strategies for conservation of *Fragaria* species will be discussed to present a holistic view of *ex situ* conservation of *Fragaria* genetic resources.


**Keywords:** biotechnology, cryopreservation, ex situ conservation, genetic stability, slow growth, tissue culture

## 1. Introduction

The genus *Fragaria* L. belongs to the family Rosaceae. The species of *Fragaria* are commonly known as strawberries. The fruit is technically known as accessory fruit (Achne) because the fleshy part is not derived from ovaries but from receptacles. The basic chromosome number is 7 (x = 7). There are around 20 species found in the genus of diploid, tetraploid, hexaploid and octoploid in nature. *Fragaria* is found in the temperate regions of the northern hemisphere as well as South America. Natural hybridization between *F. chiloensis* (octaploid) with *F. virginiana* (octaploid) gave rise to the present-day strawberry cultivars, *F.* × *ananassa* Duch. [1]. A large number of commercial varieties evolved later. *Fragaria* has *Duchesnea* and *Potentilla* as close relatives [2, 3]. Most extensively distributed species in the genus *F. vesca* L., is native to northern Eurasia, North and South America [4]. The commonly cultivated strawberry, *F.* x *ananassa* Duch., is grown in most of the arable regions of the world [5].

While strawberries are native to temperate climates, some varieties can be grown in subtropical climates. Many countries in Asia, North America, Europe and Africa produce strawberries. In 2018, the highest producer of strawberries was China (2.7 million tons) followed by US (1.2 million tons) and Mexico (0.59 million tons) [6]. Generally, strawberry is used for table and desert purposes; however various value-added products are also prepared like strawberry jam, jelly, candy and canned strawberry. A holistic approach is required to conserve such a valuable economically important species. Tissue culture technology has significantly contributed towards the propagation and conservation of *Fragaria* germplasm. The sections below will be useful for breeders, researchers, farmers, farm managers to have an overview of *Fragaria* genetic resource and its *ex situ* conservation.

### 1.1 Genetic resources of *Fragaria*

Genus *Fragaria* includes 20 species (**Table 1**) distributed in the Northern temperate and Holarctic zones [7, 8]. In *Fragaria*, four specific fertility classes are primarily associated with their ploidy level or chromosome number. Most widely distributed species *F. vesca* is diploid in nature (2n =14) and the most cultivated

| Species | Ploidy | Distribution |
|---|---|---|
| *Fragaria vesca* L. | 2x | Worldwide |
| *F. viridis* Duch. | 2x | Europe and Asia |
| *F. nilgerrensis* Schlect. | 2x | Southeastern Asia |
| *F. daltoniana* J. Gay | 2x | Himalayas |
| *F. nubicola* Lindl. | 2x | Himalayas |
| *F. iinumae* Makino | 2x | Japan |
| *F. yesoensis* Hara | 2x | Japan |
| *F. mandshurica* Staudt | 2x | North China |
| *F. nipponica* Makino | 2x | Japan |
| *F. gracilisa* A. Los | 2x | North China |
| *F. pentaphylla* Losinsk | 2x | North China |
| *F. corymbosa* Losinsk | 2x | North China |
| *F. orientalis* Losinsk | 4x | Russian Far East/ China |
| *F. moupinensis* (French.) Card | 4x | North China |
| *F. ×bringhurstii* Staudt | 5x | California |
| *F. moschata* Duch. | 6x | Euro-Siberia |
| *F. chiloensis* (L.) Miller | 8x | Western N. America and Chile |
| *F. virginiana* Miller | 8x | North America |
| *F. iturupensis* Staud | 8x | Iturup Island |
| *F. × ananassa* Duchesne ex Lamarck | 8x | Worldwide |

*Source: [3].*

**Table 1.**
*Species of Fragaria, ploidy and global distribution.*

species is *F.* × *ananassa* (2n = 56) is octoploid in nature. Diploid strawberries can be differentiated by their foliage, flower and fruits, inflorescence structure and plant habit. The *F. vesca* genome size was estimated at 197 Mb [9] and 206 Mb [10] and of *F. ananassa* 698 Mb [11]. Out of 20 species listed in **Table 1**, three species of *Fragaria* occur wild in the Himalayas in India [12–14]. These are: *1. F. nilgerrensis* Schecht. (Nilgiri strawberry): Habitat: A creeping herb found in Nilgiris and higher hills of north-eastern India. Uses: The fruit is pinkish, sub-acidic and juicy; *2. F. nubicola* Lindl. (Alpine Strawberry): Habitat: A herb distributed in the temperate Himalayas. Uses: Fruits are edible; and *3. F. vesca* L. (perpetual strawberry): Habitat: A herb found in the higher altitudes of temperate Himalayas. Uses: The red delicious fruits are edible.

## 2. *In vitro* propagation

*Fragaria* species are propagated primarily by stolons, also called runners, and at a lesser extent by seeds [15, 16]. From a Horticulture point of view, micropropagation has been practised for more than 50 years. Micropropagated strawberry plant has been introduced to prevent most of the plant and soil transmissible diseases. Over the years various protocols have been established for *in vitro* propagation of strawberry. Adventitious shoot regeneration in strawberry has been widely done using different explants such as a leaf, leaf disk, sepals, petiole and root mainly for transformation studies [17]. The *in vitro* culture has been successful in the mass propagation of true-to-type strawberry plants. A protocol that enabled strawberry micropropagation in one step, i.e., shoots multiplication and rooting in the same culture medium, emerged as a better choice for micropropagation of strawberries than shoot proliferation, (on a cytokinin supplemented medium) with subsequent rooting of shoots. In 'Bounty' strawberry, zeatin at very low levels (1–2 μM) produced 2–3 shoots per explant, averaging 88% rooting incidence on a single culture medium [18]. An efficient micropropagation protocol is a prerequisite for an *in vitro* conservation programme.

## 3. *Ex situ* conservation strategies

*Ex situ* conservation strategies ensure the conservation of plant genetic resources outside their natural habitat. The genetic material is conserved either in field, seed, *in vitro* and cryo genebank. The explants for *ex situ* conservation include seed, *in vitro* cultures, shoot tips, dormant buds, pollen, DNA, etc. Germplasm of *Fragaria* is always in demand by breeders and researchers who need germplasm with good quality traits for crop improvement purposes. Thus, conservation of available germplasm is of utmost important. As strawberries are vegetatively propagated, most common method of conservation is in the field genebank, orchards, glass house and net houses as live plants. Seeds are cross-pollinated and heterozygous thus not encouraged for genotype conservation. Germplasm conservation by conventional methods has several limitations, e.g., high inputs of cost and labour (as in field genabank), seed dormancy, seed-borne diseases, etc. The biotechnological tools, like tissue culture and cryopreservation (in liquid nitrogen), help to overcome these problems [19, 20]. Thus, strawberry germplasm is being conserved *ex situ* in the field-, *in vitro*- and cryo-genebanks in many countries [2].

### 3.1 Seed conservation

Seed conservation of any species is based on the seed storage behavior of seeds. Orthodox seeds are conserved in the seed genebank while recalcitrant and intermediate seeds are stored by other methods. Seeds of *Fragaria* exhibit orthodox seed storage behavior and are being conserved in the genebank (**Table 2**) [21]. *Fragaria* seeds are heterozygous and can be stored in the seed genebank for gene pool conservation.

### 3.2 Field genebank

Conventionally *Fragaria* species are maintained in field genebanks as live plants. For example, in the Field gene bank, Bhowali, India, 80 accessions of *Fragaria* spp. are being maintained. At NCGR, Corvallis, 1500 accessions of *Fragaria* are maintained under screen house, and other global genebanks. Conservation in field genebanks comes with limitations like: large land requirement; establishment and maintenance expenses; risk of loss due to disease and insect attacks, and loss due to natural disasters and climate change impacts.

### 3.3 *In vitro* conservation

*In vitro* conservation refers to maintaining germplasm on a defined nutrient medium under controlled environmental conditions. Three major *in vitro* conservation strategies are in practice. Conservation of germplasm at: (1) normal growth (2) slow growth and (3) cryopreservation. Slow growth techniques are for short- to medium-term-conservation storage of clonal plant material to be stored under *in vitro* conditions with extended shelf life. Slow growth methods aim for reducing the growth of *in vitro* shoots thus prolonging the subculture interval without causing any adverse effect on the plant tissue. Cryopreservation techniques are used for the long-term conservation of plant material. *In vitro* conservation strategies are discussed below:

#### *3.3.1 Normal growth*

Under this method, germplasm cultures are maintained under normal growing conditions by frequent subculturing at regular intervals. The normal growth method is advantageous as the cultures are available for immediate multiplication and distribution and it avoids the requirement of low-temperature facility (thus economical for tropical countries) or the application of stresses. *In vitro* maintenance of germplasm under normal growing conditions is the best method if the subculture interval may be extended up to a year or more [27]. About 1900 accessions of various horticultural crops including *Fragaria* are being conserved *in vitro* under normal growing conditions at National Genebank at ICAR-NBPGR, New Delhi, India. There is *in vitro* back-up of *Fragaria* germplasm conserved in the field genebank at Bhowali. Runners and suckers were collected from field genebank, Bhowali and established *in vitro*. Vegetative explants are taken from the field and established in MS media [28]. Shoot were multiplied on MS media supplemented with BAP (1 mg/l) IAA (1 mg/l) and $GA_3$ (0.1 mg/l). Thirty-five accessions of *F. vesca* are conserved in the *In vitro* genebank at 25°C/light on MS + 0.2 mg/l BAP for a conservation period of 6-months [29]. Besides 45 exotic accessions of *Fragaria* are also part of *in vitro* collection, maintained under normal growth conditions. *In vitro* shoots are rooted on IBA (1 mg/l) media. Plantlets were

| Species | Seed storage Behavior | Seed Storage Conditions |
|---|---|---|
| *Fragaria chiloensis* (L.) Mill. | Orthodox | 100% viability following drying to mc's (moisture content) in equilibrium with 15% RH and freezing for 1 month at −20°C at RBG Kew, WP |
| *F. moschata* Duchesne ex Weston | Orthodox | 92% viability following drying to mc's in equilibrium with 15% RH and freezing for 34days at −20°C at RBG Kew, WP |
| *Fragaria* spp. (strawberry) | Orthodox | No problem for long-term storage under IPGRI preferred conditions (SSLR); no loss in viability following 23 years hermetic air-dry storage at 4.5°C [22] |
| *F. vesca* L. | Orthodox | Viability is halved after 3 years open storage [23]; viability maintained after 8 years hermetic storage at −18°C [24]; long-term storage under IPGRI preferred conditions at RBG Kew, WP. Oldest collection 16 years; germination change 78 to 91.2%, 14years, 1 collection |
| *F. virginiana* Mill. | Orthodox | 75% viability following drying to mc's in equilibrium with 15% RH and freezing for 107 days at -20C at RBG Kew, WP. Longevity: When stored in sealed containers the seeds can remain viable for up to 20 years [25]. |
| *F. viridis* Weston | Uncertain | The species has been shown to form a transient soil seed bank, with seeds persisting in the soil for <1 year [26]. Although, this may suggest that seeds of the species are short-lived under ambient conditions, and perhaps recalcitrant or intermediate, several factors may have resulted in the inclusion of orthodox seeds within this category (see [26] for further detail). Further research is necessary before the storage behavior of the taxon can be reliably classified. |

*Compiled from: [21].*

**Table 2.**
*Seed storage behavior and storage conditions of Fragaria species.*

transferred in the sterilized soilrite filled pots for acclimatization. The plants raised through tissue culture exhibited normal growth, flowering and fruit setting [29].

### *3.3.2 Slow growth*

The main aim of this method is to maintain cultures undergrowth limitation conditions to reduce the requirement of frequent subculture. Some of the various approaches in practice are discussed below:

i. Low-temperature incubation

This method applies to wide range of genotypes, especially of temperate nature. Here, the *in vitro* cultures are maintained at low temperature that affects the metabolic activities which in turn restrict the growth of the plant. The storage temperature, generally, is crop-specific. *In vitro* conservation of *Fragaria* spp. at low temperature was successfully reported (**Table 3**).

ii. Use of growth retardants

Growth retardants are used to reduce the overall growth of the *in vitro* plants thereby prolonging their subculture intervals. But, the use of some of the growth retardants may

| Species | *In vitro* conservation strategy | Conservation duration | References |
|---|---|---|---|
| *Fragaria* spp. and cultivars (50) | 4°C, dark, add liquid med. Every 3 m | 6 years | [30] |
| *Fragaria* cultivars (21) | 2 °C | 9–27 m, 10–100% survival | [31] |
| *Fragaria* spp. and cultivars (96) | 4°C, dark, tissue culture bags | 9–24months; 15 months mean storage time | [32–34] |
| *Fragaria* spp. and Cultivars (9) | 4°C, 12 h photoperiod or dark, 4 BA conc., CH or not | More than 12 months, best with CH, photoperiod, low BA | [35] |
| *Fragaria* spp. | 4 °C | Up to 7 months | [36] |
| *Fragaria* (wild species) | 4 °C, five chamber bags, Jungnickel medium | 6 to 9 months | [37] |
| *Fragaria vesca* | 4 °C, dark, culture tubes | 10 months | [29] |

**Table 3.**
*In vitro conservation of* Fragaria *spp. at low temperature.*

cause mutation, owing to their mutagenic properties. These may also pose a physiological problem if used for a longer time. The most commonly used growth retardants are abscisic acid, dimethylamino succinamide, phosphon D, maleic hydrazide [27].

iii. Use of minimal growth media and restrictive growth conditions

Carbon source affects the growth rate of *in vitro* cultures. Alteration of optimum dose could reduce the growth rate of cultures in many species. The inclusion of sugar alcohol like mannitol or sorbitol in culture media is quite effective in restricting the growth of many plant species *in vitro* [27]. The use of minimal media may be more effective at low temperature.

### *3.3.3 Cryopreservation*

Cryopreservation provides a low-input method for storing base collection (long-term backup) of clonal materials. Cryopreservation techniques are based on the removal of all freeze-able water from tissues by physical or osmotic dehydration, followed by ultra-rapid freezing. Cryopreservation can be achieved through classical and new vitrification-based techniques. Classical techniques involve freeze-induced dehydration, whereas new techniques are based on vitrification. Vitrification can be defined as the transition of water directly from the liquid phase into an amorphous phase or glass, while avoiding the formation of crystalline ice [39]. The main advantages in cryopreservation are simplicity and applicability to a wide range of genotypes [40].

Literature survey showed that, *in vitro* grown shoot tips of *Fragaria* have been cryopreserved using various techniques, encapsulation-vitrification [41], encapsulation-dehydration [42], vitrification [43], droplet-vitrification [44], cold acclimation + vitrification, Encapsulation-dehydration, controlled rate cooling [45] and V-cryoplate [46]. Another cryopreservation procedure using aluminum cryo-plates, termed D-cryoplate, was successfully developed for *in vitro* mat rush (*Juncus decipiens* Nakai) basal stem buds [47]. Encapsulation-dehydration and vitrification-based techniques viz., vitrification, V-cryoplate and D-cryoplate, were applied for cryopreservation of non-cold-acclimated shoot tips of a cultivar of strawberry *F.* x *ananassa* cv. Earliglow

[48]. In comparison, the recently developed new aged technique D-cryoplate, resulted in the best with 40% recovery among the four techniques tested. Both the recently developed new-aged techniques, D-cryoplate and V-cryoplate techniques have been used in many crops due to their high efficiency and operational simplicity. Cultivars of *F.* × *ananassa* were cryopreserved by V-cryoplate method and 81% recovery of cryopreserved shoot tips was obtained [46]. The protocol included cold acclimation at 5°C for 3 weeks before cryopreservation. The importance of cold acclimation for cryopreservation has been emphasized in other studies [45, 49–52]. In case of temperate species, a cold acclimation period, which triggers cold adaptation mechanisms, is often beneficial [53].

## 4. Genetic stability

Maintenance of genetic fidelity is essential for a successful cryopreservation strategy [54] and requires tools for evaluating the genetic stability of conserved plants. Theoretically, during cryopreservation all metabolic activities stop at the ultralow temperature of LN, consequently, after rewarming from cryopreservation recovery of true-to-type plants is expected [55]. Cryopreservation protocols involve cooling in and rewarming from LN, *in vitro* culture and regeneration processes, phenotypic and genomic changes can occur due to somaclonal variation. Hence, it is necessary to verify true-to-type plants after cryopreservation [56]. The literature is overall 'positive' regarding the outcome of stability assessments from cryopreservation with studies. The various types of DNA markers detect different levels of polymorphism and different amounts of DNA change. Genetic stability is the norm in most studies of possible plant genetic variation following cryopreservation [56].

The development of molecular techniques in the recent year provides additional means for assessing genetic fidelity in plants. Single sequence repeats (SSRs) are tandemly repeated motifs of one to six bases present in coding and non-coding regions and are highly polymorphic [57]. *F.* x *ananassa* shoot tips were cryopreserved using encapsulation-dehydration and vitrification along with that new modified method cryoplate (V and D cryoplate) [48]. Plants raised through these techniques were subjected to genetic stability analysis using SSR markers. No differences were observed between *Fragaria in vitro* mother plants and *in vitro* cryopreserved plants using eight SSR primers [58]. This lack of variation suggests that there were no changes in the genetic fidelity of the plants due to cryopreservation. This was also the case in *Solanum*, in which the microsatellite sequences of plants regrown from cryopreserved apices were identical to the profiles of the parent plants and their progeny [59]. No structural changes were observed in the *in vitro* control or the Solanum plants grown from the cryopreserved germplasm, indicating stable inheritance of SSR sequences in the somatic progeny [59]. The low coverage of the genome is one criticism of molecular techniques. Despite being highly polymorphic and co-dominant, SSRs may be clustered and distributed unevenly in certain chromosome locations.

## 5. Cryotherapy and virus elimination

An important pre-requisite for the conservation of germplasm either *in vitro* or under cryopreservation is the availability of 'virus-free' starting material. The presence of viruses hinders international and national exchange and conservation of germplasm and also becomes a hurdle for utilization of germplasm in its crop improvement. The viruses cause the most serious diseases in *Fragaria.* Viruses causing serious diseases are arabis mosaic virus (ArMV), raspberry bushy dwarf virus (RBDV), strawberry mild yellow edge virus (SMYEV) and raspberry ringspot virus (RpRSV). These viruses are transmitted during traditional vegetative propagation through runners or by modern methods of *in vitro* multiplication. Thus, periodic screening for *in vitro* conserved *Fragaria* germplasm is important for viruses' status in an *in vitro* genebank [60].

It was established for the first time that cryopreservation was not only useful for germplasm conservation, but also for virus eradication in *in vitro* shoots of plum infected with plum pox virus (PPV) [61]. Cryotherapy is a novel method for virus eradication in economically important plant species [62]. In Cryotherapy, more hydrated, infected cells die because of freezing, and only small compact cells which are close to the meristem and generally virus-free survive cryopreservation [63]. Cryotherapy has been attempted for virus eradication in many important crops such as banana, grapevine, *Prunus*, raspberry, citrus and potato. In *F. vesca* post-cryotherapy 96.67% of cultures were found free from *Raspberry ringspot virus* (RpRSV) [64].

## 6. Conclusions

The development of new elite cultivars with high yield, quality, and biotic and abiotic resistance or tolerance, will improve the efficiency of *Fragaria* production. In addition, the use of virus-free plant propagules may also increase yields and limit the spread of viral diseases within and between countries. Therefore, *Fragaria* germplasm conservation and its back up is of utmost important.

For a cryopreservation protocol to be successful or to provide optimal results, experimental conditions for each of its successive steps must be optimized. The protocols stated in this study need to be optimized for better regrowth. There is scope for improvement at each step of the tested techniques, to obtain higher recovery of cryopreserved shoot tips. The modified protocols can be tested on other *Fragaria* cultivars and further can be used for cryopreservation of genetic resources of *Fragaria*.

## Acknowledgements

The author gratefully acknowledges Former and Present Directors, ICAR-NBPGR for providing research facilities, Dr. Barbara M. Reed and Dr. Kim E. Hummer for providing *Fragaria* accessions and Dr. T. Niino for providing cryo-plates to facilitate cryo-research.

## Author details

Sandhya Gupta
Indian Council of Agricultural Research-National Bureau of Plant Genetic Resources (ICAR-NBPGR), New Delhi, India

*Address all correspondence to: sandhya.gupta@icar.gov.in; sandhya_gupta87@yahoo.com

## References


[1] Sharma RM, Yamdagni R, Dubey AK, Pandey V. Strawberries: Production, Postharvest Management and Protection. Boca Raton, Florida, USA: CRC Press; 2019. p. 600

[2] Hummer KE. Global Conservation Strategy for Fragaria (Strawberry). Leuven, Belgium: ISHS Scripta Horticulturae (6); 2008. p. 87

[3] Hancock J, Sjulin TM, Lobos G. Strawberries. In: Hancock JF, editor. Temperate Fruit Crop Breeding: Germplasm to Genomics. Dordrecht, The Netherlands: Springer, Kluwer Academic Publishers; 2008. pp. 393-437. DOI: 10.1007/978-1-4020-6907-9

[4] Hummer KE. A global conservation strategy for strawberries. Acta Horticulturae. 2006;**760**:49-54

[5] Hancock JF, Maas JL, Shanks CH, Breen PJ, Luby JJ. Strawberries (Fragaria spp.). In: Moore J, Ballington J, editors. Genetic Resources in Temperate Fruit and Nut Crops International Society of Horticultural Sciences. Wageningen, The Netherlands, Leuven, Belgium: ISHS; 1990. pp. 489-546. DOI: 10.17660/ActaHortic.1991.290.11

[6] FAOSTAT. Food and Agriculture Organization Database. 2018. Available from http://www.fao.org/faostat.

[7] Staudt G. The species of *Fragaria*, their taxonomy and geographical distribution. Acta Horticulturae. 1989;**265**:23-33

[8] Staudt G. Systematics and Geographic Distribution of the American Strawberry Species: Taxonomic Studies in the Genus Fragaria (Rosaceae: Potentilleae). California, USA: University of California Publications in Botany 81; p. 1999 xii + 162

[9] Akiyama Y, Yamamoto Y, Ohmido N, Ohshima M, Fukui K. Estimation of the nuclear DNA content of strawberries (*Fragaria* spp.) compared with *Arabidopsis thaliana* by using dual-step flow cytometry. Cytologia. 2001;**66**(4):431-436

[10] Davis TM, Denoyes-Rothan B, Lerceteau-Köhler E. Strawberry. In: Fruits and Nuts. Berlin, Heidelberg: Springer; 2007. pp. 189-205

[11] Hirakawa H, Shirasawa K, Kosugi S, Tashiro K, Nakayama S, Yamada M, et al. Dissection of the octoploid strawberry genome by deep sequencing of the genomes of *Fragaria* species. DNA Research. 2014;**21**:169-181. DOI: 10.1093/dnares/dst049

[12] Arora RK, Pandey A. Wild Edible Plants of India: Diversity, Conservation and Use. New Delhi, India: National Bureau of Plant Genetic Resources; 1996. p. 294

[13] Pradheep K, Bhandari DC, Bansal KC. Wild Relatives of Cultivated Plants in India. Indian Council of Agricultural Research: New Delhi, India; 2014. p. 728

[14] Gupta S, Pradheep K. Diversity, conservation and use of underutilized and minor fruits in India: An overview. Acta Horticulturae. 2020;**1297**:51-60. DOI: 10.17660/ ActaHortic. 2020.1297.8

[15] Anderson MK, Roderick W. USDA NRCS plant guide for mountain strawberry *Fragaria virginiana* Duchesne. 2006. Available from https://era.library.ualberta.ca/items/7d998720-bd19-4847-a29c-26a38ece79b7/view/e70ae4f4-0cc9-4ab2-be19-6fb77ec29695/Fragaria-20virginiana.pdf [Accessed 2021-11-20]

[16] Gerling HS, Willoughby MG, Schoepf A, Tannas KE, Tannas CA. A guide to using native plants on disturbed lands. In: Alberta Agriculture, Food and Rural Development and Alberta Environmental Protection. Edmonton, Alberta Canada: Alberta Agriculture, Food and Rural Development Publishing Branch; 1996. 247 p. Available from https://era.library.ualberta.ca/items/7d998720-bd19-4847-a29c-26a38ece79b7/view/e70 ae4f4-0cc9-4ab2-be19-6fb77ec29695/Fragaria-20virginiana.pdf [Accessed 2021-11-20]

[17] Debnath SC. Strawberry sepal: Another explant for thidiazuron-induced adventitious shoot regeneration. *In Vitro Cellular & Developmental Biology. Plant*. 2005;**41**:671-676

[18] Debnath SC. Zeatin overcomes thidiazuron-induced inhibition of shoot elongation and promotes rooting in strawberry culture *in vitro*. The Journal of Horticultural Science and Biotechnology. 2006;**81**:349-354

[19] Thormann I, Dulloo M, Engels J. Techniques for *ex situ* plant conservation. In: Henry RJ, editor. Plant Conservation Genetics. Philadelphia, Pennsylvania, USA: Haworth Press, Inc; 2006. pp. 7-36

[20] Lynch PT, Benson EE, Harding K. Climate change: The role of economically important, vegetatively propagated plants. The Journal of Horticultural Science and Biotechnology. 2007;**82**:157-160. DOI: DOI.org/10.1080/14620316.2007.11512213

[21] Royal Botanic Gardens Kew, Fragaria. Seed Information Database. 2008. Available from: http://data.kew.org/sid/SidServlet?Clade=&Order=&Family=&APG=off&Genus= Fragaria&Species=&StorBehav=0&StFlag=on [Accessed 2021-11-20]

[22] Scott DH, Draper AD. A further note on longevity of strawbery seed in cold storage. HortScience. 1970;**5**:439

[23] Ewart AJ. On the longevity of seeds. Proceedings of the Royal Society Victoria. 1908;**21**:1-210

[24] Montezuma-De-Carvalho J, Paiva J, Pimenta MCS. Effect of cold storage on seed viability of aromatic plants from the Portuguese flora. In: EUCARPIA International Symposium on Conservation of Genetic Resources of Aromatic and Medicinal Plants. Oeiras, Portugal: EUCARPIA Genetic Resources Section; 9-11 May 1984

[25] Rose R, Chachulski CEC, Haase DL. Propagation of Pacific Northwest Native Plants. Corvallis, Oregon: Oregon State University Press; 1998. 248 p. Available from https://era.library.ualberta.ca/items/7d998720-bd19-4847-a29c-26a38ece79b7/view/e70 ae4f4-0cc9-4ab2-be19-6fb77ec29695/Fragaria-20virginiana.pdf [Accessed on 2021-11-20]

[26] Thompson K, Bakker JP, Bekker RM. The Soil Seed Banks of North West Europe: Methodology, Density and Longevity. Cambridge: Cambridge University Press; 1997. Available from: http://data.kew.org/sid/SidServlet?ID=10499&Num=rU0 [Accessed on 2021-11-20]

[27] Gupta S, Mandal BB. *In vitro* methods for PGR conservation: Principles and prospects. In: Chaudhury R, Pandey R, Mal B, editors. *In Vitro* Conservation and Cryopreservation of Tropical Fruit Species. New Delhi, India: IPGRI Office for South Asia, NBPGR; 2003. pp. 71-80

[28] Murashige T, Skoog F. A revised medium for rapid growth and bioassays with tobacco tissue cultures. Physiologia Plantarum. 1962;**15**(3):473-494. DOI: 10.1111/j.1399-3054.1962.tb08052.x

[29] Gupta S, Parakh DB, Verma SK. *Ex situ* conservation of strawberry (*Fragaria*

spp.) germplasm in ICAR-NBPGR. Acta Horticulturae. 2020;**1297**:89-98. DOI: 10.17660/ ActaHortic.2020.1297.13

[30] Mullin RH, Schlegel DE. Cold storage maintenance of strawberry meristem plantlets. HortScience. 1976;**11**:100-101

[31] Damiano C. Cold storage maintenance and resuming of propagation of *in vitro* strawberry culture. Annali dell'Istituto Sperimentale per la Frutticoltura. 1979;**10**:53-58

[32] Reed BM. Application of gas-permeable bags for *in vitro* cold storage of strawberry germplasm. Plant Cell Reports. 1991;**10**(9):431-434. DOI: org/10.1007/BF00233808

[33] Reed BM. Cold storage of strawberries *in vitro*: A comparison of three storage systems. Fruit Varieties Journal. 1992;**46**:98-102

[34] Reed BM, Hummer K. Conservation of germplasm of strawberry (Fragaria species). In: Bajaj YPS, editor. Biotechnology in Agriculture and Forestry. Vol. 32. Berlin: Springer-Verlag; 1995. pp. 354-370

[35] Reed BM. Factors affecting the *in vitro* storage of strawberry. HortScience. 1995;**30**(4):871. DOI: DOI.org/10.21273/ HORTSCI.30.4.871B

[36] Sedlak J, Zidova P, Paprstein F. Slow growth *in vitro* conservation of fruit crops. Acta Horticulturae. 2019;**1234**:119-124. DOI: org/10.17660/ ActaHortic.2019.1234.15

[37] Höfer M. Conservation strategy of genetic resources for strawberry in Germany. Acta Horticulturae. 2011;**908**:421-430. DOI: org/10.17660/ ActaHortic.2011.908.54

[38] Reed BM, Chang Y. Medium- and long-term storage of *in vitro* cultures of temperate fruit and nut crops. In: Razdan MK, Cocking EC, editors. Conservation of Plant Genetic Resources *In Vitro*. Vol. 1. Enfield, NH, USA: Science Publishers, Inc.; 1997. pp. 67-105

[39] Fahy GM, MacFarlane DR, Angell CA, Meryman HT. Vitrification as an approach to cryopreservation. Cryobiology. 1984;**21**:407-426

[40] Engelmann F. Plant cryopreservation: Progress and prospects. *In Vitro Cellular & Developmental Biology. Plant*. 2004;**40**(5):427-433

[41] Hirai D, Shirai K, Shirai S, Sakai A. Cryopreservation of *in vitro*-grown meristems of strawberry (*Fragaria* x *ananassa* Duch.) by encapsulation-vitrification. Euphytica. 1998;**101**(1): 109-115. DOI: 10.1023/A:1018386 211577

[42] Halmagyi A, Deliu C. Cryopreservation of strawberry shoot tips by encapsulation-dehydration. Notulae Botanicae Horti Agrobotanici Cluj-Napoca. 2006;**34**(1):28-33. DOI: 10.15835/nbha341260

[43] Zhao M, Wang Z, Su J, Qian Y. Cryopreservation on *in vitro* cultured shoot tips of strawberry. Acta Horticulturae. 2006;**708**:265-270. DOI: 10.17660/ActaHortic.2006.708.45

[44] Pinker I, Halmagyi A, Olbricht K, Ehrig F. Cryopreservation of strawberry cv. 'Senga Sengana' by droplet vitrification. Acta Horticulturae. 2009;**839**:253-260. DOI: 10.17660/ ActaHortic.2009.839.31

[45] Höfer M, Reed BM. Cryopreservation of strawberry genetic resources in Germany. Acta Horticulturae. 2011;**918**:139-146. DOI: 10.17660/ ActaHortic.2011.918.16

[46] Yamamoto S, Fukui K, Rafique T, Khan NI, Martinez CRC, Sekizawa K, et al. Cryopreservation of *in vitro*-grown shoot tips of strawberry by the vitrification method using aluminium cryoplates. Plant Genetic Resources. 2012;**10**(1):14-19. DOI: 10.1017/S1479262111000906

[47] Niino T, Yamamoto SI, Fukui K, Castillo-Martinez CR, Arizaga MV, Matsumoto T, et al. Dehydration improves cryopreservation of mat rush (*Juncus decipiens* Nakai) basal stem buds on cryo-plates. CryoLetters. 2013;**34**(6):549-560

[48] Gupta S, Tewari P. Cryopreservation of Fragaria × ananassa using different techniques. Acta Horticulturae. 2020;**1298**:161-166. DOI: 10.17660/ActaHortic.2020.1298.23

[49] Reed BM. Cold acclimation as a method to improve survival of cryopreserved Rubus meristems. CryoLetters. 1988;**9**:166-171

[50] Chang Y, Reed BM. Extended cold acclimation and recovery medium alteration improve regrowth of Rubus shoot tips following cryopreservation. CryoLetters. 1999;**20**(6):371-376

[51] Gupta S, Reed BM. Cryopreservation of shoot tips of blackberry and raspberry by encapsulation-dehydration and vitrification. CryoLetters. 2006;**27**(1):29-42

[52] Höfer M. Cryopreservation of *in vitro* shoot tips of strawberry by the vitrification method - establishment of a duplicate collection of *Fragaria* germplasm. CryoLetters. 2016;**37**(3):163-172

[53] Engelmann F. Cryopreservation of clonal crops: A review of key parameters. Acta Horticulturae. 2014;**1039**:31-39. DOI: DOI.org/10.17660/ActaHortic.2014.1039.2

[54] Bajaj YPS. Tuberization in potato plants regenerated from freeze-preserved meristems. Crop Improvement. 1978;**5**:137-141

[55] Panis B, Swennen R, Engelmann F. Cryopreservation of plant germplasm. Acta Horticulturae. 2001;**560**:79-86

[56] Harding K. Genetic integrity of cryopreserved plant cells: A review. CryoLetters. 2004;**25**:3-22

[57] Zane L, Bargelloni L, Patarnello T. Strategies for microsatellite isolation: A review. Molecular Ecology. 2002;**11**:1-16

[58] Gupta S, Tewari P, Mishra M, Singh R, Nerwal DK. Cryopreservation and assessment of genetic stability of cryopreserved *Fragaria* spp. using SSR markers. In: Thammasiri K, Kongsawadworakul P, Swangpol SC, editors. Book of Abstracts: The IX International Scientific and Practical Conference on Biotechnology as an Instrument for Plant Biodiversity Conservation (Physiological, Biochemical, Embryological, Genetic and Legal Aspects) (Biotech 2021). Bangkok, Thailand: Department of Plant Science, Faculty of Science, Mahidol University; 2021. p. 90

[59] Harding K, Benson EE. The use of microsatellite analysis in Solanum tuberosum L. *in vitro* plantlets derived from cryopreserved germplasm. CryoLetters. 2001;**22**:199-208

[60] Gupta S. Cryopreservation of shoot tips of *Fragaria* spp. and virus elimination through cryotherapy. Cryobiology. 2016;**3**(73):40

[61] Brison M, de Boucaud MT, Pierronnet A, Dosba F. Effect of

cryopreservation on the sanitary state of a cv *prunus* rootstock experimentally contaminated with plum pox potyvirus. Plant Science. 1997;**123**:189-196

[62] Wang MR, Chen L, Zhang Z, Blystad DR, Wang QC. Cryotherapy: A novel method for virus eradication in economically important plant species. In: Plant Cell Culture Protocols. New York, NY: Humana Press; 2018. pp. 257-268

[63] Helliot B, Panis B, Poumay Y, Swennen R, Lepoivre P, Frison E. Cryopreservation for the elimination of cucumber mosaic and banana streak viruses from banana (*Musa* spp.). Plant Cell Reports. 2002;**20**(12):1117-1122

[64] Gumachanamardi P. Cryopreservation of in vitro Fragaria vesca L. shoot tips and virus elimination through cryotherapy [thesis]. India: Division of Plant Genetic Resources ICAR-Indian Agricultural Research Institute New Delhi; 2021. p. 57

Chapter 5

# Chemical Fertilizer Medium on Tissue Culture of Strawberries

*Sumana Neera*


## Abstract

The strawberry is a nutrient-rich fruit with high concentrations of vitamins A, C, $B_1$, $B_2$, protein, calcium, potassium, copper, iron, and especially antioxidants. Strawberries are an extremely popular fruit because the antioxidants in them are much more than in tomatoes, up to 7 times. Strawberry propagation now requires a tissue culture-derived cultivar. Infection from the mother plant causes low yield due to diseases, and the chemicals utilized as bait in vitro are expensive. The purpose of this experiment was to find a replacement for the chemical fertilizer mix that was affecting the strawberry's in vitro growth. The experiment was conducted with the strawberry cultivar Royal 80 and the 5 × 4 Factorial in CRD. Explants were cultivated in media containing five types of chemical fertilizer: 21-21-21 ($N-P_2O_5-K_2O$), 15-30-15 ($N-P_2O_5-K_2O$), 36-5-5 ($N-P_2O_5-K_2O$), 15-30-15 ($N-P_2O_5-K_2O$) and 10-52-17 ($N-P_2O_5-K_2O$) at concentrations of 0.5, 1.5, 2.5 and 3.5 g $L^{-1}$, compared to MS medium, which provided control for each type of fertilizer. The plantlets were cultured in 0.5 g/l of 21-21-21 showed the highest number of plantlets was 4.30 plantlets/explant, which was better than MS medium (3.44 plantlets/explant), 15-30-15 and 36-5-5 media (3.48, 3.87 plantlets/explant, respectively).


**Keywords:** strawberry, antioxidants, chemical fertilizers, plantlet, *in vitro*

## 1. Introduction

The strawberry (*Fragaria spp.*) is one of the most popular berry fruits in the world. Strawberries contain high levels of vitamins C and K, as well as fiber, folic acid, manganese, and potassium [1]. They are also rich in phytonutrients and flavonoids, which give them their bright red color. They have been used medicinally for centuries to help with digestive problems, tooth whitening, and skin irritations. Their high fiber and fructose content may help control blood sugar levels by slowing digestion, and the fiber is thought to have a satiating effect. Leaves can be eaten raw, cooked, or used to make tea [2–5].

In Thailand, strawberries have been grown for many years in the north of the country and have become economically important since the early 1980s. The Royal Project Foundation's research resulted in the creation of a popular cultivar known as "Royal 80". This cultivar has a pleasant flavor, is early to bear fruit, and is of good quality. The absence of good mother plants for runner plant production is an issue in strawberry farming in Thailand. There are very few virus-free plants available, and plant vigor is low [6, 7]. The propagation of strawberries using tissue culture technique is one way to solve this problem. However, production costs are high due to the expensive chemicals used for the propagation recipes [8]. To reduce the high expenses of the tissue culture technique, we are interested in using chemical fertilizers, which farmers used for strawberry production instead of MS medium. Farmers for strawberry production use the strawberry fertilizers such as 21-21-21 (N-$P_2O_5$-$K_2O$), 15-30-15 (N-$P_2O_5$-$K_2O$), 36-5-5 (N-$P_2O_5$-$K_2O$), 15-30-15 (N-$P_2O_5$-$K_2O$), and 10-52-17 (N-$P_2O_5$-$K_2O$). Therefore, developing recipes to reduce production costs is necessary in conducting this research. So far, there is no report on the use of fertilizers to substitute strawberry production medium. The aim of this study was to investigate the effect of types and concentration of chemical fertilizers by using five types of replacement mediums on *Fragaria* spp. to reduce production costs.

## 2. Materials and methods

### 2.1 Explant preparation

Strawberries, cultivar Royal 80 stolons were obtained and used for explant from a strawberry farm in Petchaboon province, Thailand. Stolon (**Figure 1**) was chopped to 3–4 cm and washed in a 10% detergent solution with running tap water. It was washed in regular tap water. The stolons were surface-sterilized for 7 min with 10% sodium hypochlorite and then rinsed three times in sterile distilled water [9]. The stolon was cut to a length of 2–3 cm and placed in MS medium [10], which contained sucrose 30 g $L^{-1}$ and phytagel 2.0 g $L^{-1}$. Prior to autoclaving at 121°C, the pH was adjusted to 5.7 with 1 N NaOH or 1 N HCl. The cultures were maintained in a culture room with a temperature of 25 ± 2°C, a light intensity of 100 μmol $m^{-2}$ $s^{-1}$, and a photoperiod of 16 h [8]. After 2 months of culture, the young shoot was induced in MS medium.

### 2.2 Statistical analysis

The experiment carried out in 5 × 4 Factorial in CRD (Completely Randomized Design) consisted of two factors as factor A was 5 types of chemical fertilizers as 21-21-21 (N-$P_2O_5$-$K_2O$: Procon), 36-5-5 (N-$P_2O_5$-$K_2O$: Superspeed), 30-20-10 (N-$P_2O_5$-$K_2O$: Plalungchang pink), 15-30-15 (N-$P_2O_5$-$K_2O$: Megafor), 10-52-17 (N-$P_2O_5$-$K_2O$:

**Figure 1.**
*Stolon of strawberry cultivar royal 80 was used to be an explant.*

Plalungchang blue) and MS as a control, factor B was 4 levels of chemical fertilizer concentration as 0.5, 1.5, 2.5, and 3.5 g $L^{-1}$. The data was analyzed statistically by Statistix 8 program and the significance of differences among means was conducted using LSD at $p \leq 0.05$ and $p \leq 0.01$. Data was recorded after 4 weeks of culture.

### 2.3 Multiplication

The young shoots of strawberries from MS medium [10] were transferred to the experiment, consisting of 21 treatments as above. All treatments had 10 replicates and were supplemented with sucrose 30 g $L^{-1}$, phytagel 2.5 g $L^{-1}$, and adjusted pH to 5.7. All cultures were maintained in a culture room kept at 25 ± 2°C with a light intensity of 100 mol $m^{-2}$ $s^{1}$ and a photoperiod of 16 h [9]. The following data were corrected after 4 weeks of culture: number of plantlets (plants/explant), fresh weight of plantlets (g/explant), number of roots (root/explant), and root length (root/explant).

## 3. Results and discussions

### 3.1 Fresh weight

From **Table 1**, the comparison with five types of chemical fertilizers (factor A), the concentration of chemical fertilizer (factor B), and the interaction between types and the concentration of chemical fertilizer (factor AxB) were highly significant. If compared with the fresh weight of strawberry plantlets, five types of chemical fertilizers as 21-21-21 ($N-P_2O_5-K_2O$), 36-5-5 ($N-P_2O_5-K_2O$), 30-20-10 ($N-P_2O_5-K_2O$), 15-30-15($N-P_2O_5-K_2O$) and 10-52-17 ($N-P_2O_5-K_2O$) showed an average amount of

| Type of chemical fertilizers (A) | Fresh weight of plantlets (g/explant) | | | | |
|---|---|---|---|---|---|
| | Concentration of chemical fertilizers (g $L^{-1}$) (B) | | | | |
| | 0.5 | 1.5 | 2.5 | 3.5 | Average |
| MS (control) | | | | | 1.72 a |
| 21-21-21 ($N-P_2O_5-K_2O$) | 1.31 | 1.32 | 1.35 | 1.33 | 1.33 bc |
| 36-5-5 ($N-P_2O_5-K_2O$) | 1.56 | 1.25 | 1.31 | 1.25 | 1.34 b |
| 30-20-10 ($N-P_2O_5-K_2O$) | 1.34 | 1.25 | 1.25 | 1.25 | 1.28 c |
| 10-52-17 ($N-P_2O_5-K_2O$) | 1.31 | 1.27 | 1.26 | 1.31 | 1.29 bc |
| 15-30-15 ($N-P_2O_5-K_2O$) | 1.29 | 1.32 | 1.46 | 1.33 | 1.35 b |
| Average | 1.42 a | 1.36 b | 1.39 ab | 1.36 b | |
| A | ** | | | | |
| B | ** | | | | |
| A × B | ** | | | | |

*Different alphabets in the same column and row are significantly different at level 0.05 compared to LSD.*
***Significant difference at the level of 0.01.*

**Table 1.**
*The effect of different chemical fertilizer types and concentrations on the fresh weight of strawberry plantlets after cultured for 4 weeks.*

| Type of chemical fertilizers (A) | Number of plantlets (plantlet/explant) | | | | |
|---|---|---|---|---|---|
| | Concentration of chemical fertilizers (g $L^{-1}$) (B) | | | | |
| | 0.5 | 1.5 | 2.5 | 3.5 | Average |
| MS (control) | | | | | 3.44 b |
| 21-21-21 ($N-P_2O_5-K_2O$) | 4.30 | 3.60 | 3.66 | 3.92 | 3.87 a |
| 36-5-5 ($N-P_2O_5-K_2O$) | 3.66 | 2.98 | 1.25 | 1.25 | 2.29 c |
| 30-20-10 ($N-P_2O_5-K_2O$) | 3.97 | 1.25 | 1.25 | 1.25 | 1.93 c |
| 10-52-17 ($N-P_2O_5-K_2O$) | 2.17 | 2.17 | 2.30 | 2.17 | 2.20 c |
| 15-30-15 ($N-P_2O_5-K_2O$) | 2.96 | 2.84 | 4.30 | 3.81 | 3.48 ab |
| Average | 3.42 a | 2.71 b | 2.70 b | 2.64 b | |
| A | ** | | | | |
| B | ** | | | | |
| A × B | ** | | | | |

*Different alphabets in the same column and row are significantly different at level 0.05 compared to LSD.*
*$^{**}$Significant difference at the level of 0.01.*

**Table 2.**
*The effect of chemical fertilizer types and concentrations on the number of strawberry plantlets after 4 weeks of culture.*

fresh weight of 1.28–1.35 $gL^{-1}$. According to the experiment, MS medium was the best medium to induce the fresh weight of plantlets (1.72 $gL^{-1}$) because it contained macronutrients, micronutrients, and many kinds of vitamins [9, 11].

### 3.2 Number of plantlets

In **Table 2**, to compare five types of chemical fertilizers (factor A), the concentration of chemical fertilizer (factor B) and interaction between types and concentration of chemical fertilizer (AxB) were all highly significant at level 0.01. MS medium, 21-21-21 ($N-P_2O_5-K_2O$), and 15-30-15 ($N-P_2O_5-K_2O$) chemical fertilizers showed a similar number of plantlets (3.44, 3.87, and 3.48 plantlets/explant, respectively). At the concentration of chemical fertilizer, 0.5 g $L^{-1}$ was the best concentration to induce plantlets of strawberries. The medium containing 0.5 g $L^{-1}$ of 21-21-21 ($N-P_2O_5-K_2O$) chemical fertilizer was the best medium to produce plantlets of strawberries (**Figure 2**) [11].

### 3.3 Number of roots

To study the effects of five types of chemical fertilizer, the concentration of chemical fertilizer and the interaction between type of chemical fertilizer and concentration of chemical fertilizer on root induction of strawberries were shown in **Table 3**. The type of chemical fertilizer (factor A) was significantly different at level 0.01. The concentration of chemical fertilizer (factor B) was shown to be significantly different at level 0.05. Also, the interaction between factors A and B was significantly different at level 0.05 too. Similar to [8], MS medium is shown to be the best medium to induce root growth in strawberries (5.09 root/explant). On the media, 21-21-21 ($N-P_2O_5-K_2O$) and 15-30-15 ($N-P_2O_5-K_2O$) were also good media to induce strawberry root (3.97 and 3.64 root/explant, respectively).

**Figure 2.**
*Plantlet was cultured on medium containing 0.5 g $L^{-1}$ of 21-21-21 ($N$-$P_2O_5$-$K_2O$) chemical fertilizer.*

| Type of chemical fertilizer (A) | Number of roots (root/explant) | | | | |
|---|---|---|---|---|---|
| | Concentration of chemical fertilizer (g $L^{-1}$) (B) | | | | |
| | 0.5 | 1.5 | 2.5 | 3.5 | Average |
| MS (control) | | | | | 5.09 a |
| 21-21-21 ($N$-$P_2O_5$-$K_2O$) | 4.18 | 5.63 | 1.91 | 4.16 | 3.97 b |
| 36-5-5 ($N$-$P_2O_5$-$K_2O$) | 4.74 | 1.25 | 1.56 | 1.25 | 2.20 c |
| 30-20-10 ($N$-$P_2O_5$-$K_2O$) | 4.45 | 1.44 | 1.25 | 1.25 | 2.10 c |
| 10-52-17 ($N$-$P_2O_5$-$K_2O$) | 2.17 | 1.25 | 2.12 | 1.31 | 1.98 c |
| 15-30-15 ($N$-$P_2O_5$-$K_2O$) | 2.02 | 3.35 | 5.11 | 1.33 | 3.64 b |
| Average | 1.42 a | 1.36 b | 1.39 b | 1.36 b | |
| A | ** | | | | |
| B | * | | | | |
| A × B | ** | | | | |

*Different alphabets in the same column and row are significantly different at level 0.05 compared to LSD.*
**Significantly different at level 0.05.*
***Significantly different at level 0.01.*

**Table 3.**
*The effect of chemical fertilizer types and concentrations on a number of roots of strawberry plantlet after being cultured for 4 weeks.*

## 3.4 Root length

Five types of chemical fertilizer (factor A), the concentration of chemical fertilizer (factor B), and interactions between types of chemical fertilizer and concentration of chemical fertilizer (A × B) on the root length of strawberries were investigated (**Table 4**). Factor A, factor B, and factor A × B were shown to be significantly different at level 0.01. The MS medium and the 21-21-21 ($N$-$P_2O_5$-$K_2O$) chemical fertilizer medium were the best media to induce root length (2.42 and 2.21 cm/explant, respectively). Because higher fertilizer concentrations caused the plant to be unable

| Type of chemical fertilizers (A) | Root lengths (cm/explant) | | | | |
|---|---|---|---|---|---|
| | Concentration of chemical fertilizers (g $L^{-1}$) (B) | | | | |
| | 0.5 | 1.5 | 2.5 | 3.5 | Average |
| MS (control) | | | | | 2.42 a |
| 21-21-21 ($N$-$P_2O_5$-$K_2O$) | 2.65 | 2.41 | 2.10 | 1.68 | 2.21 ab |
| 36-5-5 ($N$-$P_2O_5$-$K_2O$) | 4.06 | 1.25 | 1.41 | 1.25 | 1.99 b |
| 30-20-10 ($N$-$P_2O_5$-$K_2O$) | 1.96 | 1.36 | 1.25 | 1.25 | 1.46 c |
| 10-52-17 ($N$-$P_2O_5$-$K_2O$) | 1.67 | 1.25 | 1.72 | 2.09 | 1.69 c |
| 15-30-15 ($N$-$P_2O_5$-$K_2O$) | 1.61 | 1.95 | 2.44 | 1.91 | 1.98 c |
| Average | 2.39 a | 1.77 b | 1.89 b | 1.77 b | |
| A | ** | | | | |
| B | ** | | | | |
| A × B | ** | | | | |

*Different alphabets in the same column and row are significantly different at level 0.05 compared to LSD.*
*[**]Significant difference at the level of 0.01.*

**Table 4.**
*The effect of chemical fertilizer types and concentrations on the root length of strawberry plantlets after being cultured for 4 weeks.*

to properly absorb nutrients for growth, the optimal concentration to promote root length (2.39 cm/explant) was 0.5 g $L^{-1}$ [12].

## 4. Conclusion

This experiment was conducted to study the effect of chemical fertilizer instead of MS medium on induced plantlets of strawberry (*F.* spp.) cultivar Royal 80. The findings revealed that the type and concentration of chemical fertilizer had an effect on the fresh weight of plantlets, the number of plantlets, the number of roots, and the root length. The strawberry was cultured on 0.5 g$L^{-1}$ of 21-21-21 medium, which was the best medium.

## Acknowledgements

The author gratefully acknowledges Khon Kaen University and the Faculty of Agriculture for financial support and for preparing the manuscript.

## Author details

Sumana Neera
Khon Kaen University, Khon Kaen, Thailand

*Address all correspondence to: sumana@kku.ac.th

## References

[1] Voltolina V. 14 Proven Health Benefits of Strawberries 18 Nutritional Benefits of Strawberries [Internet]. 2019. Available from: https://www.organicfacts.net/health-benefits/fruit/strawberries.html [Accessed: 16 July 2021]

[2] Devlieghere F, Vermelen A, Debevere J. Chitisan: Antimicrobial activity, interactions with food components and applicability as a coating on fruit and vegetables. Food Microbiology. 2004;**21**(6):703-714

[3] Food and Agricultural Organization (FAO). The State of Food Insecurity in the Word [Internet]. USA: FAO; 2008. Available from: http://www.fao.org/3/i0291e00.htm [ISBN: 978-92-5-106049-0]

[4] Karhu S, Hakala K. Micropropagated strawberries on the field. Acta Horticulturae. 2002;**567**:321-324. DOI: 10.17660/ACTAHORTIC.2002.567.68

[5] Lewin J. The Health Benefits of Strawberries [Internet]. 2018. Available from: http://www.bbcgoodfood.com

[6] Chien YK, Al-Jowid AM, Baiz A. An effective disinfection protocol for plant regeneration from shoot tip cultures of strawberry. American Journal of Biotechnology. 2009;**8**(11):2611-2615. DOI: 10.5897/AJB09.228

[7] Pipattanawong N. Strawberry production and the Royal Project Foundation, Thailand. Journal of Developments in Sustainable Agriculture. 2015;**10**:15-18

[8] Sutan AN, Popescu A, Isac V. *In vitro* culture medium and explant type effect on callogenesis and shoot regeneration in two genotypes of ornamental strawberry. Romanian Biotechnological Letters. 2010;**15**(2):12-18

[9] Neera S, Boungphengphanh D. Potential of common chemical fertilizers for micropropagation of *Rhynchostylis gigantea* (Lindl) Ridl. Acta Horticulturae. 2016;**1113**:73-78. DOI: 10.17660/ActaHortic.2016.1113.10

[10] Murashige T, Skoog F. A revised medium for rapid growth and bioassays with tobacco tissue cultures. Plant Physiology. 1962;**15**(3):473-479

[11] Li H, Huang R, Li T, Hu K. Ability of nitrogen and phosphorus assimilation of seven strawberry cultivars in a northern Atlantic coastal soil. In: 19th Congress of Soil Science, Soil Solution for Changing World. 1-6 August 2010; Brisbane, Australia

[12] Somboon T. Plant Physiology. Bangkok, Thailand: Kasetsart University; 2001. 237p

Section 3

# Strawberry Breeding and Genetic Diversity

Chapter 6

# The Genetic Diversity of Strawberry Species, the Underutilized Gene Pool and the Need for Cultivars with New Quality and Agronomic Attributes

*Pedro Antonio Dávalos-González, Ramón Aguilar-García, Alejandro Rodríguez-Guillén and Alba Estela Jofre-y-Garfias*

## Abstract

The growing increase in global strawberry production and consumption has been spectacular during this century. In 2019, 396,401 ha were planted, and 8.9 million tons of fruit were produced globally, and more than 50% of that volume was in the subtropical climate. The problems and losses caused by diseases and pests are of global importance, particularly with root and crown diseases, the severity and spread of which has been magnified by the cancelation of certain soil fumigants, and by the susceptibility to one or more of the parasites of the group of cultivars currently planted. The use of the genetic reservoir available both in the cultivated species, as in the 26 wild species, is a formidable wealth of genes, partially collected, and characterized, which can be of fundamental importance to introduce new genetic combinations into modern commercial cultivars and to redesign them, so that they have a greater adaptation to stresses caused by biotic and abiotic factors, in addition to an important improvement in the nutraceutical quality of the fruit. This chapter documents the importance of this gene pool in the development of elite cultivars with these qualities.

**Keywords:** *Fragaria spp.*, resistance to biotic agents, resistance to abiotic agents, nutraceuticals, breeding

## 1. Introduction

Global strawberry production grew at a rate close to 5% per year in the first two decades of the twenty-first century. At the beginning of this century, 4.57 million tons were produced annually, on an area of 40,000 ha versus 8.9 million tons and 396,401 ha in 2019 [1]. The origin of the production was given as follows: Asia and America are the continents with the highest contribution, where, in decreasing order,

China, USA, Mexico, Turkey and Egypt are the five largest producers in the world. The statistics of the last 40 years stand out several factors (1) The cultivation spread from 53 countries in 1980 to 77 in 2000 and 79 in 2019; (2) More than half of the current fruit production is in the subtropical climate; (3) Emerging countries such as: Turkey, Egypt and Morocco, became important production poles; and, (4) The high altitude tropics whose typical case is Mexico, showed its climatic benignity, which placed Mexico as the world's leading producer of fresh strawberries in autumn-winter, a period in which there is a deficit in the global market.

Other factors that are changing the role in the production-demand binomial are the cancelation of methyl-bromide [2], the promotion of organic cultivation, the interest in developing cultivars rich in bioactive and nutraceutical compounds [3], and the increasing importance of day-neutral cultivars [4]. These global trends are changing the profile of the strawberry industry, ultimately creating new technological demands of all kinds, especially for the main component, which are cultivars. In a holistic context, broadening the genetic base for new attributes and the formation of elite cultivars could have a major impact on better use of water, fertilizers, and adaptation to various stresses such as: alkaline pH, excessive heat, tolerance to frost damage, etc., furthermore, to help mitigate and/or eliminate future demand for synthetic pesticides.

Developing elite genotypes will imply a greater exploration, collection, and characterization of wild strawberry germplasm to face global problems [5, 6], a deep scientific knowledge of the genetic complexities to use it, especially in the case of those with ploidy levels other than octoploid. Nevertheless, molecular biology is currently advancing rapidly, and must be an ally of classical improvement, to advance more quickly in the objective of enriching the genetic base of the crop and achieving the development of cultivars with new characteristics. This chapter will present a review of contemporary problems of this crop, the use of current genetic resources as the main strategy to design their management, the factors that affect the under-utilization of the genetic reservoir, the demands for elite cultivars, with genetic resistance to biotic and abiotic factors, and better nutraceutical qualities, and the limitations of this approach.

## 2. Global context

The predominant plantation system in the subtropical environment is that developed by California, USA during the 20th century, it was adopted and/or adapted with certain variations in other countries from the equator to 42°latitude in both hemispheres [7]. Its technological support was the disinfection of soils with methyl-bromide + chloropicrin to eliminate soil diseases [8], the development of cultivars with high productivity and sensory quality of the fruit [9], and the optimization of the technological package for cultivation, fertigation, pest, disease and weed management practices [10]. The reproducibility of the previous production model, and the adoption of the macro-tunnel, located in Spain, Mexico, and other Mediterranean countries, among the largest producers of fresh strawberries.

When the use of methyl bromide ended in 2005 and 2015 in developed and developing countries, respectively, ended the relatively simple Era to eradicate biotic agents from the soil, since to date substitutes or alternatives are being investigated to replace it, being chemical, physical, microbiological agents, or a combination of them, that exerts action on a wide spectrum of biological entities [2].

Strawberries are grown in a wide variety of environments. In terms of latitude, it can be said that, from the equator to the polar zone [11], mainly in the northern hemisphere. Regarding altitude, from sea level to altitudes above 2000 meters above sea level [12]. These macroenvironments, with their different photoperiod, temperature, and rainfall regimes, as well as different pH and soil texture, are the genesis of an infinite series of microenvironments, and give rise to the so-called geographical and regional adaptation, a situation that affects cultivars. They can be adapted to a better or lesser degree to a certain environment [11].

The strawberry industry is experiencing a continuous varietal change. Except for China, where cultivars from Japan predominate [7], in the nine main strawberry producing countries, perhaps no more than 15 cultivars, generated by the Universities of California and Florida and, a few others from private companies are used. However, these genotypes share a close relationship since they descend from common or related parentals. Modern cultivars stand out for their productive qualities, good adaptation, and high sensory quality. The risky facet is associated with genetic uniformity and genetic erosion for traits that can confer tolerance and adaptation to biotic and abiotic factors, and their clonal spread, which is a risk of transmission of infectious agents.

The genetic vulnerability was shown since the end of the last century, both for nuclear genes [13] and for the cytoplasm [14] and becomes more valid in contemporary times, before the first signs of the globalization of phytosanitary problems of strawberry. During the twentieth century, biotic problems were caused by 20 pests, 108 diseases, and eight nematodes, in addition to five abiotic agents [15]. It was anticipated that others could arise [16] and this was the case in this century with *Drosophila suzukii* [17], and the diseases caused by *Colletotrichum* spp. [18], *Fusarium oxysporum* f. sp. *fragariae*, *Macrophomina phaseolina*, and probably *Neopestalotiopsis* spp.

## 3. Global challenges

Based on the available information, experiences of classical genetic improvement for the development of cultivars tolerant or resistant to diseases, pests, abiotic factors and recently to improve the nutraceutical quality of the fruit, will be addressed. An important aspect is that, for each goal of incorporating tolerance or resistance to a certain problem, the required sequence is to search for sources of genetic resistance [19], or of the richness of nutraceutical compounds [3], and then transfer it to the new cultivars. The commercial strawberry is octoploid, and wild plant populations of the 26 species known to date are found in nature [20], including their ancestors, and the newly discovered *F. emeiensis* [21]. The availability and use of this genetic wealth in the formation of cultivars were derived from the international literature. The ease of use of these genes depends on the chromosomal level of the species, the nuclear and cytoplasmic genetic compatibility between them, the type of inheritance of the resistance (qualitative or quantitative) and the availability of an appropriate technique to identify the resistant individuals.

### 3.1 Disease resistance

The development of cultivars with tolerance or resistance to certain diseases has been an approach of limited use in strawberries. Root and crown diseases are the group of parasites that cause the most economic damage. In the last century and the current one, the presence of at least seven important diseases has been reported: *Phytophthora fragariae*, *Verticillium* spp., *Phytophthora cactorum* [15]. In the latter, *Colletotrichum* spp. [18], *Fusarium oxysporum* f. sp. *fragariae* in Chile [22]; China [23]; Spain [24]; California, USA [25]; Iraq [26]; Serbia [27]; Turkey [28]; Iran [29]; Bangladesh [30]; Ecuador [31]; *Macrophomina phaseolina* in Florida, USA [32]; Israel [33]; California, USA [34]; Spain [35]; Argentina [36]; Iran [37]; Australia [38]; Chile [39]; Tunisia [40]; Italy [41]; and probably *Neopestalotiopsis* spp., emerging parasite whose presence has been reported in 17 countries during this century in Brazil [42], Egypt [43]; Morocco [44]; Spain [45]; Iran [46]; Vietnam [47]; Belgium [48]; Argentina [49]; India [50]; Korea [51]; Uruguay [52]; Italy [53]; Mexico [54]; China [55]; Ecuador [56]; Finland [57]; Taiwan [58], and that it could acquire global importance if the damage persists in future years. It is important to highlight that *Colletotrichum* spp. [59], *Fusarium oxysporum* f. sp. *fragariae* (FOF), *Macrophomina phaseolina* [60–62], and *Neopestalotiopsis* spp. [63–65] were reported since the twentieth century, but only FOF was important in Australia [66], Japan [67], Argentina [68], Korea [69], and Mexico [70].

Genetic resistance in strawberries has a history dating back to the last century, and valuable experiences that confirm the goodness of this strategy. In this sense, the United States Department of Agriculture, released a multitude of cultivars resistant to various races of *P. fragariae*, adapted to the cold climate of various countries in the world, whose commercial use and/or as sources of resistance, were used to obtain cultivars in other countries. The resistance genes used by the USDA were transferred from *F. chiloensis* clones Yaquina and Del Norte; and *Fragaria* x *ananassa* through the cultivars 'Md 683' and 'Aberdeen' (**Table 1**) [71, 72]. In populations of *F. chiloensis* from California, USA there is also resistance to this fungus [73, 74].

Other diseases of the twentieth century that justified the development of resistant cultivars were *Verticillium* spp., and *P. cactorum*. For the first disease, in the last century sources of genetic resistance were detected in *Fragaria* x *ananassa* [75], and *F. chiloensis* [76], but more recent studies also found resistance in *F. virginiana* and three diploid species [77]. The same situation occurred for *P. cactorum*, where resistance genes have been tracked back in commercial cultivars from the USA and Germany [78–81], and in two diploid species [82] (**Table 1**).

Root diseases that acquired global importance from the XXI century, have been the subject of research that allowed to strengthen the efforts made regionally during the twentieth century, such was the case of anthracnose. The disease can be caused by the species *C. fragariae*, (CF), *C. gloesporioides* (CG) and *C. acutatum* (CA), also affecting all organs, among them flower and fruit rot [18]. The following cultivars resistant to CF were developed: 'Florida Belle' [83]; to CF and CG 'Treasure' [84]; and resistant to CF and CA 'Pelican' [18] and at least one race of CG [85]; and resistant to CA, 'Flavorfest' [86]. Subsequent studies identified sources of resistance to CG in octoploid species [86, 87].

The same happened for FOF, resistant cultivars were detected in Korea [69], Japan [88]. Genetic resistance to FOF strains was detected in Mexico in cultivars from the United States and, also in *F. chiloensis* ssp. *pacifica* [89, 90] (**Figure 1**). The 'Ventana' cv is FOF resistant in California, USA [91]. In Mexico, there are selections in an intermediate stage of advance, which carry genes from both the cultivated species and *F. chiloensis* for resistance to FOF, adapted to the tropical high-altitude climate in Mexico [92] (**Table 1**; **Figure 2**). Recent investigations detected genotypes resistant to *Macrophomina phaseolina* in Australia and Egypt, although some are out of date cultivars [93, 94].

| Disease-causing agent | Species/resistance genes | Ploidy level | Genes used in cultivars | References |
|---|---|---|---|---|
| *Phytophthora fragariae* | *F.* x *ananassa* | 8x | Yes | [11, 71] |
| | *Fragaria chiloensis* Oregon | 8x | Yes | [11, 71] |
| | *F. chiloensis* California | 8x | Yes | [73, 74] |
| | *F. virginiana* | 8x | Yes | [11, 71] |
| *Verticillium dahliae* | *F.* x *ananassa* | 8x | Yes | [75, 77] |
| | *F. chiloensis* California | 8x | ? | [76] |
| | *F. virginiana* | 8x | ? | [77] |
| | *F. vesca* | 2x | No | [77] |
| | *F. nipponica* | 2x | No | [77] |
| | *F. iinumae* | 2x | No | [77] |
| *Phytophthora cactorum* | *F.* x *ananassa* | 8x | Yes | [78–81] |
| | *F. vesca* | 2x | No | [82] |
| | *F. nilgerrensis* | 2x | No | [82] |
| *Colletotrichum* spp. | *F.* x *ananassa* | 8x | Yes | [83, 85] |
| | *F. chiloensis* | 8x | ? | [86, 87] |
| | *F. virginiana* | 8x | ? | [86, 87] |
| *Fusarium oxysporum* f. sp. *fragariae* | *F.* x *ananassa* | 8x | Yes | [69, 88, 89] |
| | *F. chiloensis* spp. *pacifica* | 8x | In progress | [89, 90] |
| *Macrophomina phaseolina* | *Fragaria* x *ananassa* | 8x | Yes | [93, 94] |
| *Botrytis cinerea* | *F.* x *ananassa* | 8x | Yes | [95, 96] |
| | *Fragaria chiloensis ssp.* | 8x | ? | [97] |
| | *chiloensis* f. *chiloensis* | 8x | No | [98] |
| | *F. virginiana* | 2x | No | [98] |
| | *Fragaria vesca ssp. bracteata* | | | |
| *Xanthomonas fragariae* | *F.* x *ananassa* | 8x | Si | [99] |
| | *F. virginiana* | 8x | In progress | [99, 100] |
| | *F. moschata* | 6x | No | [101, 102] |
| | *F. pentaphylla* | 2x | No | [101] |
| | *F. vesca* | 2x | No | [102] |
| | *F. nilgerrensis* | 2x | No | [102] |
| *Podosphaeraaphanis* | *F. x ananassa* | 8x | Yes | [103] |
| | *F. virginiana* ssps. *Grayana and virginiana* | 8x | ? | [104] |
| Other foliar diseases* | *F. x ananassa* | 8x | Yes | [11] |
| | *F. virginiana* | 8x | No | [105] |
| | *F. viridis* | 2x | No | [106] |
| | *F. pentaphylla* | 2x | No | [107] |
| | *F. nilgerrensis* | 2x | No | [108] |
| Virus | *F.* x *ananassa* | 8x | Yes | [11, 109] |
| | *F. chiloensis* ssps. *Lucida*, and *pacifica* | 8x | Yes | [73, 74, 90] |

**Refer to Mycosphaerella fragariae and Diplocarpon earlianum.*

**Table 1.**
*Sources of disease resistance in F. x ananassa and other species of wild strawberry.*

Other diseases that attack foliage, flower, and fruit, caused mainly by fungi and a bacterium, are documented in **Table 1**. The damage due to *Botrytis cinerea* is globally important [15], and there have been detected although commercial cultivars are only moderately susceptible [95, 96], and apparently, the diversity and genetic variation for tolerance or resistance to the fungus is absent in the commercial species, which has

**Figure 1.**
*Root system from clones of Fragaria chiloensis ssp. pacifica showing different degrees of resistance, 60 days after being inoculated with fusarium oxysporum f. sp. fragariae.*

**Figure 2.**
*Comparison of an experimental clone carrying genes of F. chiloensis resistant to fusarium oxysporum f. sp. fragariae, and the viral complex of Irapuato, Gto. (central furrow), and two susceptible genotypes (left and right furrows).*

been an impediment to develop tolerant cultivars. However, in this century sources of genetic resistance were identified in progenitor species of cultivated strawberry and in a diploid species [97, 98] (**Table 1**).

The case of the bacterium *Xanthomonas fragariae* is quite similar to that of *Botrytis*, except that until recently resistance was detected in *F. virginiana*, since no sources of resistance were found in the commercial species [99, 100]. Additionally, evidence of immunity genes found in *F. moschata* [100–102], and resistance genes in the diploids *F. nilgerrensis* and *F. vesca* f. *alba,* and 'Illa Martin' and in the diploid *F. pentaphylla* [102], but not in *F. nilgerrensis*, *F. daltoniana*, *F. iinumae*, *F. vesca*, *F. viridis*, *F. gracilis*; *F. nubicola* and *F. orientalis* [101].

For other foliage diseases such as powdery mildew [103, 104], *Mycosphaerella fragariae* and *Diplocarpon earliana* [11, 105]; since the previous century, resistance genes were found in cultivated species and in *F. virginiana*, and tolerant or resistant cultivars were developed. Subsequently, knowledge has been enriched with the detection of resistance genes in other octoploid and diploid species [106–108] (**Table 1**).

A scientifically important and economically transcendental case, for the strawberry industry in California, during the twentieth century, was the practical demonstration that genetic tolerance was the best alternative to avoid economic losses, caused by the yellowing viral complex [109]. Around 1945, the University of California released the cultivars 'Shasta' and 'Lassen', tolerant to the viral complex. This event marked the beginning of the Era of the formation of cultivars with high yield potential and sensory quality of strawberry, adapted to the subtropical climate of California. The tolerance genes introduced in these cultivars were derived from a cultivar called 'Ettersburg 121', which had within its ancestors' genes from *Fragaria chiloensis* of the central coast of California, USA (**Table 1**) [73, 74]. In Irapuato, Gto., Mexico, there are advanced clones with tolerance to the local viral complex, which carry genes of *F. chiloensis* ssp. *pacifica* from California USA, [92] (**Figure 3**).

### 3.2 Pest resistance

Pests of greatest global importance and causing major economic damage, are the two-spotted spider mite (*Tetranychus urticae*), lygus bug (*L. lineolaris*, *L. hesperus*) and possibly other genus of bugs, in addition to thrips (*Frankliniella occidentalis*). The aphid *Chaetosiphon* spp. is a pest that transmits at least five viruses and for that reason, it is also important [15].

The genetic improvement in strawberries for tolerance or resistance to some of these pests was almost null in the previous century, for several reasons. Partly because of the availability of synthetic pesticides, which at first allowed easy control. Also due to the technical difficulty, time invested and economic cost of maintaining a genetic improvement program to achieve this objective, and in another, because there was a lot of pressure to develop cultivars with high yield potential and good sensory quality, even if they were susceptible to the most important pests of the crop.

Despite this unfavorable environment, there were pioneering scientists in spider mite and aphid resistance research. By far, the two-spotted spider mite has always been the main pest of strawberries and for this reason, the first studies evaluated the reaction of cultivars of the time to the mite. Experience showed that it developed larger populations on certain genotypes, which confirmed the presence of genetic variation in the host, with various degrees of damage, from tolerant to susceptible [110].

**Figure 3.**
*Comparison between a resistant clone (upper furrow), and a susceptible one to the viral complex present in Irapuato's region.*

A survey with a greater number of cultivars and clones of octoploid species, allowed us to locate sources of resistance in the cultivated species, in *F. chiloensis* from North America [111, 112] and from South America and in *F. virginiana* [113]. However, an important aspect was that in the wild clones of both octoploid species, a higher level of resistance was identified than in the cultivars. In addition, and very important, is that some *F. chiloensis* clones, that were resistant to the spider mite, were also resistant to the aphid *C. fragaefolii*, which is a vector of at least five economically important viruses (**Table 2**).

For the other pests of global importance, genetic variation is generally mentioned at the cultivar level, and this is the case of *Lygus* spp. [114] and *Frankliniella occidentalis* [115], although the information available for both pests is still very limited, and many aspects of the host–parasite relationship are unknown.

An outstanding case is a problem with the oriental fly *Drosophila suzukii*, a pest that has spread throughout the main strawberry-producing countries, which parasitizes other small fruits as well, in which the damage could be considered in the future. A study carried out in Germany, with octoploid and diploid genotypes, identified that in the diploid *F. vesca* certain clones had a low mosquito emergence [17] (**Table 2**).

### 3.3 Outstanding characters of adaptation and resistance to abiotic factors

In this section, a series of agronomic attributes are presented and discussed, which allow the plant a better adaptation to the environment and/or mitigate its adverse effect on it and eventually result in higher productivity and quality of the fruit and therefore are attributed with high economic importance.

| Pest | Species/resistance genes | Ploidy level | Genes used in cultivars | References |
|---|---|---|---|---|
| *Tetranychus urticae* | *F.* x *ananassa* | 8x | Yes | [110] |
| | *Fragaria chiloensis* ssps. *Lucida, and pacifica* | 8x | No | [111, 112] |
| | | 8x | No | [112] |
| | *Fragaria chiloensis ssp. chiloensis* | 8x | No | [113] |
| | *F. virginiana* | | | |
| *Chaetosiphon fragaefolii* | *F. chiloensis* | 8x | ? | [112] |
| *Lygus spp.* | *F.* x *ananassa* | 8x | Yes | [114] |
| *Frankliniella occidentalis* | *F. x ananassa* | 8x | ? | [115] |
| *Drosophila suzukii* | *F. x ananassa* | 8x | ? | [17] |
| | *F. vesca* | 2x | Yes | |

**Table 2.**
*Sources of resistance to pests of global importance in F. x ananassa and other species of wild strawberry.*

Among the 11 listed attributes, those related to wide adaptation, low chilling requirements, resistance to low temperatures and versatility for different photoperiod regimes, have had primary importance with the evolution under cultivation of the octoploid strawberry and consequently with the already cited adaptation to environments as contrasting as the duration of the photoperiod, temperature regime, cold-chilling needs, rainfall, soil texture, etc. [11, 74, 116–120].

Previous characters present in one, or both, octoploid parental species of cultivated strawberry were surely transferred to it during the synthesis of both species in Europe in the seventeenth century, as well as with the introduction of these ancient European cultivars into the USA, and its numerous introgressions of *F. chiloensis* and *F. virginiana* genes, by amateur breeders in the United States during the eighteenth century, which originated a multitude of cultivars more rustic and adaptable to the environments of that country, where the availability at first of short-day cultivars, over the years and due to this introgression, originated long-day octoploid cultivars [11] and later, as explained below, the day-neutral cultivars. Other sources of cyclical flowering have been documented in *F. virginiana* [118, 119].

One of the classic examples of the impact on the strawberry industry is the incorporation of genes that confer the day-neutral character and allow continuous flowering in the subtropical environment. The original source of the day-neutral character was found in *F. virginiana* ssp. *glauca,* in a plant collected in Utah [74]. These genes were introduced through the cultivar 'Shasta' and by means of a carefully modified backcross program, after three cycles of crossing and selection, the first cultivars with the day-neutral trait were released in 1979. In contemporary times most day-neutral cultivars carry those genes transferred by Bringhurst [4, 9].

Resistance of the plant and its different organs to low temperatures is another attribute, which allows minimizing the damage with temperatures below 0°C and is crucial to mitigate the damage of these organs. In the tropical climate at altitudes above 1500 meters above sea level, temperatures below the mentioned threshold can cause large yield losses during autumn and winter in cultivation without a macro-tunnel. However, the fore effect is probably maximized by the sudden increase in temperature up to 25°C in three hours, so this wide thermal oscillation could be the cause of the damage to the plant, flower, and fruit. Among the genotypes sown in Mexico, it has been

| Factor | Species/resistance genes | Ploidy level | Genes used in cultivars | References |
|---|---|---|---|---|
| Wide adaptation | *F. x ananassa* | 8x | Yes | [11] |
| | *Fragaria chiloensis* | 8x | ? | [116] |
| | *F. vesca* | 2x | No | [117] |
| Low chilling requirements | *F. chiloensis* ssps. *Pacifica* and *lucida* | 8x | Yes | [73] |
| | *Fragaria chiloensis ssp. chiloensis* | 8x | Yes | [118] |
| | *F. virginiana* | 8x | ? | [118] |
| Photoperiod response | *F. virginiana* | 8x | Yes | [74] |
| | Day neutral (DN) | 8x | In progress? | [4, 119] |
| | Other sources of DN | 8x | ? | [11] |
| | Short day and long day | 2x | No | [119] |
| | *F. chiloensis* ssps. *Pacifica and. Lucida* | | | |
| | *F. vesca* | | | |
| Cold resistance | *F. virginiana* | 8x | ? | [11] |
| | *F. chiloensis* ssp. *chiloensis* f. | 8x | No | [6, 11] |
| | *patagonica* | 6x | No | [117] |
| | *F. moschata* | 4x | No | [108, 117] |
| | *F. orientalis* | 4x | No | [117] |
| | *F. moupinensis* | 2x | No | [117] |
| | *F. viridis* | 2x | No | [11, 117] |
| | *F. daltoniana* | 2x | No | [120] |
| | *F. iinumae* | 2x | No | [117] |
| | *F. nipponica* | 2x | No | [117] |
| | *Fraxinus mandshurica* | | | |
| Heat resistance | *F. virginiana* | 8x | ? | [4, 11] |
| Drought resistance | *F. chiloensis* North America | 8x | ? | [76] |
| | *F. chiloensis* South America | 8x | ? | [6] |
| | *F. virginiana* | 8x | ? | [11] |
| | *F. orientalis* | 4x | No | [117] |
| Resistance to waterlogging | *F. viridis* | 2x | No | [108] |
| | *F. nilgerrensis* | 2x | No | [108] |
| | *F. pentaphylla* | 2x | No | [108] |
| Resistance to salinity | *F. chiloensis* ssps. *Pacifica* and *lucida* | 8x | Yes | [76] |
| | *F. chiloensis* ssp. *chiloensis* | 8x | ¿ | [6] |
| Resistance to alkaline pH | *F.* x *ananassa* | 8x | Yes | [11] |
| | *F. virginiana* | 8x | ? | [117] |
| | *F. chiloensis* | 8x | ? | [117] |
| | *F. viridis* | 2x | No | [117] |
| Iron deficiency resistance | *F. x ananassa* | 8x | Yes | [122] |
| | *Fragaria chiloensis ssp. pacifica* | 8x | In progress | [122, 123] |
| Low nutrient requirements | *F. chiloensis* ssps. *Chiloensis and pacifica* | 8x | No | [74, 124] |

**Table 3.**
*Genetic diversity for traits associated with abiotic factors of global importance in F. x ananassa and other species of wild strawberry.*

observed that the most susceptible to this thermal shock are the day-neutral cultivars of California, compared to the short-day cultivars of California and Florida, respectively.

There are reports about sources of resistance to low temperatures in the progenitor species of the cultivated strawberry [6, 11], and also in some diplo, tetra and hexaploid species, [11, 108, 117, 120] as can be seen in the summary of **Table 3**. This desirable quality could acquire more relevance as more strawberry is grown in the tropics, as macro trends in crop expansion suggest for the near future. The opposite character, which is resistant to high temperatures during summer, could be important in certain latitudes, although except for Darrow [11], there was only recently interest in this stress as a cause of inhibition of flowering in short-day and day-neutral cultivars, when daytime temperatures are around 26°C [4].

Adaptation to deficits and excesses of moisture is documented for some strawberry species [6, 11, 108, 117, 121] (**Table 3**), and considering the current and future growing environments, both characteristics can be valuable, particularly a gradient related to the efficient use of water, or in other words, cultivars that require the fewer amount of water per kg of fruit produced, since most of the strawberries are grown under irrigation, and this is an input whose availability for agricultural use is less and less.

Other qualities that are found in wild species and are of great economic and environmental importance are those related to resistance to alkaline pH, salinity, and efficient use of iron in those soils [6, 11, 117]. In many of the countries where strawberries are grown, there are problems of iron deficiencies (**Figure 4**), induced by the alkaline pH of the soils, a problem that is partially solved with the application of iron in different forms. It has been observed that there is genetic variation for the efficient use of iron by certain cultivars and octoploid species of strawberry [122, 123], but unfortunately, on many occasions, there are no genotypes available that have this quality and are also adapted and productive to cultivation environments, where these nutritional deficiencies are manifested [122] (**Figure 5**).

**Figure 4.**
*Iron deficiency in a commercial plantation planted in soil with alkaline pH in Irapuato, Gto., Mexico.*

**Figure 5.**
*Clones with genes of Fragaria chiloensis resistant to iron deficiencies planted in an alkaline pH soil of Irapuato, Gto., Mexico.*

On the other hand, non-renewable inputs such as the use of synthetic fertilizers, could be better used by incorporating in modern cultivars the genes that confer a more efficient use of them, qualities present in certain wild species [73, 124, 125] and that until now have not been used (**Table 3**).

### 3.4 Characters associated with sensory and nutraceutical quality

Strawberry has a long history of genetic improvement for traits associated with the sensory quality of the fruit. Certainly, since ancient times, the aborigines of the new world [12, 120, 125, 126], practiced selection for some organoleptic characteristics such as fruit weight, color, firmness and flavor, outstanding attributes that have been reported in the landrace's varieties of Chile [124–127]. These qualities, which are under genetic control, have been incorporated into commercial cultivars of

| Character | Species | Ploidy level | Used genes in cultivars | References |
|---|---|---|---|---|
| Big fruit | *Fragaria chiloensis* | 8x | Yes | [11] |
| | *F. daltoniana* | 2x | No | [11, 117] |
| | *F. nilgerrensis* | 2x | No | [11, 117] |
| Fruit firmness | *Fragaria chiloensis ssp. chiloensis* | 8x | Yes | [124, 126] |
| | *F. pentaphylla* | 2x | No | [107] |
| | *F. viridis* | 2x | No | [117] |
| Aromatic fruit | *F. moschata* | 6x | No | [11, 117] |
| | *F. nilgerrensis* | 2x | Yes | [128] |
| | *F. viridis* | 2x | No | [117] |
| Fruit color<br>Almost white to red<br>Unusual bright red | *F. chiloensis* ssp. *chiloensis*<br>*F. pentaphylla* | 8x<br>2x | Yes<br>No | [11]<br>[107] |
| High nutraceutical content | *Fragaria virginiana ssp. glauca*<br>*Fragaria chiloensis ssp. lucida*<br>(cyanidin)<br>*F. vesca* | 8x<br>8x<br>2x | In process<br>No<br>No | [3]<br>[134]<br>[133] |

**Table 4.**
*Characteristics of sensory and nutraceutical qualities of strawberry in F. x ananassa, and other wild species of strawberry.*

*F. xananassa*. The large fruit size and firmness undoubtedly came from *F. chiloensis* ssp. *chiloensis* f. *chiloensis* [11]. Clones with large fruit have been identified in other species such as *F. daltoniana* and *F. nilgerrensis* [11, 117]. The firmness of fruit, which is a highly appreciated quality in strawberries, derived from *F. chiloensis* [125, 126] has also been found in certain diploid species [107, 117] (**Table 4**).

Color, flavor and aroma are attributes of the fruit that influence consumer acceptance [128]. The genetic diversity for these traits is partially documented. For example, for color it is possible to find a range of tones from albino to red in some species [11, 129, 130], the same happens with the flavor where outstands certain octoploids and diploids, while aroma *F. moschata* is recognized as a species that is above all [117] (**Table 4**).

Nutritional qualities of strawberries were documented since the previous century for the high content of vitamin C, as much or more than some citrus fruits, and Hansen and Waldo demonstrated in 1944 its genetic control in commercial strawberry cultivars [75]. Evaluations of California's cultivars showed a range of 50 to 100 mg of vitamin C per 100 g fresh weight, with 'Tufts' standing out [131] (**Table 4**).

With the medical recognition of the benefits for human health of certain bioactive compounds such as flavonoids and polyphenols [132], in addition to the already known properties of vitamins and minerals, and with the confirmation that the strawberry belongs to the group of fruits with high content of these substances, its consumption increased and there was an interest in increasing the nutraceutical properties of strawberries, through genetic improvement [3].

Research groups of some prestigious institutions have identified some important compounds, their presence in cultivars [133], and certain cases, which are the strawberry species whose contents are higher and can be the appropriate genetic source for these traits to be transferred to new cultivars. For example, wild plants with a high content of cyanidin, a type of anthocyanin, that helps to reduce risks of type 2 diabetes, certain types of cancer and heart problems, were identified in *F. chiloensis* [134] (**Table 4**).

Diamanti et al. [3], identified certain wild clones of *F. virginiana* ssp. *glauca* with high antioxidant content, and through interspecific hybridization, followed by three cycles of backcrosses, managed to recombine strawberry genotypes with high potential yield and with a higher content of anthocyanins, polyphenols and greater antioxidant capacity, than the cultivar 'Romina'. This is a cutting-edge genetic improvement approach, since the results suggest that it is possible to reconcile in this particular case, a high biomass yield with a higher content of bioactive compounds.

## 4. Expanding the use of the genetic pool

As a corollary of the information gathered from the available literature on the subject under discussion, it was evident that genetic improvement in strawberries was fundamentally aimed at increasing productivity, the sensory quality of strawberries, and adaptation to various environments [9], and in certain countries, there were also relevant experiences in the formation of varieties with resistance to root diseases [71, 135], foliar diseases [11] and to viral complexes [74]. Among the wide genetic diversity existing in the 26 species, only a part of the reservoir has been used, basically from *Fragaria* x *ananassa* [11], infrequently, from other octoploid species [74, 136], and exceptionally, with a lower of ploidy level [128].

Several avant-garde approaches have been besought and applied to use the genetic richness of wild species, to expand the genetic base of cultivated strawberries. One is the use of synthetic octoploids to take advantage of the genes of different levels of ploidy and bring them to the octoploid level [137], another is to perform the synthesis of *Fragaria* x *ananassa* again, but using a select group of progenitors of both, *F. chiloensis* as well as *F. virginiana*, which have been chosen for many characters such as: yield, quality and resistance to adverse factors [4, 6, 105, 138], and one more, is to form synthetic decaploids to introduce the complete genome of both, octoploid and diploid species [128, 139]. The introgression of genes of octoploid species towards the cultivated one [74] has been practiced successfully and with economically impressive results, but its contribution to broadening the genetic base is minimal. The first three day-neutral cultivars released, contributed only 6.25% of *F. virginiana* ssp. *glauca* genes.

The collected germplasm of *F. chiloensis* [117, 140–143] and *F. virginiana* [142, 144, 145], are available in the gene banks of certain countries [140], they have been characterized for certain attributes of cyclic flowering [4, 146], horticultural [5], resistance to diseases [89], resistance to pests [111, 112, 147], and at the molecular level [148–150]. However, it is important to characterize them to deal with global problems like pests, diseases, adaptation, and tolerance to abiotic factors, as well as for their nutraceutical properties. Among the global problems that the strawberry industry is facing, the best cultivars are susceptible to the most important pests and diseases, and regarding genetics, it stands out that there is little information on the sources of resistance to pests. The same situation occurs, towards resistance to low temperatures and there is also a great deficit of information on the richness of nutraceuticals in wild species.

Researchers from Michigan University [4] evaluated 2500 *F. virginiana* clones and 6000 *F. chiloensis* clones collected in California, the Pacific coast of North America and Chile. Among them, 38 elite parents were identified, and used to synthesized populations of *F.* x *ananassa*. If this germplasm were available, it could be characterized and used to deal with the global problems mentioned. If no sources of resistance are found for all the problems mentioned, the collection of octoploid species should be expanded. The populations of *F. chiloensis* from Alaska, whose genotypes withstand temperatures of −10°C and whose usefulness for that purpose was mentioned last century [11].

Little documented is the gene pool in species other than the octoploid level. There are likely valuable genes of economic importance that do not exist in octoploid species, that are currently underutilized, and that could contribute to solve emerging problems in the strawberry industry. As an example, the immunity reported to *Xanthomonas fragariae*, in *F. moschata* [102], and also, about the resistance in *F. pentaphylla* [101], and *F. vesca* [102], for the probable tolerance to *Drosophila suzukii* in *F. vesca* [17], to name just a few cases already published.

Under the above-mentioned needs, it is important to continue with the germplasm collections of Asian species [108, 151] since, to date, it is the region with the highest number of reported species, where all levels of ploidy are found, except the hexa and, octoploids. Due to the contrasting environments where they are found, and the molecular genetic diversity existing in regions such as Tibet [152]; the presence of genes for resistance to low temperatures is potentially suspected, and certain indications reinforce this hypothesis. Luo et al. [153] demonstrated the possibility of transferring resistance at low temperatures from a wild pentaploid parent from China. It is also possible that, in Chinese species, there is resistance to moisture deficits and excesses, and resistance to foliar diseases [102, 108]. Tetraploid species are particularly interesting and hypothetically important, since if there were genes for outstanding traits absent in the octoploids [154], their transfer to these in some evolutionarily related species would be relatively less genetically complicated [138]. There is a lack of knowledge of the degree of genetic affinity between the five tetraploid and octoploid species, for the use of the possible genetic richness of the tetraploids at the octoploid level. Classic breeding methods for crosses between species of the same and different ploidy levels have been widely described [11, 74, 75, 137, 139], and they should be surely complemented with recently developed biotechnological techniques [155].

## 5. Conclusions

During the present century, except for explorations in China, strawberry germplasm collections have ceased. In attention to the serious phytosanitary problems of strawberries, international collaboration is important to take advantage of the germplasm collections and populations derived from them, deposited in different repositories, and characterize them for those cases of global problems. Simultaneously with the above, evaluation techniques must be developed to rigorously characterize the germplasm reaction and identify the sources of valuable genes.

Faced with the need to mitigate global phytosanitary problems, genetic resources must provide part of the solution, and use those underused genes to form more rustic strawberry genotypes for biotic and abiotic factors, and with better nutraceutical quality of the fruit.

## Acknowledgements

The authors acknowledge the generosity of Dr. R.S. Bringhurst (RIP) for donating the germplasm of *F. chiloensis* for the detection of sources of genetic resistance to *Fusarium oxysporum* f. sp. *Fragaria*e.

## Author details

Pedro Antonio Dávalos-González[1], Ramón Aguilar-García[1],
Alejandro Rodríguez-Guillén[1] and Alba Estela Jofre-y-Garfias[2]*

1 National Institute of Forestry, Agricultural and Livestock Research (INIFAP), Celaya, Guanajuato, Mexico

2 National School of Higher Studies León Unit, National Autonomous University of Mexico (ENES-León, U.N.A.M). León, Guanajuato, Mexico

*Address all correspondence to: ajofregarfias@gmail.com

## References

[1] FAOSTAT. 2021. Available from: http://faostat.fao.org/site/339/default.aspx [Accessed: November 17, 2021]

[2] Daugovish O, Knapp S, Gordon T, Fennimore S, Muramoto J, Bolda M. Soil pest management in current California strawberry production: A review. Acta Horticulturae. 2021;**2021**(1309):701-709. DOI: 10.17660/ActaHortic.2021.1309.101

[3] Diamanti J, Mazzoni L, Balducci F, Capelletti R, Capocasa F, Battino M, et al. Use of wild genotypes in breeding program increases strawberry fruit sensorial and nutritional quality. Journal of Agricultural and Food Chemistry. 2014;**62**:3944-3953. DOI: 10.1021/jf500708x

[4] Hancock JF, Edger PP, Callow PW, Herlache T, Finn CE. Generating a unique germplasm base for the breeding of day-neutral strawberry cultivars. HortScience. 2018;**53**:1069-1071. DOI: 10.21273/HORTSCI12840-18

[5] Hancock JF, Callow PW, Serce S, Son PQ. Variation in horticultural characteristics of native *F. virginiana* and *F. chiloensis* of north and South America. Journal of the American Society for Horticultural Science. 2003;**128**:201-208. DOI: 10.21273/JASHS.128.2.0201

[6] Hancock JF, Hokanson SC, Finn CE, Hummer KE. Introducing a supercore collection of wild octoploid strawberries. Acta Horticulturae. 2002;**567**:77-79. DOI: 10.17660/ActaHortic.2002.567.6

[7] López-Aranda JM, Soria C, Santos BM, Miranda L, Domínguez P, Medina-MínguezJJ.Strawberryproduction in mild climates of the world: A review of current cultivar use. International Journal of Fruit Science. 2011;**11**:232-244. DOI: 10.1080/15538362.2011.608294

[8] Wilhelm S, Paulus AO. How soil fumigation benefits the California strawberry industry. Plant Disease. 1980;**64**:264-270

[9] Bringhurst RS, Voth V, Shaw DV. University of California strawberry breeding. HortScience. 1990;**25**(834):998-999. DOI: 10.21273/HORTSCI.25.8.834

[10] Welch N. Strawberry production in California. California: University of California. Cooperative extension leaflet.; 1989:No. 2959.

[11] Darrow GM. The Strawberry: History, Breeding and Physiology. 1rd ed. New York: Holt, Rinehart and Winston; 1966. p. 447

[12] Finn CE, Hancock J, Heider C. Notes on the strawberry of Ecuador: Ancient land races, the community of farmers, and modern production. HortScience. 1998;**33**:583-587. DOI: 10.21273/HORTSCI.33.4.583

[13] Sjulin TM, Dale A. Genetic diversity of north American strawberry cultivars. Journal of the American Society for Horticultural Science. 1987;**112**:375-385

[14] Dale A, Sjulin TM. Few cytoplasms contribute to north American strawberry cultivars. HortScience. 1990;**25**:1341-1342. DOI: 10.21273/HORTSCI.25.11.1341

[15] Maas JL. Compendium of Strawberry Diseases. 2nd. ed. St. Paul Mineapolis; 1998, 1998. p. 98. DOI: 10.2307/3793273

[16] Maas JL, Galletta GJ. Recent progress in strawberry disease research. Acta

Horticulturae. 1997;**439**:769-779. DOI: 10.17660/ACTAHORTIC.1997.439.128

[17] Gong X, Bracker L, Bolke N, Plata C, Zeitlmayr S, Metzler D, et al. Strawberry accessions with reduced *Drosophila suzukii* emergence from fruits. Frontiers in Plant Science. 2016;**7**:1880. DOI: 10.3389/fpls.2016.01880

[18] Smith BJ. Epidemiology and pathology of strawberry anthracnose: A north American perspective. HortScience. 2008;**43**:69-73. DOI: 10.21273/HORTSCI.43.1.69

[19] Harlan JR. Genetic resources in wild relatives of crops. Crops Science. 1976;**16**:329-333. DOI: 10.2135/CROPSCI1976.0011183X001600030004X

[20] Hummer KE, Bassil N, Njuguna W. Fragaria. In: Kole C, editor. Wild Crop Relatives: Genomic and Breeding Resources Temperate Fruits. Vol. 2011. Berlin, Heidelberg: Springer-Verlag; 2011. pp. 17-44. DOI: 10.1017/S0014479711000950

[21] Quiao Q, Edger PP, Xue L, Quiong L, Lu J, Zhang Y, et al. Evolutionary history and pan-genome dynamics of strawberry (*Fragaria* spp.). PNAS. 2021;**118**(45):e2105431118. DOI: 10.1073/pnas.2105431118

[22] González FJ, Walls S, Mancilla M. Detection by PCR-RFLP of *fusarium oxysporum* f. sp. *Fragariae* causal agent of fusariosis in strawberry. Mycological Bulletin. 2005;**20**:63-72. DOI: 10.22370/BOLMICOL.2005.20.0.278

[23] Huang Y, Zhen W, Zhang L, Zhang G, Tian T, Cheng H, et al. Separation of causal organisms and biocontrol of strawberry (*Fragaria ananassa* Duch.) replant disease. Biotechnology. 2005;**15**:74-76

[24] Arroyo FT, Llergo Y, Aguado A, Romero F. First report of *Fusarium* wilt caused by *fusarium oxysporum* on strawberry in Spain. Plant Disease. 2009;**93**:323. DOI: 10.1094/PDIS-93-3-0323B

[25] Koike ST, Kirpatrick SC, Gordon TR. *Fusarium* wilt of strawberry caused by *fusarium oxysporum* in California. Plant Disease. 2009;**93**:1077. DOI: 10.1094/PDIS-93-10-1077A

[26] Juber KS, Al-Juboory HH, Al-Juboory SB. *Fusarium* wilt disease of strawberry caused by *fusarium oxysporum* f.sp. *fragariae* in Iraq and its control. Journal of Experimental Biology and Agricultural Sciences. 2014;**2**:419-427

[27] Stankovic I, Ristic D, Vucurovic A, Milojevic K, Nicolic D, Krstic B, et al. First report of *fusarium* wilt of strawberry caused by *fusarium oxysporum* in Serbia. Plant Disease. 2014;**98**:1435. DOI: 10.1094/PDIS-04-14-0346-PDN

[28] Dinler H, Benlioglu S, Benlioglu K. Occurrence of *fusarium* wilt caused by *fusarium oxysporum* on strawberry transplants in Aydın Province in Turkey. Australasian Plant Disease Notes. 2016;**11**:10. DOI: 10.1007/s13314-016-0196-3

[29] Ayoubi N, Soleimani MJ. Morphologicalandmolecularidentification of pathogenic *fusarium* spp. on strawberry in Iran. Sydowia. 2016;**68**:163-171. DOI: 10.12905/0380.sydowia68-2016-0163

[30] Surovy MZ, Kabir MK, Gupta DR, Hassan O, Mahmud NU, Sabir AA, et al. First report of *fusarium* wilt caused by *fusarium oxysporum* on strawberry in Bangladesh. Plant Disease. 2018;**103**:367. DOI: 10.1094/PDIS-07-18-1121-PDN

[31] Salazar S, Bayas-Morejón F, Niño Z, Guevara J, Ruilova M, Merino O, et al. Morphological and biochemical indicators of *fusarium oxysporum* f. sp. *fragariae* in strawberry crops (*Fragaria* x *ananassa* Duch) in the province of Pichincha, Ecuador. Agronomy Research. 2018;**16**:1854-1861. DOI: 10.15159/ar.18.206

[32] Mertely J, Seijo T, Peres N. First report of *Macrophomina phaseolina* causing a crown rot of strawberry in Florida. Plant Disease. 2005;**89**:434. DOI: 10.1094/PD-89-0434A

[33] Zveibil A, Freeman S. First report of crown and root rot in strawberry caused by *Macrophomina phaseolina* in Israel. Plant Disease. 2005;**89**(4):434. DOI: 10.1094/PD-89-1014C

[34] Koike ST. Crown rot of strawberry caused by *Macrophomina phaseolina* in California. Plant Disease. 2008;**92**:1253. DOI: 10.1094/PDIS-92-8-1253B

[35] Avilés M, Castillo S, Bascon J, Zea-Bonilla T, Martín-Sánchez PM, Pérez-Jiménez RM. First report of *Macrophomina phaseolina* causing crown and root rot of strawberry in Spain. Plant Pathology. 2008;**57**:382. DOI: 10.1111/J.1365-3059.2007.01717.X

[36] Baino OM, Salazar SM, Ramallo AC, Kirschbaum DS. First report *of Macrophomina phaseolina* causing strawberry crown and root rot in northwestern Argentina. Plant Disease. 2011;**95**:1477. DOI: 10.1094/PDIS-03-11-0193

[37] Sharifi K, Mahdavi M. First report of strawberry crown and root rot caused by *Macrophomina phaseolina* in Iran. Iranian Journal of Plant Pathology. 2012;**47**:161

[38] Hutton DG, Gomez AO, Mattner SW. *Macrophomina phaseolina* and its association with strawberry crown rot in Australia. International Journal of Fruit Science. 2013;**13**:149-155. DOI: 10.1080/15538362.2012.698143

[39] Sánchez S, Gambardella M, Enriquez JL, Diaz I. First report of crown rot of strawberry caused by *Macrophomina phaseolina* in Chile. Plant Disease. 2013;**97**:996. DOI: 10.1094/PDIS-12-12-1121-PDN

[40] Hajloui MR, Mnari-Hattab M, Sayeh M, Zarrouh I, Jemmali A, Koike ST. First report of *Macrophomina phaseolina* causing charcoal rot of strawberry in Tunisia. New Disease Reports. 2015;**32**:14. DOI: 10.5197/J.2044-0588.2015.032.014

[41] Gerin D, Dongiovanni C, De Miccolis Angelini RM, Pollastro S, Faretra F. First report of *Macrophomina phaseolina* causing crown and root rot on strawberry in Italy. Plant Disease. 2018;**102**:1857. DOI: 10.1094/PDIS-01-18-0191-PDN

[42] Camili EC, Carbonari M, de SouzaNL. Characterization of *Pestalotiopsis longisetula* and its pathogenicity in strawberry. Summa Phytopathologica. 2002;**28**:213-214

[43] Embaby E. *Pestalotia* fruit rot on strawberry plants in Egypt. Egypt Journal of Phytopathology. 2007;**35**:99-110

[44] Mouden N, Benkirane R, OuazzaniTouhami A, Douira A. Mycoflore de quelques varieties deu fraisier (*Fragaria ananassa* L.), cultivées dans la regíón du Gharb et le Loukkos (Maroc). Journal of Applied Biosciences. 2013;**61**:4490-4514. DOI: 10.4314/JAB.V61I0.85597

[45] Chamorro M, Aguado A, De los Santos B. First report of root and crown rot caused by *Pestalotiopsis clavispora* (*Neopestalotiopsis clavispora*)

on strawberry in Spain. Plant Disease. 2015;**100**:1495. DOI: 10.1094/PDIS-11-15-1308-PDN

[46] Ayoubi N, Soleimani MJ. Strawberry fruit rot caused by *Neopestalotiopsis iranensis* sp. nov., and *N. mesopotamica*. Current Microbiology. 2016;**72**:329-336. DOI: 10.1007/s00284-015-0955-y

[47] Dung L, Thi N, Hoang P, Ngoc P. The first report of *Pestalotiopsis* sp. causing crown rot disease on strawberries in Dalat. Dalat University Journal Science. 2016;**6**:364-367. DOI: 10.37569/dalatuniversity.6.3.81(2016)

[48] Van Hemelrijck W, Ceustermans A, Van Campenhout J, Lieten P, Bylemans D. Crown rot in strawberry caused by *Pestalotiopsis*. Acta Horticulturae. 2017;**1156**:781-786. DOI: 10.17660/ACTAHORTIC.2017.1156.115

[49] Obregon VG, Meneguzzi NG, Ibañez JM, Lattar TE, Kirschbaum DS. First report of *Neopestalotiopsis clavispora* causing root and crown rot on strawberry plants in Argentina. Plant Disease. 2018;**102**:1856. DOI: 10.1094/PDIS-02-18-0330-PDN

[50] Amrutha P, Vijayaraghavan R. Evaluation of fungicides and biocontrol agents against *Neopestalotiopsis clavispora* causing leaf blight of strawberry (*Fragaria* x *ananassa* Duch.). International Journal of Current Microbiology and Applied Sciences. 2018;**7**:622-628. DOI: 10.20546/ijcmas.2018.708.067

[51] Park K, Han I, Lee SM, Choi SL, Kim MC, Lee H. Crown and root rot of strawberry caused by *Neopestalotiopsis clavispora* in Korea. Korean Journal of Mycology. 2019;**47**:427-435. DOI: 10.4489/KJM.20190047

[52] Machín A, González P, Vicente E, Sánchez M, Estelda C, Ghelfi J, et al. First report of root and crown rot caused by *Neopestalotiopsis clavispora* on strawberry in Uruguay. Plant Disease. 2019;**103**:2946. DOI: 10.1094/PDIS-05-19-0948-PDN

[53] Di Sigillo L, Ruocco M, Gualtieri L, Pane C, Zaccardelli M. First report of *Neopestalotiopsis clavispora* causing crown rot in strawberry in Italy. Journal of Plant Pathology. 2020;**102**:281. DOI: 10.1007/s42161-019-00415-2

[54] Rebollar-Alviter A, Silva-Rojas HV, Fuentes-Aragón D, Acosta-González U, Martínez-Ruiz M, Parra-Robles BE. An emerging strawberry fungal disease associated with root rot, crown rot and leaf spot caused by *Neopestalotiopsis rosae* in Mexico. Plant Disease. 2020;**104**:2054-2059. DOI: 10.1094/pdis-11-19-2493-sc

[55] Sun Q, Harishchandra D, Jia J, Zuo Q, Zhang G, Wang Q, et al. Role of *Neopestalotiopsis rosae* in causing root rot of strawberry in Beijing, China. Crop Protection. 2021;**147**:105710. DOI: 10.1016/j.cropro.2021.105710

[56] Intiago-Reyna HO, Rivas-Figueroa FJ, Rivera-Casignia AM, Álvarez-Romero PI, Ferreira-e Ferreira AFTA. Outbreaks of crown rot in *Fragaria* x *ananassa* caused by *Neopestalotiopsis mesopotamica* in Ecuador. Emirates Journal of Food and Agriculture. 2021;**33**:520-527. DOI: 10.9755/ejfa.2021.v33.i6.2718

[57] Parikka P, Latvala S. Soil-borne diseases increasing in Finnish strawberry production. Acta Horticulturae. 2021;**1309**:725-730. DOI: 10.17660/ACTAHORTIC.2021.1309.104

[58] Wu HY, Tsai CY, Wu YM, Ariyawansa HA, Chung CL, Chung PC. First report of *Neopestalotiopsis rosae* causing leaf blight and crown rot on strawberry in Taiwan. Plant Disease. 2021;**105**:487. DOI: 10.1094/PDIS-05-20-1045-PDN

[59] Howard CM, Maas JL, Chandler CK, Albregts EE. Anthracnose of strawberry caused by the *Colletotrichum* complex in Florida. Plant Disease. 1992;**76**:976-981. DOI: 10.1094/PD-76-0976

[60] Tweedy B, Powell D. Charcoal rot on strawberry in Illinois. Plant Disease Report. 1958;**42**:107

[61] Madkour MA, Aly MH. Cell wall degrading enzymes produced during pathogenesis of *Macrophomina phaseolina* on strawberry plants. Phytopathologische Zeitschrift. 1981;**100**:36-43. DOI: 10.1111/J.1439-0434.1981.TB03288.X

[62] Baudry AA, Morzières JP. First report of charcoal rot of strawberry in France. Acta Horticulturae. 1993;**348**:485-488. DOI: 10.17660/ACTAHORTIC.1993.348.99

[63] Kenneth RG, Barkai-Golan R, Netzer D. A *Pestalotia* fruit rot of strawberry in Israel. Plant Disease Report. 1968;**52**:472-474

[64] Howard CM, Albregts EE. A strawberry fruit rot caused by *Pestalotia longisetula*. Phytopathology. 1973;**63**:862-863. DOI: 10.1094/PHYTO-63-862

[65] Royse DJ, Ries SM. A cortical decay of petioles and stolons of strawberry caused by *Pestalotia longisetula*. Plant Disease Report. 1976;**60**:901-902

[66] Winks BL, Williams YN. A wilt of strawberry caused by a new form of *fusarium oxysporum*. Queensland Journal of Agricultural and Animal Sciences. 1965;**22**:475-479

[67] Okamoto H, Fuji S, Kato K, Yoshioka A. A new strawberry disease '*fusarium* wilt'. Plant Protection. 1970;**24**:231-235

[68] Mena AJ, Palacios de García ME, González MA. Root diseases of strawberry caused by *fusarium oxysporum* Sch. f. Sp. *fragariae* Winks et will and *Rhizoctonia fragariae* Hussain et Mckeen. Revista Agronómica del Noroeste Argentino. 1975;**12**:299-307

[69] Kim CH, Seo HD, Cho WD, Kim SB. Studies on varietal resistance and chemical control to the wilt of strawberry caused by *fusarium oxysporum*. Korean Journal of Plant Protection. 1982;**21**:61-67

[70] Castro-Franco J, Dávalos-González PA. Etiología de la "secadera" o pudrición de raíz y corona de la fresa en Irapuato. Revista Mexicana de Fitopatología. 1990;**8**:80-86

[71] Scott DH, Draper AD, Galleta GJ. Breeding strawberries for red stele resistance. Plant Breeding Reviews. 1984;**1984**(2):195-214. DOI: 10.1002/9781118060995.CH6

[72] Lawrence FJ, Galleta GJ, Scott DH. Strawberry breeding work of the U.S. Department of Agriculture. HortScience. 1990;**25**:895-896. DOI: 10.21273/HORTSCI.25.8.895

[73] Bringhurst RS, Hancock JF, Voth V. The beach strawberry, an important natural resource. California Agriculture. 1977;**31**:10

[74] Bringhurst RS, Voth V. Breeding octoploid strawberries. Iowa State Journal of Research. 1984;**58**:371-381

[75] Scott DH, Lawrence FJ. Strawberries. In: Janick J, Moore JN, editors. Advances in Fruit Breeding. West Lafayette, Indiana: Purdue University Press; 1975, 1975. pp. 71-97

[76] Bringhurst RS, Wilhelm S, Voth V. Verticillium wilt resistance in natural

populations of *Fragaria chiloensis* in California. Phytopathology. 1966;**56**:219-222

[77] Vining KJ, Davis TM, Jamieson AR, Mahoney LL. Germplasm resources for Verticillium wilt resistance breeding and genetics in strawberry (*Fragaria*). Journal of Berry Research. 2015;**5**:183-195. DOI: 10.3233/JBR-150096

[78] Roudeillac P. Objectives and methods of the strawberry breeding program sponsored by the CIREF. Acta Horticulturae. 1993;**348**:117-123. DOI: 10.17660/ACTAHORTIC.1993.348.12

[79] Simpson DW, Bell JA, Hammond KJ, Whitehouse AB. The latest strawberry cultivars from horticultural research international. Acta Horticulturae. 2002;**567**:165-168. DOI: 10.17660/ACTAHORTIC.2002.567.29

[80] Parikka P. Screening of strawberry resistance to *Phytophthora cactorum*. Scientific Works of the Lithuanian Institute of Horticulture and Lituanian University of Agriculture. 2007;**26**:23-30

[81] Pérez-Jiménez RM, De Cal A, Melgarejo P, Cubero J, Soria C, Zea-Bonilla T, et al. Resistance of several strawberry cultivars against three different pathogens. Spanish Journal of Agricultural Research. 2012;**10**:502-512. DOI: 10.5424/SJAR/2012102-345-11

[82] Eikemo H, Brurberg MB, Davik J. Resistance to *Phytophthora cactorum* in diploid *Fragaria* species. HortScience. 2010;**45**:193-197. DOI: 10.21273/HORTSCI.45.2.193

[83] Howard CM, Albregts EE. 'Florida Belle', an anthracnose resistant strawberry for Florida. Florida Agricultural Experiment Station. 1975;**233**:6

[84] Mackenzie SJ, Legard D, Timmer LW, Chandler CK, Peres NA. Resistance of strawberry cultivars to crown rot caused by *Colletotrichum gloeosporioides* isolates from Florida is nonspecific. Plant Disease. 2006;**90**:1091-1097. DOI: 10.1094/PD-90-1091

[85] Lewers KS, Castro PR, Enns JM, Hokanson SC, Galleta G, Handley DT, et al. 'Flavorfest' strawberry. HortScience. 2017;**52**:1627-1632. DOI: 10.21273/HORTSCI11893-17

[86] Lewers KS, Turecheck WW, Hokanson SC, Maas JL, Hancock JF, Serce S, et al. Evaluation of elite native strawberry germplasm for resistance to anthracnose crown rot disease caused by *Colletotrichum* species. Journal of the American Society for Horticultural Science. 2007;**132**:842-849. DOI: 10.21273/JASHS.132.6.842

[87] Mangandi J, Peres NA, Whitaker VM. Identifying resistance to crown rot caused by *Colletotrichum gloeosporioides* in strawberry. Plant Disease. 2015;**99**:954-961. DOI: 10.1094/PDIS-09-14-0907-RE

[88] Kodama T. Characters of strawberry yellows caused by *fusarium* and difference of its effect on the grown varieties. Bulletin of the Nara Agricultural Experiment Station (Japan). 1974;**6**:68-75

[89] Dávalos-González PA. Respuesta de la fresa a dos métodos de inoculación de *Fusarium oxysporum* f. sp. *fragariae* e identificación de fuentes de resistencia. [thesis MSc]. Saltillo, Coah: Universidad Autónoma Agraria Antonio Narro; 1990

[90] Dávalos-González PA, Jofre-Garfias AE, Hernández-Razo AR, Narro-Sánchez J, Castro-Franco J, Vázquez-Sánchez N, et al. Strawberry breeding for the central plateau of México. Acta Horticulturae.

2006;**708**:547-552. DOI: 10.17660/ACTAHORTIC.2006.708.97

[91] Koike ST, Gordon TR. Management of *Fusarium* wilt of strawberry. Crop Protection. 2015;**73**:67-72. DOI: 10.1016/J.CROPRO.2015.02.003

[92] Dávalos-González PA, Jofre-Garfias AE, Díaz-Espino LF, Mariscal-Amaro LA, López-Pérez MG. Advances in strawberry breeding in México: Genotypes selection with high productivity, good fruit quality, and tolerance to the local viral complex and *fusarium oxysporum*. Acta Horticulturae. 2017;**1156**:159-165. DOI: 10.17660/ACTAHORTIC.2017.1156.23

[93] Gomez AO, De Faveri J, Neal JM, Aitken EAB, Herrington ME. Response of strawberry cultivars inoculated with *Macrophomina phaseolina* in Australia. International Journal of Fruit Science. 2020;**20**:sup 2. DOI: 10.1080/15538362.2019.1709114

[94] Youssef S, Shalaby AA, Pickel B, Maymon M, Borenstein M, Dai N, et al. Selecting tolerant strawberry germplasm to the fungus *Macrophomina phaseolina*. Acta Horticulturae. 2021;**1309**:751-758. DOI: 10.17660/ActaHortic.2021.1309.107

[95] Maas JL, Smith WL Jr. 'Earliglow', a possible source of resistance to Botrytis fruit rot in strawberry. HortScience. 1978;**13**:275-276

[96] Barritt BH. Resistance of strawberry clones to Botrytis fruit rot. Journal of the American Society for Horticultural Science. 1980;**105**:160-164

[97] González G, Moyá M, Sandoval C, Herrera R. Genetic diversity in Chilean strawberry (*Fragaria chiloensis*): Differential response to *Botrytis cinerea* infection. Spanish Journal of Agricultural Research. 2009;**7**:886-895. DOI: 10.5424/SJAR/2009074-1102

[98] Bestfleisch M, Pflimpfl ML, Höfer M, Schulte E, Wünsche JN, Hanke MV, et al. Evaluation of strawberry (*Fragaria* L.) genetic resources for resistance to *Botrytis cinerea*. Plant Pathology. 2015;**64**:396-405. DOI: 10.1111/PPA.12278

[99] Maas JL, Behe CG, Hartung JS, Hokanson SC. Sources of resistance for two differentially pathogenic strains of *Xanthomonas fragariae* in *Fragaria* genotypes. HortScience. 2000;**35**:128-131. DOI: 10.21273/HORTSCI.35.1.128

[100] Maas JL, Behe CG, Hartung JS, Hokanson SC. Strawberry parent clones US 4808 and US 4809 resistant to bacterial angular leafspot disease caused by *Xanthomonas fragariae*. HortScience. 2002;**37**:716-717. DOI: 10.21273/HORTSCI.37.4.716

[101] Xue S, Bors RH, Strelkov SE. Resistance sources to *Xanthomonas fragariae* in non-octoploid strawberry species. HortScience. 2005;**40**:1653-1656. DOI: 10.21273/HORTSCI.40.6.1653

[102] Bestfleisch M, Richter K, Wensing A, Wünsche JN, Hanke MV, Höfer M, et al. Resistance and systemic dispersal of *Xanthomonas fragariae* in strawberry germplasm (*Fragaria* L.). Plant Pathology. 2015;**64**:71-80. DOI: 10.1111/PPA.12232

[103] Nelson MD, Gubler WD, Shaw DV. Relative resistance of 47 strawberry cultivars to powdery mildew in California greenhouse and field environments. Plant Disease. 1996;**80**:326-328. DOI: 10.1094/PD-80-0326

[104] Kennedy C, Hasing TN, Perez NA, Whitaker VM. Evaluation of strawberry species and cultivars for powdery mildew resistance in open-field and high tunnel

production systems. HortScience. 2013;**48**:1125-1129. DOI: 10.21273/HORTSCI.48.9.1125

[105] Hancock JF, Dale A, Luby JJ. Should we reconstitute the strawberry? Acta Horticultura. 1993;**348**:86-93. DOI: 10.17660/ACTAHORTIC.1993.348.8

[106] Govorova G. Strawberry breeding in Russia. Acta Horticulturae. 1993;**348**:45-55. DOI: 10.17660/ActaHortic.1993.348.2

[107] Hancock JF, Sjulin TM, Lobos GA. Strawberries. In: Hancock JF, editor. Temperate Fruit Crop Breeding: Germplasm to Genomics. Dordrechts: Springer; 1999. pp. 393-438. DOI: 10.1007/978-1-4020-6907-9

[108] Guo RX, Hue L, Luo GJ, Zhang T, Lei JJ. Investigation and taxonomy of wild *Fragaria* resources in Tibet, China. Genetic Resources and Crop Evolution. 2018;**65**:405-417. DOI: 10.1007/s10722-017-0541-1

[109] Wilhelm S, Sagen JE. A History of the Strawberry from Ancient Gardens to Modern Markets. Berkeley: University of California; 1974. p. 298

[110] Schuster DJ, Price JF, Martin FG, Howard CM, Albregts EE. Tolerance of strawberry cultivars to two-spotted spider mites in Florida. Journal of Economic Entomology. 1980;**73**:52-54. DOI: 10.1093/JEE/73.1.52

[111] Shanks CH Jr, Barritt BH. Resistance of *Fragaria chiloensis* clones to the two-spotted spider mite. HortScience. 1984;**19**:640-641

[112] Shanks CH Jr, Moore PP. Resistance to two-spotted spider mite and strawberry aphid in *Fragaria chiloensis, F. virginiana* and *F.* x *ananassa* clones. HortScience. 1995;**30**:596-599. DOI: 10.21273/HORTSCI.30.3.596

[113] Serce S, Hancock JF. Screening of strawberry germplasm for resistance to the two-spotted spider mite. HortScience. 2002;**37**:593-594. DOI: 10.21273/HORTSCI.37.3.593

[114] Handley DT, Dill JF, Pollard JE. Field susceptibility of twenty strawberry cultivars to tarnished plant bug injury. Fruit Varieties Journal. 1991;**45**:166-169

[115] Kitamura T, Kashio T. Varietal differences in the susceptibility of strawberry to western flowers trips *Frankliniella occidentalis* (Pergande). Kyushu Plant Protection Research. 2004;**50**:55-61

[116] Clausen RE. Ettersburg strawberries. Journal of Heredity. 1915;**VI**:324-331

[117] Luby JJ, Hancock JF, Cameron JS. Expansion of the strawberry germplasm base in North America. In: Dale A, Luby JJ, editors. The Strawberry into the 21st Century. Vol. 1991. Portland: Timber Press; 1991. pp. 66-75

[118] Kennedy C, Hasing TN, Whitaker V. Characterization of *Fragaria virginiana* and *F. chiloensis* in a minimal-chill, winter annual production system. HortScience. 2014;**49**:848-855. DOI: 10.21273/HORTSCI.49.7.848

[119] Cai W, Zurn JD, Bassil NV, Hummer KE. Perpetual flowering in strawberry species. HorScience. 2017;**52**:1496-1500. DOI: 10.21273/HORTSCI12025-17

[120] Hancock JF, Lavín A, Retamales JB. Our Southern strawberry heritage: *Fragaria chiloensis* of Chile. HortScience. 1999;(34):814-816. DOI: 10.21273/HORTSCI.34.5.814

[121] Darrow GM. Strawberries in Mexico, Central America, Colombia and Ecuador. Ceiba. 1953;**3**:179-185

[122] Aguilar-García R, Dávalos-González PA, Rodríguez-Guillén A. La clorosis Férrico-Manganésica en fresa y su manejo. Technical Book (in Review). Instituto Nacional de Investigaciones, Forestales, Agrícolas y Pecuarias. Celaya, Gto. México: Campo Experimental Bajío; 2021. p. 48

[123] Torum AA, Kakar YA, Bicen B, Erdem N, Serce S. In vitro screening of octoploid *Fragaria chiloensis* and *Fragaria virginiana* genotypes against iron deficiency. Turkish Journal of Agriculture and Forestry. 2014;**38**: 169-179. DOI: 10.3906/TAR-1305-83

[124] Popenoe W. The frutilla, or Chilean strawberry. The Journal of Heredity. 1921;**12**(457):466. DOI: 10.1093/OXFORDJOURNALS.JHERED.A102048

[125] Darrow GM. Exploration in South America for strawberries and other small fruits. Fruit Varieties and Horticultural Digest. 1957;**12**:5-7

[126] Lavín A. Caracterización botánica, fisiológica y agronómica de ecotipos chilenos de *Fragaria chiloensis* (L) Duch. recolectados en las X y XI regiones de Chile. Informe final. Proyecto Fondecyt. 1997;**1940083**

[127] Finn CE, Retamales JB, Lobos GA, Hancock JF. The Chilean strawberry (*Fragaria chiloensis*): Over 1000 years of domestication. HortScience. 2013;**48**:418-421. DOI: 10.21273/HORTSCI.48.4.418

[128] Noguchi Y, Mochizuki T, Sone K. Breeding of a new aromatic strawberry by interspecific hybridization *Fragaria* x *ananassa* x *F. nilgerrensis*. Journal of the Japanese Society for Horticultural Science. 2002;**71**:208-213. DOI: 10.2503/JJSHS.71.208

[129] Lei J, Xue L, Guo RX, Dai HP. The *Fragaria* species native to China and their geographical distribution. Acta Horticulturae. 2016;**1156**:37-46. DOI: 10.17660/ACTAHORTIC.2017.1156.5

[130] Lei J, Hue L, Dai HP, Deng MQ. The taxonomy of Chinese *Fragaria* species. Acta Horticulturae. 2014;**1049**:289-294. DOI: 10.17660/ACTAHORTIC.2014.1049.37

[131] Bringhurst RS, Rae H, Voth V. Ascorbic acid in California strawberries. 1974 SNB Pomology Report # 9. Strawberry News Bulletin. 1975;XXI Number 5. Watsonville, California: California Strawberry Advisory Board;

[132] Wang SY. Antioxidants and health benefits of strawberries. Acta Horticulturae. 2014;**1049**:49-62. DOI: 10.17660/ACTAHORTIC.2014.1049.1

[133] Yildiz H, Ercisli S, Hegedus A, Akbulut M, Topdas EF, Aliman J. Bioactive content and antioxidant characteristics of wild (*Fragaria vesca* L.) and cultivated strawberry (*Fragaria* x *ananassa*) fruits from Turkey. Journal of Applied Botany and Food Quality. 2014;**87**:274-278. DOI: 10.5073/JABFQ.2014.087.038

[134] Mahoney L, Davis TM, Curran-Celentano J. Investigation of cyanidin and pelargonidin contents in the genus *Fragaria* L. Acta Horticulturae. 2009;**842**:915-918. DOI: 10.17660/ACTAHORTIC.2009.842.202

[135] Galleta GJ, Draper AD, Maas JL. Combining disease resistance, plant adaptation and fruit quality in breeding short day and day-neutral strawberries. Acta Horticulturae. 1989;**265**:43-51. DOI: 10.17660/ACTAHORTIC.1989.265.3

[136] Olbricht K, Ulrich D, Waurich V, Wagner H, Bicking D, Gerischer U, et al. Breeding potential of underutilized *Fragaria* species. Acta Horticulturae.

2021;**2021**(1309):139-146. DOI: 10.17660/ActaHortic.2021.1309.20

[137] Evans WD. The production of multispecific octoploids from *Fragaria* species and the cultivated strawberry. Euphytica. 1982;**31**:901-907. DOI: 10.1007/BF00039230

[138] Hancock JF, Finn CE, Luby JJ, Dale A, Callow PW, Serce S. Reconstruction of the strawberry, *Fragaria* x *ananassa*, using genotypes of *F. virginiana* and *F. chiloensis*. HortScience. 2010;**45**:1006-1013. DOI: 10.21273/HORTSCI.45.7.1006

[139] Ahmadi H, Bringhurst RS. Breeding strawberries at the decaploid level. Journal of the American Society for Horticultural Science. 1992;**117**:856-862. DOI: 10.21273/JASHS.117.5.856

[140] Cameron JS, Shanks CH, Sjulin TM, Muñoz CE. Collection of *Fragaria chiloensis* in central and southern Chile. In: Dale A, Luby JJ, editors. The Strawberry Into the 21st Century. Portland: Timber Press; 1991. pp. 108-110. DOI: 10.21273/HORTSCI.25.9.1141F

[141] Cameron JS, Sjulin TM, Ballington JR, Shanks CH, Muñoz C, Lavín A. Exploration, collection and evaluation of Chilean *Fragaria*: Summary of 1990 and 1992 expeditions. Acta Horticulturae. 1993;**348**:65-74. DOI: 10.17660/ACTAHORTIC.1993.348.5

[142] Hancock JF, Callow PW, Dale A, Luby JJ, Finn CE, Hokanson SC, et al. From the Andes to the Rockies: Native strawberry collection and utilization. HortScience. 2001;**36**:221-225. DOI: 10.21273/HORTSCI.36.2.221

[143] Gambardella M, Infante R, López Aranda JM, Faedi W, Roudeillac P. Collection of wild and cultivated native *Fragaria* in southern Chile. Acta Horticulturae. 2002;**567**:61-63. DOI: 10.17660/ACTAHORTIC.2002.567.2

[144] Luby JJ, Hancock JF, jr., Ballington JR. Collection of native strawberry germplasm in the Pacific northwest and northern Rocky Mountains of the United States. HortScience. 1992;**27**:12-17. DOI: 10.21273/HORTSCI.27.1.12

[145] Dale A, Daubeny HA, Luffman M, Sullivan JA. Development of *Fragaria* germplasm in Canada. Acta Horticulturae. 1993;**348**:75-80. DOI: 10.17660/ACTAHORTIC.1993.348.6

[146] Sakin M, Hancock JF, Luby JJ. Identifying new sources of genes that determine cyclic flowering in Rocky Mountains populations of *Fragaria virginiana* spp. *glauca* Staudt. Journal of the American Society for Horticultural Science. 1997;**122**:205-210. DOI: 10.21273/JASHS.122.2.205

[147] Crock JE, Shanks CH, Barritt BH. Resistance in *Fragaria chiloensis* and *F.* x *ananassa* to the aphids *Chaetosiphon fragaefolii* and *C. thomasi*. HortScience. 1982;**17**:959-960

[148] Becerra V, Paredes M, Lavín A. Biochemical and molecular diversity in the Chilean strawberry and its implications for plant breeding. HortScience. 2005;**40**:1642-1643. DOI: 10.21273/HORTSCI.40.6.1642

[149] Gambardella M, Cadavid A, Díaz V, Pertuzé R. Molecular and morphological characterization of wild and cultivated native *Fragaria* in Southern Chile. HortScience. 2005;**40**:1640-1641. DOI: 10.21273/HORTSCI.40.6.1640

[150] Carrasco B, Garcés M, Rojas P, Saud G, Herrera R, Retamales JB, et al. The Chilean strawberry [*Fragaria chiloensis* (L) Duch.]: Genetic diversity

and structure. Journal of the American Society for Horticultural Science. 2007;**132**:501-506. DOI: 10.21273/JASHS.132.4.501

[151] Zhao MH, Wu WM, Lei JJ, Quian YM, Wuang ZG, Yuan J. Geographical distribution and morphological diversity of wild strawberry germplasm resource in China. Acta Horticulturae. 2009;**842**:593-596. DOI: 10.17660/ACTAHORTIC.2009.842.126

[152] Peng M, Zong X, Wang C, Peng F. Genetic diversity of strawberry (*Fragaria ananassa* Duch.) from the Motuo county of the Tibet plateau determined by AFLP markers. Biotechnology and Biotechnological Equipment. 2015;**29**:876-881. DOI: 10.1080/13102818.2015.1050968

[153] Luo G, Xue L, Guo R, Xu W, Lei J. Creating cold resistant strawberry via interploidy hybridization between octoploid and dodecaploid. Euphytica. 2017;**213**:194. DOI: 10.1007/s10681-017-1980-7

[154] Zhao J, Xue L, Che XD, Jiang S, Lei JJ, Dai HP. Investigation and classification of the five tetraploid *Fragaria* species native to China. Acta Horticulturae. 2021;**1309**:51-56. DOI: 10.17660/ACTAHORTIC.2021.1309.8

[155] Pérez de Castro AM, Vilanova S, Cañizares J, Pascual L, Blanca JM, Díez MJ, et al. Application of genomic tools in plant breeding. Current Genomics. 2012;**13**:179-195. DOI: 10.2174/138920212800543084

Chapter 7

# Hermaphroditism in *Fragaria moschata*, a Cultivated Strawberry Species Preceding the Evolution and Growing of *F.* ×*ananassa* in Europe

*Hannu Ahokas*


## Abstract

Two cases of hermaphroditism in *Fragaria moschata* Weston, (2n = 42), are described from Finland. One of them is a mutation in a clone known as a female since 1941 but appearing as a somatic segregant in 2013. The clone also carries a recessive gene for yellow-green leaf variegation. The mutations probably date back to the airborne radioactive fallout from the fire of the Chernobyl nuclear power plant in the Soviet Union in 1986. By interspecific crosses with the diploid (2n = 14) strawberry species, *F.* ×*bifera* Duch., *Fragaria vesca* L., and *Fragaria*. *viridis* Weston, the genetic determination of sex in *F. moschata* was studied and shown to depend on different genetic mechanisms in the two hermaphroditic musk strawberries. Selected tetraploid hybrids from crosses between musk and diploid strawberry species provide a wide variation of berry flavor and may have used as home garden strawberries. Two spontaneous hybrids between *F. moschata* and *F. vesca* are shown.


**Keywords:** Chernobyl 1986, edaphic requirements, green strawberry, musk strawberry, mutant manifestation, redomestication, wood strawberry

## 1. Introduction

The musk strawberry, *Fragaria moschata* is originally a Central European species having spread as an escape from gardens in many areas [1–5]. In the Tortona municipality in North Italy, cultivation of the musk strawberry, 'Profumata di Tortona' dates historically back to the year 1411 [6 cited in 7], when in a register of the *Pieve di Garbagna*, *magiostre* or *fravole* are indicated among the local fruits and vegetables [8]. The strawberry of Tortona is a local, dioecious ecotype, and the name 'Profumata' appeared for the first time in the mid-19th century [8]. The first known cultivar name of musk strawberry was 'Le Chapiron' in Europe in 1576; the cultivar name was derived further and apparently called 'Capron' by Jean de la Quintinie in 1672 [9]. In 1773, 14 varieties of the musk strawberry were listed in England by the botanist



IntechOpen

Richard Weston [10]. The binominal name *Fragaria moschata* valid nowadays was ascribed to Weston [11]. In 1788, the Frenchman A. N. Duchesne described the hermaphroditic musk strawberry cultivar 'Le Caperonnier royal', the origin of which was known from Brussels in 1770 [12]. With the name 'Capiton' the variant was probably introduced to cultivation in Wallonia by about the middle of the 16th century [11]. A paper recognizing a hermaphroditic musk strawberry was red in England in 1817 by Keens [13] who, however, preferred dioecious varieties. Of the 'Le Caperonnier royal' variety was probably the hermaphroditic musk strawberries with name 'Capron royal' in St. Petersburg, Russia since 1869 [14, 15]. 'Capron' was studied by Ahmadi & Bringhurst [16] in USA in 1991. The use of the cultivar names of musk strawberries was confusing in the past [11, 17].

Hegi [1] considered the distribution of *F. moschata* in the north to reach Ingria (see [18, 19]), i.e., the territory on the eastern coast of the Gulf of Finland. The archaeobotanist Terttu Lempiäinen [20] thought *F. moschata* (erroneously under the name *F. muricata*) to have occurred in the 17th on Karelian Isthmus, in the territory of Finland of the Kingdom of Sweden. The growing or semicultural choosing of *F. moschata* preceded the evolution and spread of *F.* ×*ananassa* Duch., which occurred in Europe between the latter half of the 18th and early half of the 19th century [11, 21–23]. In Swedish texts of the Nordic literature, *F. moschata* was usually called *jordgubbe*, e.g., by Kalm [24] in 1759, which term later started to mean the garden strawberry, *F.* ×*ananassa* [25]. In Finland, the growing of *F.* ×*ananassa* began in the 1840s with a few early European cultivars [26]. Growing strawberries from seeds were prompted in the past [27], which may have resulted in the development of local varieties.

The earliest introductions to Finland of cultivated *F. moschata* occurred to gardens of manor houses in the 18th century principally from Sweden, less likely from Germany, and later on, from Russia. Stocks of musk strawberries were subsequently transferred from garden to garden. To the Valamo (Валаам, Valaam in Russian) islands on the Lake Ladoga, then of the territory of Finland, *F. moschata* varieties were introduced from Dr. E. Regel's garden nursery in St. Petersburg as written in 1870 by Damaskin [28], the igumene of the Valamo monastery. Regel" [29] recommended *F. moschata* varieties 'Belle Bordelaise', 'Bloch Hautbois', 'Capron royal', 'Capron framboisé' and 'Large flat' in 1866. He listed elsewhere altogether 11 varieties of *F. moschata* in 1869 [14]. *F. moschata* also thrived in the wilderness on Valamo, in a grove, where berry forming clones were found by Dr. Med. T. Sælan in 1881 [30] shown by Sælan's specimens H 371514 – H 371516, in Botancal Museum, Luomus, Helsinki. For berries, the growing of clones of both female and male, or a hermaphroditic clone are needed.

In the latter half of the 19th century, the supply of the garden nurseries in Europe turned to sell cultivars of *F.* ×*ananassa*. As examples, the catalog of Regel" & Kessel'ring" [15] in St. Petersburg, Russia in 1879 had 11 named cultivars of *F. vesca* s.l., 10 named cultivars of *F. moschata*, and 168 cultivars of *F.* ×*ananassa* (by far the most), with *F. virginiana* Miller and *F. chiloensis* (L.) Miller cultivars pooled together. The catalog of Vilmorin-Andrieux et C$^{ie}$ [31] in Paris, France in 1891 listed eight named cultivars of *F. vesca* s.l., one of *F. viridis*, three of *F. moschata*, one of *F. virginiana*, one of *F. chiloensis* and 46 of *F.* ×*ananassa*.

In the now ceased Soviet Union before the year 1930, musk strawberry was grown for industrial purposes, though less than *F.* ×*ananassa*, at the most, on tens of hectares in different regions assessed then in the country recovering from its inner wars. There the cultivar names 'Španka' ('Шпанка'), 'Russkaya' ('Русская') and 'Kapron' ('Капрон') were recognized [32]. The 'Capron' cultivar usually meant a hermaphroditic variety. Soviet Union exported strawberry products [32], and their ingredients probably also contained musk strawberries.

The Austrian organization Arche Noah has maintained a hermaphroditic cultivar 'Oke' of *F. moschata* [33]. Among 108 *F. moschata* plants collected in Ziegenbusch, near Dresden, Germany, six were found to be hermaphroditic [5]. *F. moschata* is clearly a trioecious species also having hermaphroditic genotypes beside females and males.

Prof. Günter Staudt (*1926 †2008) did not believe that hermaphroditic *F. moschata* really exists [4, 12 p207]. He evidently held the concept of his teacher, Prof. Elisabeth Schiemann [34] that hermaphroditic *F. moschata*-like plants are tetraploid hybrids between *F. moschata* and *F. viridis*. Nowadays, the ancestry of the hexaploid *F. moschata* is thought to trace to three diploid species, *F. viridis*, *F. mandshurica* Staudt, and *F. vesca* [35] which are monoecious.

## 2. Find of two hermaphroditic *Fragaria moschata* clones in Finland

In the summer of 2012, I visited a private experimental farm in Loimaa town, SW Finland, where I to my surprise saw berry carrying plants of *F. moschata*, without any indication of separate females or males in the stand. The land owner's family had moved to the farm in 1954, when the strawberry was already there. The preceding history of the strawberry is unknown. I received a living sample of the musk strawberry, hereafter called 'Loimaa' (**Figures 1** and **2**), and deposited its herbarium specimen to Botanical Museum, Luomus (H 831495). Studying in the Museum specimens of *F. moschata* with berries, revealed a sample (H 223584) with well-formed anthers and pollen collected 13.5 km away from the experimental farm in Loimaa in 1971. This specimen and 'Loimaa' may be of the same gendre.

A surprise to a land owner in Kotka town, on the S coast of Finland, was the occurrence of berries in 2013 in an area of about 4 m$^2$ among about 3 ares of large strawberry plants, which had never set any berry during the observed period since

**Figure 1.**
*The hermaphroditic 'Loimaa' of* Fragaria moschata *at anthesis. June 5, 2013.*

**Figure 2.**
*The hermaphroditic 'Loimaa' of* Fragaria moschata *at maturity. June 19, 2014.*

1941. I got a message about the strange berry-set through conversations between two elderly people who met in a clinic waiting room, one being the landowner's relative, and the other my father. Visiting the farm in Kotka showed that an apparent mutation had occurred in the female clone of *F. moschata* to hermaphroditism (**Figures 3** and **4**).

**Figure 3.**
*Occurrence of berries on a patch with flowers carrying fertile anthers in the female clone of* Fragaria moschata *in Kotka, South Finland. Before the 2013 season, the musk strawberry occurrence has not made any berry since 1941, the year of the start of observation. The occurrence in the yard of the farm house has been isolated spatially from other clones of musk strawberry. July 4, 2013.*

**Figure 4.**
*A dessicated late flower carrying fertile anthers in the anew risen hermaphrodite of* Fragaria moschata *from the female clone in Kotka. The anthers have dehiscing stomia with pollen grains. See* ***Figure 3*** *for the stand already advanced to the berry maturing period on the day of the sampling, July 4, 2013.*

This mutated clone is called 'Kotka' hereafter. Its level of male fertility has appeared variable, being inferior in part of the flowers, or flowers opening at a certain point of the flowering season.

I performed crossing between 'Kotka' and 'Loimaa', and interspecific crosses of the hexaploid *F. moschata* with the diploid *F.* ×*bifera*, *F. vesca* and *F. viridis*, in order to reveal the inheritance of the sex-determining genotype in *F. moschata* in the resulting tetraploid hybrids and to study the possible use of the hybrids as cultural berries. *F.* ×*bifera* is the variably fertile hybrid between *F. vesca* and *F. viridis* [36]. Two spontaneous hybrids between *F. moschata* and *F. vesca* occurred.

The hexaploid chromosome number (2n = 42) in *F. moschata* in material from Finland has been determined in a female clone [37] and, in a male clone from Viitasaari, Central Finland (H 335838), by Ahokas (unpublished).

## 3. Crossing methods, seed germination and testing of self-pollination

I performed emasculation under magnification glasses in unopened flower-buds. The emasculated flowers were isolated in glassine crossing bags. Unopened flowers of the pollen parents were usually isolated in glassine crossing bags in advance to eliminated contaminating pollen. Such isolated flowers at the anthesis were directly used for pollination after the emasculated seed parent flowers were checked for the absence of anthers under magnification.

Cleaned mature seeds (achenes) were usually germinated on agar gel, where axenic conditions were tried to be attained. The seeds were surface sterilized with 0.001% $HgCl_2$ in water in tubes for 20 min, at times vortexed, rinsed with two changes of sterile water, and placed individually on sterile 1% gel of Bacto-agar (Difco, Detroit, USA) made in tap-water, with added 0.01% $CaCO_3$. After germination, the seedlings with one or two vegetative leaves were transferred to soil (**Figure 5**). The commercial organic growing soil Puutarhan Musta multa (Biolan, Eura, Finland) was usually used

**Figure 5.**
*Axenic germination of hybrid seeds of* Fragaria moschata *'Kotka'* × F. viridis *on water agar gel. The germinats are at a suitable size to be transferred to soil. Scale digits millimeters.*

for growing. The experimental cultivation was mostly done in Kirkkonummi Municipality, on the southern coast of Finland.

Inflorescences before the opening of the flower buds were isolated in glassine crossing bags to test the self-pollination ability of the flowers. If the earliest flowers had already opened, they were removed before the isolation. Crossing bags were cut to a suitable length, closed around the inflorescence stem with a paper-clip, and were often needed to be tied with a thread to a bamboo-stick for support. The opened flowers were also scored for the sex organs once or twice during the flowering period.

## 4. Cross between the two hermaphrodites of *Fragaria moschata*

The $F_1$ progeny of the cross of the hermaphrodites, *F. moschata* 'Kotka' × 'Loimaa' comprised 21 females, 10 hermaphrodites and 37 males. Pooling females and hermaphrodites together, the ratio 31: 37 would fit to a 1: 1 segregation, ($\chi^2$ = 0.529, df = 1, p = 0.467). The offspring of the reciprocal cross has not been studied. The genetic bases of the hermaphroditism in 'Kotka' and 'Loimaa' are different.

Three berries around the 'Kotka' hermaphrodite were taken in 2015. These were evidently from female flowers open-pollinated by the near-by hermaphrodite, and as appeared later, one by *F. vesca*. The offspring of the three berries were much alike. Along with typical musk strawberries, two of the berry-series produced also variegating, morphologically distinct and weak seedlings (**Table 1**, **Figure 6**) and one exceptional seedling with 2n = 28 chromosomes, the *F. vesca* hybrid (see next sections).

## 5. Two spontaneous tetraploid hybrids, *Fragaria moschata* × *F. vesca*

Berries of the tetraploid strawberry found in Jokioinen, SW Finland [38] have the fragrance of *F. moschata* or *F. viridis*. At the time I found the tetraploid strawberry in

| Berry no. | Females | Males | Sex unknown | Variegated leaves | Hybrid with *F. vesca* |
|---|---|---|---|---|---|
| 1/3 | 14 | 13 | 8 | 3* | 1 |
| 2/3 | 5 | 5 | 1 | 0 | 0 |
| 3/3 | 3 | 5 | 5 | 2 | 0 |
| Total | 22 | 23 | 14 | 5** | 1 |

**One with flowers could be ranked to be female, all the other plants with variegated leaves remained unknown for the flowers.*
***The gene for variegation must be recessive, but the viability of the homozygotes is reduced and the number is less than expected, or 1/4.*

**Table 1.**
*$F_1$ progenies from three berries taken from the* Fragaria moschata *of the Kotka population. No hermaphrodite was observed. Before discarding, the $F_1$ plants were grown for two seasons during which the flowers were ranked using isolation of florescence and observing the anther development grade.*

**Figure 6.**
*A variegating chlorotic, deformed and weak descendent grown from the seeds of an open-pollinated flower of the Kotka occurrence of* Fragaria moschata *(**Table 1**). The occurrence of such mutant seedlings suggests inbreeding of heterozygosity of the allele or instability of the gene. The fact that there appeared significantly less than one forth (1: 3) of the mutants suggests them to have been deleterious with reduced viability in any stages between zygote and seedling. June 3, 2019.*

Jokioinen, explanations for its origin were open, but as the first obvious reason was the hybrid of *F. moschata* and *F. vesca*. Additional information has confirmed it to be an obvious hybrid between *F. moschata* and *F. vesca* along with the knowledge published from the genomic constitution of *F. moschata* since that. Two of its tetraploid seedling derivatives [39] or further spontaneous seedlings form the present sample of the Jokioinen hybrid (**Figures 7** and **8**). The original habitat in Jokioinen has been destroyed.

The other hybrid with *F. vesca* originates from open-pollinated female clone of *F. moschata* in Kotka. It has 28 chromosomes. Its phenotype is stout. The central leaflet overlaps partly the lateral leaflets. The leaflets are roundish. Apparent due to meiotic

**Figure 7.**
*Under open-pollination nearly completely fertile, tetraploid hybrid of* Fragaria moschata × F. vesca *originating from Jokioinen, Finland [38]. Evidently seedling of the original clone. July 5, 2021.*

irregularities it has low partial fertility of female organs under open-pollination, where there are various tetraploid hybrids around (**Figure 9**). The Kotka and Jokioinen hybrids crossed with difficulty resulting in an $F_1$ plant distinct in morphology from the parents (**Figure 10**).

## 6. Artificial hybrids, *Fragaria moschata* × *F. vesca*

An artificial hybrid of the hermaphroditic *F. moschata* 'Loimaa' × *F. vesca*. In advance isolated flowers of the *F. vesca* pollen parfent were of the spontaneous local Käpykallio clone in Kirkkonummi. *F. vesca* of Käpykallio is self-fertile, pollinating itself in isolated inflorescence. The cross gave in $F_1$ progeny the segregation of the hermaphroditic to male ratio of 6: 7 which is the best fit to a 1: 1 segregation among 13 examples (**Table 2**). The hermaphroditic flowers often showed partial fertility when selfed in isolation bags, which can be ascribed to meiotic irregularities of the interspecific hybrid.

## 7. Artificial hybrids, *Fragaria moschata* × *F. viridis*

Hybrids of *F. moschata* as the seed parent with *F. viridis* were fairly easily obtained, which I first tried with the clone of musk strawberries around the Jokioinen Manor,

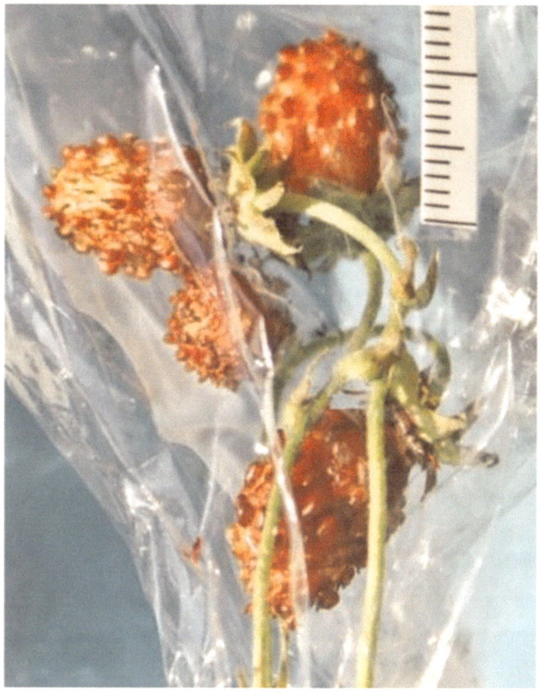

**Figure 8.**
*Under isolation, a nearly perfect set of berries in the tetraploid hybrid of* Fragaria moschata × F. vesca *originating from Jokioinen, Finland [38]. Evidently from a seedling of the original clone. The achenes are superficial and larger than those of* F. vesca. *The berries have a mild fragrance resembling that of* F. viridis *and* F. moschata. *Scale digits millimeters. July 19, 2020.*

**Figure 9.**
*Under open-pollination, slightly fertile, tetraploid, spontaneous $F_1$ hybrid of* Fragaria moschata *(female in Kotka)* × F. vesca. *Note the roundish leaflets with the margins of the central leaflet overlapping the laterals and the stout structure. Senescence of leaves has started along with the set of berries. July 13, 2021.*

**Figure 10.**
*A plant of the first generation between the spontaneous hybrid of* Fragaria moschata × F. vesca *from Kotka and Jokioinen. The leaf morphology does not resemble those of the parents (cf.* ***Figures 7*** *and* ***9****). Only slightly fertile under open-pollination. June 25, 2021.*

| Seed parent (*F. moschata*) | Number of flowering $F_1$ phenotypes | | |
|---|---|---|---|
| | Females | Hermaphrodites | Males |
| 'Loimaa', hermaphrodite | 0 | 6 | 7 |

**Table 2.**
*$F_1$ progeny of the crosses of the hermaphroditic* Fragaria moschata *'Loimaa' with the hermaphroditic pollen parent* F. vesca.

where this species had evidently been introduced in the 18th century. I obtained two plants from a single pollinated flower with a clone of *F. viridis* obtained from the native occurrence in Lemland, Alandia, Finland (Dr. Carl-Adam Hæggström's collection no. 7451, H 258115). A female and male $F_1$ plant resulted and proved to be interfertile (**Figure 11**). An octoploid derivative of it was nearly sterile (**Figure 12**). The genetic determining hermaphroditism in 'Loimaa' and 'Kotka' appear different with the $F_1$ segregation numbers (**Table 3**). The $F_1$ fertility with the 'Kotka' hermaphrodite was in general good both as selfed and open-pollinated (**Figures 13** and **14**). While *F. viridis* has monopodial branching pattern of the stolon, the *F. moschata* × *F. viridis* hybrids have sympodial stolon as also the *F. moschata* parent species.

The phenotype distributions of the crosses with the hermaphrodites 'Loimaa' and 'Kotka' are highly significantly different, $\chi^2 = 57.008$, df = 2, $p < 0.001$. In the 'Kotka' pedigree, the pooled female + hermaphrodite, 13 ratio to 6 males does not differ from the 67: 26 ratio of the 'Loimaa' pedigree, $\chi^2 = 0.0016$, df = 1, p = 0.968.

**Figure 11.**
*The $F_1$ male and female plants of an artificial hybridization of* Fragaria moschata *(female)* × F. viridis. *The whitered male flowers (yellow circle) have pollinated those of the female. August, 1998. The blue-framed insert shows a female flower of the hybrid on June 30, 1997. The female and male offspring of the hybrid with the hermaphroditic* F. viridis *as the pollen parent corroborate the female heterogamety in* F. moschata.

**Figure 12.**
*An octoploid derivative of the tetraploid $F_1$ female plants of the artificial hybrid of* Fragaria moschata × F. viridis. *Open-pollinated. Practically also female sterile. The polyploidy causes more roundish leaflets (cf.* ***Figure 11****) and leaves with five leaflets are frequent. The polyploidy was induced with the method described [40]. Scale 1 × 18 cm. July 19, 2021.*

| Seed parent (*F. moschata*), and its flower type | Number of flowering $F_1$ phenotypes | | |
|---|---|---|---|
| | Females | Hermaphrodites | Males |
| Female from Jokioinen | 1 | 0 | 1 |
| 'Loimaa', hermaphrodite | 0 | 67 | 26 |
| 'Kotka', hermaphrodite | 10 | 3 | 6 |

**Table 3.**
*$F_1$ progenies of the crosses of the female clone from Jokioinen, Finland and the hermaphroditic* Fragaria moschata *'Loimaa' and 'Kotka' with the hermaphroditic pollen parent* F. viridis.

**Figure 13.**
*Fertile, female $F_1$ hybrid plant of* Fragaria moschata *'Kotka'* × F. viridis *forming plenty of berries by open-pollination. Under the isolation-bag the antherless female flowers did not set any berries. Scale 1 × 18 cm. July 7, 2021.*

**Figure 14.**
*Fertile, hermaphroditic $F_1$ hybrid of* Fragaria moschata *'Kotka'* × F. viridis *forming berries under isolation-bag. July 16, 2021.*

## 8. Artificial hybrids, *Fragaria moschata* × *F.* ×*bifera*

The hybrid species *F.* ×*bifera* (syn. *F. hagenbachiana*) is a morphologically variable, intermediate of its parental species, and variably fertile (**Figure 15**). The hybrid species may also include backcross derivatives with the parental species. Principally, the $F_1$ hybrids from the cross *F. moschata* × *F.* ×*bifera* in my experiments resembled

**Figure 15.**
*A fairly fertile* Fragaria ×bifera *(interspecific hybrid of* F. vesca *and* F. viridis*) was used as the pollen parents in crosses with* F. moschata*. The loose berries fell accidentally from the plant, while removing the metal network (mesh 12 mm) sheltering from local white tail deers, roes and moles. July 13, 2019.*

those of the *F. moschata* × *F. viridis* hybrids. The berry flavor of the hybrids is usually strong having taste of both the parents and varies from sweet to sour. The fragrance and sweetness of the delicate berries (**Figure 16**) also attract flying insects, a disadvantage in free-land growing. In *F. moschata* and its hybrids the senescence of leaves begins after flowering (**Figures 17** and **18**). Only hermaphrodites and males were obtained in $F_1$ (**Table 4**).

**Figure 16.**
*Sweet and fragrant berries set with self-pollination under the tight isolation bag in the $F_1$ hybrid of* Fragaria moschata × F. ×bifera*. Good fertility is revealed by the high seed set. Note the mode of the detachment and pinkish inner of the berry. Total fresh mass from the single inflorescence 33.5 g. Scale digits millimeters. June 19, 2020.*

**Figure 17.**
*An $F_1$ hybrid of* Fragaria moschata *('Loimaa')* × F. ×bifera *at anthesis. The leaves stay dark-green up to flowering. June 16, 2021.*

**Figure 18.**
*The $F_1$ hybrid of* Fragaria moschata × F. ×bifera *of* ***Figure 17*** *at maturity. The senescence of the leaves begins at the berry set. July 13, 2021.*

| Seed parent (*F. moschata*) | Number of flowering $F_1$ phenotypes | | |
|---|---|---|---|
| | Females | Hermaphrodites | Males |
| 'Loimaa', hermaphrodite | 0 | 31* | 18 |

**Including three partially fertile inflorescences in isolation bags.*

**Table 4.**
*$F_1$ progenies of the crosses of the hermaphroditic* Fragaria moschata *'Loimaa' with the hermaphroditic pollen parent* F. ×bifera.

The 31: 18 ratio has a poor fit to 1: 1, $\chi^2 = 3.449$, df = 1, p = 0.063. The 31: 18 ratio does not differ significant from the 67: 26 ratio of the *F. moschata* 'Loimaa' × *F. viridis* progeny (**Table 3**), $\chi^2 = 0.782$, df = 1, p = 0.376.

## 9. Edaphic requirements of *Fragaria moschata* in natural habitat

The soil analysis of the habitat of *F. moschata*, from the site where the species has survived over 220 years in Hämeenkylä, Kouvola, SE Finland (**Figure 19**) shows a relatively high Ca content about the K and Mg contents (**Table 5**). The survival of *F. viridis* and the hybrid species *F.* ×*bifera* on their northern limit of distribution in southernmost Finland are highly dependent on a high Ca content of the soil in the habitat, where *F. vesca* is indifferent, but *F. vesca* also turns dependent on high Ca in its northern limit of distribution [42]. Another important issue is the high Ca: K ration of the soil in the habitats of *F. viridis*, while the *F.* ×*bifera* habitats are intermediate between those of *F. vesca* and *F. viridis* [42].

In garden strawberry (*F.* ×*ananassa*) a deficiency of Ca in emerging leaves causes leaf tip-burn [43]. A high availability of K in the soil may decrease Ca mobility in the strawberry plant, and result in tip-burn [44]. A further reason for tip-burn can be boron deficiency [45]. In the populations surviving in the wildness, I have not observed tip-burn disorder in *F. moschata* in the natural populations, but in cultivation for years in limited soil bags, tip-burn may occur. The incipient tip-burn in *F. moschata* may deform the leaflet shape from the typical rhomboid to that of *F.* ×*ananassa* (**Figure 20**).

**Figure 19.**
*A female clone of* Fragaria moschata *which has survived* in situ *over 220 years, see Munsterhjelm 1799–1801 [41].* F. moschata *is practically the only field layer species on the site in the highly shadowed deciduous grove in Hämeenkylä, Kouvola, Finland. Photography needed the use of flash on a sunny day. June 11, 2020.*

| | |
|---|---|
| Soil type | Fine-sandy moraine Rich in organic |
| Conductivity | 0.7 mS/cm |
| pH | 5.4 |
| Ca | 1100 mg/l |
| P | 8.8 mg/l |
| K | 160 mg/l |
| Mg | 170 mg/l |
| S | 11 mg/l |
| B | 0.6 mg/l |
| Cu | 2.9 mg/l |
| Mn | 120 mg/l |
| Zn | 16 mg/l |
| Cation exchange capacity | 13 cmol/kg |
| Ca/cat. exch. cap. | 42% |
| K/cat. exch. cap. | 3% |
| Mg/cat. exch. cap. | 11% |
| Na/cat. exch. cap. | 2% |

**Table 5.**
*Results of the topsoil analysis of the Hämeenkylä site in Kouvola, Finland, where* F. moschata *has survived over 220 years, see Munsterhjelm 1799–1801 [41]. Sampling date Aug. 14, 2020. The analysis was purchased from Viljavuuspalvelu oy Eurofins (Mikkeli, Finland); analysis no. 20–00079100 named Mansikka-2.*

**Figure 20.**
*Grown in limited soil bags, an incipient tip-burn disorder in leaflets of* Fragaria moschata *with healthy leaves of the neighboring plant on the left. The affected leaves start to resemble those of a garden strawberry* F. ×ananassa. *A male descendent grown from the seeds of an open pollinated flower of the Kotka occurrence (**Table 1**). The seed parent plant must have been a female clone pollinated by the 'Kotka' hermaphroditic mutant. June 13, 2020.*

The foliage of *F. moschata* and its hybrids start to senesce when the berries grow. These plants have a tendency in the autumn to make over-wintering, dark-green leaves characterized with short petioles about half those of the summer leaves.

## 10. Discussion

Ahmadi and Bringhurst [16] reported having received only hermaphroditic descendants (24 plants) from the selfing of the hermaphroditic cultivar 'Capron' (*F. moschata*), while the cross between the female 'Perfumata' × hermaphroditic 'Capron' resulted in 12 females and 10 males and no hermaphroditic. This 1: 1 ratio was confirmed in the 9-ploid hybrids, 74 females and 72 males. The female *F. moschata* (Jokioinen) × hermaphroditic *F. viridis* cross (**Table 3**) also fits the heterogametic nature of female plants.

The hermaphroditism seems to lower yield in octoploid strawberry species compared with dioecious female clones [46, 47]. With three endemics American octoploid strawberries, also the sex of the plant was found to significantly affect the microbiomes on the flowers [48]. In environments, where both female and male clones of *F. moschata* are frequent enough for cross pollination, the fitness of female clones may be higher than that of hermaphrodites. In the margins of the sparse distribution of *F. moschata* with only one sex attainable on the place, this species relays on vegetative spread. The vegetative spread has also been promoted by human activities. The hermaphroditic *F. moschata* (syn. *F. elatior*) variety 'La Grange' could have sterile flowers from 7 to 90%, with an average of 52% of fertile flowers as written by Longworth in 1846 [49 cited in 46]. The recently arisen hermaphroditic mutant 'Kotka' of musk strawberry resembles the properties of 'La Grange' having variably perfect male fertility.

In most octoploid *Fragaria* taxa, a 13 kb sequence that occurred in all females and never in males was called sex-determining region (SDR). The SDR cassettes revealed a history of repeated translocation [50]. This also raises the hypothesis that the sex mutation from female to hermaphroditic ('Kotka') could be caused by a transposition. The probability for an occurrence of a transposon in the Kotka female clone is high, because transposable element sequences make up about 36% of the total genome assembly in the octoploid strawberry [51], and variation in the location of the sex-determining genes within the homoeologous chromosome group VI was shown in the North-American octoploid *F. virginiana* [52].

Thinly variegating leaves (**Figure 6**) occurred in five descendants of the open-pollinated *F. moschata* in Kotka, where the female clone was evidently back-crossed by its male-fertile hermaphroditic mutant. The variegation can be an expression of an unstable gene and could be caused by the movement of a transposing element, which may have been activated by an environmental mutagen. Already in 1955, Darrow [53] concluded that "The evidence is that variegation is due to a frequently mutating or unstable gene and that the tendency to mutate is inherited as recessive" and "A similar variegation has been obtained when seeds have been irradiated". The Kotka site was subjected to high doses of radioactive airborne fallout from the fire of the Chernobyl nuclear power plant in the Soviet Union in April–May of 1986. In Kotka, on June 6, 1986, a hot or high-activity particle apparently caused dysfunction in a monitor of radioactivity [54]. About five months after the beginning of the Chernobyl fallout, on October 1, 1986, in the Kotka area, the external dose rate was up to 0.159 μSv/h and the estimated $^{137}Cs$ surface activity up to 45 $kBq/m^2$ [55]. The mutating rate was high in parts of Finland after airborne fallout radioactivity from Chernobyl since the 1986 April, because one could make autoradiograms on Kodak X-OMAT AR films with fresh grass [54], and because of the occurrences of the hot particles [56]. On some sites, tens of such hot particles occurred on a 1-$dm^2$ area of land [54]. In a highly radioactively polluted area 32 km NW of the site of the 'Kotka' strawberry [55], three mutants occurred in less than 200 wheat plants grown for other research purposes in the field soil in the 1986 season, and were detected in subsequent seasons and generations [57]. On a tested site, radioactivity dose rate (mR/h) was 50 higher on the soil surfaces than at the 1-m level still on May 28, 1986 [54].

The 1986 radioactive airborne fallout from Chernobyl may have induced genetic changes in vegetative, on land surface lying strawberry stolon tissue, later somatically segregating into the phenotype like the hermaphroditism or, after inbreeding, occurring in seedlings like the variegation. Somatic segregation of polyploidy from stolon was shown in the polyploidization technique derived for strawberries ([40], figure 1d). There under the action of colchicine on cell division, polyploidized cells and cell-groups are numerous. Instead, a radiation-induced mutation on a given gene per cell is extremely rare, and it takes years that a viable mutated cell-line forms a mutant-carrying strawberry stolon, if ever. A recessive mutant needs a meiotic division to manifest, like the variegation gene (**Figure 6**).

The hybrids between *F. moschata* and *F.* ×*bifera*, *F. vesca* or *F. viridis* provide a wide variety of berry flavors and sweetness. They could be added as flavors mixed to other berries and berry products. The musk strawberries 'Capron Royal' and 'Profumata di Tortona' were among the highest aromatic 17 cultivars, 15 of which were modern *F.* ×*ananassa* cultivars [58]. Studying the aroma of two sources of musk strawberries, Ferreira et al. [59] detected more than 100 distinctive volatile compounds by GC–MS.

During some seasons, berries of musk strawberry may be infected with the fungus *Spherotheca macularis* f. sp. *fragariae*. Environmental factors like drought, high proportion of sunshine and amount of UV may control the occurrence of the infection. There may be hope to recombine resistance towards the fungus from *F. viridis* or *F. vesca* strains.

*F. viridis* is self-incompatible [60] and needs another genotype of incompatibility to pollinate. The self-incompatibility apparently increases chance of *F. viridis* to be pollinated with *F. vesca* where *F. viridis* is marginal and rare. The self-incompatibility is not inherited to the *F. moschata* × *F. viridis* hybrids. The self-incompatibility also gives up in the autotetraploid form of *F. viridis* studied with the polyploidization technique [40] which permits to test the single genotype on diploid and tetraploid level (unpublished).

## 11. Hybrids referred to as *Fragaria* ×*neglecta* and *F.* ×*intermedia*

The *F. neglecta* (*F.* ×*neglecta* Lindem.) hybrids were reported from Austria in 1896 [61]. *F. neglecta* was, however, described as a variant of *F. viridis* in 1865 [1, 62]. The dioecious *F. moschata* female pollinated with the hermaphroditic *F. viridis* gave tetraploid female and male descendants in a 1: 1 ratio [63]. Bors & Sullivan [64] obtained mostly tetraploids, some hexaploids and one mixoploid (4x/6x) among the offspring from the crosses between *F. moschata* and *F. viridis*. Their tetraploid hybrids were evidently fertile.

In 1841 v. Bach [65] described *β. intermedia* as a variant of *F. moschata* (syn. *F. elatior*). *F.* ×*intermedia* (Bach) Beck has later been considered as the hybrid between *F. moschata* and *F. vesca* [1, 61]. *F.* ×*intermedia* hybrids were collected by V. Feráková and their tetraploidy determined by A. Murin in Slovakia [66]. The unexpected chromosome numbers published for *F. moschata* by Lippert [67] might have been counted from interspecific hybrids with the diploid strawberries and their derivatives, though the plants were morphologically of the *F. moschata* type. Odd polyploidy levels, penta- and heptaploidy have been detected in spontaneous *F. moschata* × *F. vesca* hybrids in Europe [68, 69]. Evolutionary trials are ongoing with the trioecious *F. moschata*.

## Author details

Hannu Ahokas
Independent Scientist, Helsinki, Finland

*Address all correspondence to: hannu.ahokas0@saunalahti.fi

## References


[1] Hegi G. *Fragária* L. Erdbeere. Illustrierte Flora von Mittel-Europa. 1923;**4**(2):895-907

[2] Lee DV. Early history of the strawberry. In: Darrow GM, editor. The Strawberry. New York, Chicago, San Francisco: Holt, Rinehart, Winston; 1966. pp. 15-23. Also available from http://specialcollections.nal.usda.gov//speccoll/collectionsguide/darrow/Darrow_TheStrawberry.pdf

[3] Hultén E, Fries M. Atlas of North European Vascular Plants. Vol II-Vol III. Königstein: Koeltz Scientific Books; 1986

[4] Staudt G. Strawberry biogeography, genetics and systematics. Acta Horticulturae. 2009;**842**:71-83

[5] Lesemann SS, Bölke N, Tejeda Fernández CE, Buschmann S, Gerischer U, Neinhuis C, et al. Fruit diversity in a natural European habitat of *Fragaria moschata* Weston. Acta Horticulturae. 2017;**1156**: 215-222

[6] Bergaglio A. Il porto della fragole. Amministrazione Comunade di Tortona. 2007

[7] Negri AS, Allegra D, Simoni L, Rusconi F, Tonelli C, Espen L, et al. Comparative analysis of fruit aroma patterns in the domesticated wild strawberries "Profumata di Tortona" (*F. moschata*) and "Regina delle Valli" (*F. vesca*). Frontiers in Plant Science. 2015;**6**(56):1. fpls-06-00056.pdf

[8] Fragola Profumata di Tortona. L'Azienda Agricola Biologica "La Carcassola". https://www.carcassola.it/fragola_profumata_di_tortona.html [accessed on September 24, 2021]

[9] Reich L, Arlein VH. Uncommon Fruits for every Garden. 2nd ed. Portland, London: Timber Press; 2008

[10] Weston R. Tracts on Practical Agriculture and Gardening. 2nd ed. London: S. Hooper; 1773. https://books.google.com

[11] Wilhelm S, Sagen JE. A History of the Strawberry from Ancient Gardens to Modern Markets. Berkeley: University of California, Division of Agricultural Sciences; 1974

[12] Staudt G, Duchesne AN, Les dessins d AN. Duchesne pour son Histoire naturelle des fraisiers. A. N. Duchesne's drawings for his Histoire naturelle des fraisiers. Paris: Publications Scientifiques du Museum; 2003

[13] Keens M. On the cultivation of strawberries, in the open ground. Transactions of the Horticultural Society of London. 1818;**2**:392-397

[14] Regel E. Versuche über die Erdbeersorten im petersburger Klima. Gartenflora. 1869;**18**:172-179

[15] Регелъ ЭЛ, Кессельрингъ Я. 1880 пятнадцатый списокъ сортамъ плодовыхъ, древесныхъ, кустарныхъ, и многолътнихъ растенiй. Catalog von Obstsorten, Ziersträuchern und Stauden des Pomologischen Gartens und der Baumschulen. S.-Peterburg'; 1879

[16] Ahmadi H, Bringhurst RS. Genetics of sex expression in *Fragaria* species. American Journal of Botany. 1991;**78**: 504-514

[17] Goeschke F. Das Buch der Erdbeeren. 2nd ed. Berlin: Paul Parey; 1888

[18] Ruprecht FJ. Flora Ingrica. Vol. 1. Petropol, Riga, Lipsia: Eggers et soc., Sam. Schmidt, Leopold Voss; 1860

[19] Meinshausen KF. Flora Ingrica. St. Petersburg; 1878

[20] Lempiäinen T. Notes on the plants of Johannes municipality in the Karelian isthmus (Russia). Iskos. 2008; **16**:212-214

[21] Solms-Laubach Hz. Ueber unsere Erdbeeren und ihre Geschichte. Botanische Zeitung. 1907;**65**(1 Abt):46-76

[22] Staudt G. Die Entstehung und Geschichte der groβfrüchtigen Gartenerdbeeren *Fragaria ×ananassa* Duch. Der Züchter. 1962;**31**:212-218

[23] Lee DV. The strawberry from Chile. Duchesne and his work. Early breeding in Europe. In: Darrow GM, editor. The Strawberry. New York, Chicago, San Francisco: Holt, Rinehart, Winston; 1966. pp. 24-84. Also available from: http://specialcollections.nal.usda.gov// speccoll/collectionsguide/darrow/ Darrow_The_Strawberry.pdf

[24] Kalm P. Kortt berättelse om de publique planteringar jag skiött och hwad wid de samma blifuit tilgiort, sedan till riksens höglofl. ständers manufactur-contoir min sidsta ödmiuka berättelse blef ingifwen. In: Leinberg KG, editor. Bidrag till kännedomen af vårt land. Vol. 4. Jyväskylä: J. Länkeläs förlag; 1888. pp. 73-90

[25] Hylander N. Linné, Duchesne och smultronen. Linnæus, Duchesne and strawberries. Svenska Linné-Sällskapets Årsskrift. 1945;**28**:17-40 – 95-97

[26] Anonymous. Om Huwitus. Åbo Tidningar. 1850;**73**(72):1-3

[27] Heikel BW. Hedelmä- ja marjaviljelyn opas. Helsinki: Otava; 1908

[28] Дамаскин. Садовство и огородничество на о. Валаамъ. Вѣстникъ Императорскаго Россiйкаго Общества Садоводства. 1870;**10**:224-234

[29] Регелъ Э. Земляника. Вѣстникъ Императорскаго Россiйкаго Общества Садоводства. 1866;**6**:104-126

[30] Hjelt H. Conspectus florae fennicae. Acta Societatis pro Fauna et Flora Fennica. 1915-1919;**41**(1):1-502

[31] Vilmorin-Andrieux et Cie. Les plantes potagères. 2nd ed. Paris: Vilmorin-Andrieux et Cie marchands grainiers; 1891

[32] Шарина Н. Промышленная культура земляники и клубники в СССР. Bulletin of Applied Botany, of Genetics and Plant-Breeding. 1930-1931;**25**(5):197-229

[33] Wachsmuth B. Von Monats-, Wald- und Moschus-Erdbeeren. Gartenpraxis. 2009;**35**(4):20-28

[34] Schiemann E. Artkreuzungen bei *Fragaria*. II. Zeitschrift für Induktive Abstammungs- und Vererbungslehre. 1937;**73**:375-390

[35] Kamneva OK, Syring J, Liston A, Rosenberg NA. Evaluating allopolyploid origins in strawberries (*Fragaria*) using haplotypes generated from target capture sequencing. BMC Evolutionary Biology. 2017;**17**(180):1-19 additional files

[36] Staudt G, DiMeglio LM, Davis TM, Gerstberger P. *Fragaria ×bifera* Duch.: Origin and taxonomy. Botanische Jahrbücher für Systematik, Pflanzengeschichte und Pflanzengeographie. 2003;**125**:53-72

[37] Evans WD. Guelph SO1 synthetic octoploid strawberry breeding clone. HortScience. 1982;**17**:833-834

[38] Ahokas H. Spontaneous tetraploidy in strawberry (*Fragaria* sp., Rosaceae). Nordic Journal of Botany. 1999;**19**:227-234

[39] Antonius K, Ahokas H. Flow-cytometric determination of polyploidy level in spontaneous clones of strawberries. Hereditas. 1996;**124**:285

[40] Ahokas H. A method to induce polyploids from adult strawberry plants, *Fragaria* spp., avoiding tissue culture *in vitro*. Plant Breeding. 1998;**117**: 500-502

[41] Munsterhjelm JC. Jacobina Charlotta Munsterhjelms dagböcker 1799-1801. In: Lönnqvist B, editor. Helsinki (Helsingfors) Svenska litteratursällskapet i Finland: Folklivsstudier 7. 1970. pp. 1-149

[42] Ahokas H. Factors controlling the north-edge distribution of *Fragaria viridis* and its *F. vesca* hybrids in South Finland. Acta Horticulturae. 2002;**567**: 385-388

[43] Guttridge CG, Bradfield EG, Holder R. Dependence of calcium transport into strawberry leaves on positive pressure in the xylem. Annals of Botany. 1981;**48**:473-480

[44] Chiu TF, Bould C. Effects of calcium and potassium on $^{45}$Ca mobility, growth and nutritional disorders of strawberry plants (*Fragaria* spp.). Journal of Horticultural Science. 1976;**51**:525-531

[45] Maas JL, editor. Compendium of Strawberry Diseases. 2nd ed. St. Paul: APS Press; 1987

[46] Darrow GM. The importance of sex in the strawberry. Journal of Heredity. 1925;**16**:193-204

[47] Stahler MM, Luby JJ, Ascher PD. Comparative yield of female and hermaphroditic *Fragaria virgniana* germplasm collected in Minnesota and Wisconsin. In: Dale A, Luby JJ, editors. The Strawberry into the 21st Century. Portland: Timber Press; 1991. pp. 104-105

[48] Wei N, Ashman T-L. The effects of host species and sexual dimorphism differ among root, leaf and flower microbiomes of wild strawberries *in situ*. Scientific Reports. 2018;**8**:5195. DOI: 10.1038/s41598-018-23518-9

[49] Longworth N. Report on the cultivation of the strawberry. Report of Transactions of the Cincinnati Horticultural Society. 1846;**4**:59-61

[50] Tennessen JA, Wei N, Straub SCK, Govindarajulu R, Liston A, Ashman T-L. Repeated translocation of a gene cassette drives sex-chromosome turnover in strawberries. PLoS Biology. 2018;**16**(8): 1-22, e2006062, supporting material. DOI: 10.1371/journal.pbio.2006062

[51] Edger PP, Poorten TJ, Van Buren R, Hardigan MA, Colle M, McKain MR, et al. Origin and evolution of the octoploid strawberry genome. Nature Genetics. 2019;**51**:541-547. www.nature.com/naturegenetics

[52] Wei N, Govindarajulu R, Tennessen JA, Liston A, Ashman T-L. Genetic mapping and phylogenetic analysis reveal interspecific variation in sex chromosomes of the Virginian strawberry. Journal of Heredity. 2017; **108**:731-739. DOI: 10.1093/jhered/esx077

[53] Darrow GM. Leaf variegation in strawberry - a review. Plant Disease Reporter. 1955;**39**:363-370

[54] Rytömaa T, Servomaa K, Toivonen H. Tšernobylin ydinvoimalaturmassa syntyneet kuumat hiukkaset: Mahdollinen vaara

terveydelle Suomessa. Chernobyl reactor accident and radioactive particles found in Finland. Duodecim. 1986;**102**: 1190-1196

[55] Arvela H, Blomqvist L, Lemmelä H, Savolainen AL, Sarkkula S. Environmental gamma radiation measurements in Finland and the influence of meteorological conditions after the Chernobyl accident in 1986. In: STUK-A65. Helsinki: Finnish Centre for Radiation and Nuclear Safety; 1987. pp. 1-40

[56] Toivonen H, Servomaa K, Rytömaa T. Aerosols from Chernobyl: Particle characteristics and health implications. In: Philipsborn H, Steinhäusler F, editors. Hot Particles from the Chernobyl Fallout. Vol. 16. Theuern: Schriftenreihe des Bergbau- und Industriemuseum; 1988. pp. 97-105

[57] Ahokas H. Vehnämutantteja syntymävuosi 1986, Tsernobyl-vuosi. Wheat mutants, generated in 1986, the year of Chernobyl. Luonnon Tutkija. 1989;**93**:26-27

[58] Urruty L, Giraudel J-L, Lek S, Roudeillac P, Montury M. Assessment of strawberry aroma through SPME/GC and ANN methods. Classification and discrimination of varieties. Journal of Agricultural and Food Chemistry. 2002; **50**:3129-3136

[59] Ferreira V, Pet'ka J, Leitner E, Parameswaran B. Musk strawberries: The flavour of a formerly famous fruit reassessed. Flavour and Fragrance Journal. 2012;**27**:273-279

[60] Staudt G. Zur Selbststerilität von *Fragaria viridis* Duch. Die Naturwissenschaften. 1952;**39**:572

[61] Halácsy Ev. Flora von Niederösterreich. Prag, Wien, Leipzig: F. Tempsky, G. Freytag; 1896

[62] Lindemann Ev. Ueber eine sehr verbreitete und bisher verkannte Erdbeereart, *Fragaria neglecta*. Bulletin de la Société Impériale des Naturalistes de Moscou. 1865;**38**:218-220

[63] Schiemann E. Über die Geschlechtsvererbung bei *Fragaria*. Berichte der Deutschen Botanischen Gesellschaft. 1938;**56**(5):161-162

[64] Bors RH, Sullivan JA. Interspecific hybridization of *Fragaria moschata* with two diploid species, *F. nubicola* and *F. viridis*. Euphytica. 2005; **143**:201-207. DOI: 10.1007/s10681-005-3496-9

[65] Bach Mv. Bemerkungen über einige der selteneren Pflanzen des Mittelrheines. Flora. 1841;**45**:715-720

[66] Májovský J. Index of chromosome numbers of Slovakian flora. Acta Facultatis rerum naturalium Universitatis Comenianae - Botanica. 1976;**25**:1-18

[67] Lippert W. Morphologische und zytologische Untersuchung an *Fragaria*, insbesondere an *Fragaria moschata* in Bayern. Botanische Jahrbücher für Systematik, Pflanzengeschichte und Pflanzengeographie. 1985;**107**:195-202

[68] Nosrati H, Price AH, Gerstberger P, Wilcock CC. Identification of a natural allopentaploid hybrid *Fragaria* (Rosaceae), new to Europe. New Journal of Botany. 2011;**1**:88-92

[69] Nosrati H, Price AH, Gerstberger P, Wilcock CC. The first report of an alloheptaploid from the genus *Fragaria* (Rosaceae). New Journal of Botany. 2013;**3**:205-209

Chapter 8

# Quantitative Trait Loci Associated with Agronomical Traits in Strawberry

*Harun Karci, Habibullah Tevfik, Nesibe Ebru Kafkas and Salih Kafkas*


## Abstract

The cultivated strawberry (*Fragaria* x *ananassa*) is derived from *Fragaria chiloensis* and *Fragaria virginiana* species a few centuries ago, and it is one of the most preferred and consumed berries all over the world because of its a good source in terms of many nutritional elements. Strawberry has high genetic variability and adaptation to different environmental conditions due to its highly heterozygous nature. In the last decades, many farmers, breeders, researchers even consumers have started to focus on berry quality traits such as large fruit, uniform shape, high fruit firmness, high fruit sensorial quality (aroma contents), color, gloss, and resistance to pathogens. Thus, the development of novel strawberry cultivars or genotypes with high nutritionally quality traits has become one of the main aims in strawberry breeding programs. Biotechnological tools such as the identification of quantitative trait loci (QTL) and marker-assisted selection (MAS) are the most widely used technologies in fruit breeding programs for shortening the breeding period. Identification of QTLs in agnomical important traits are very valuable tools for early selection in strawberry breeding programs. This chapter is focused on QTL and marker assisted breeding studies in strawberry to date and provides new perspectives on molecular breeding in strawberry breeding.


**Keywords:** strawberry, QTL, NGS, SNP

## 1. Introduction

Strawberry is one of the temperate climate fruit species which can be grown almost anywhere in the world from Ecuador to Siberia within wide ecological limits due to its high adaptability. Strawberry is loved by everyone and has a great market advantage in fresh and industrial consumption [1]. Strawberry, which is one of 23 species in the Rosaceae family, including apple, cherry, and rose species, is in the *Fragaria* genus [2]. The most common commercially grown strawberries are octoploid ($2n = 8x = 56$) and haploid chromosome number ($n = 7$) [3] and constitute a large part of the world's production [1]. The estimated genome size of the Camasora strawberry cultivar which belongs to the *Fragaria* x *ananassa* Duchesne species has been reported as 813.4 Mb [4]. Strawberry production in the world is increasing day by day. Strawberry contains important mineral substances and vitamins for human health and nutrition, such as salicylic acid, calcium, iron, and phosphorus. In addition, strawberry is rich in antioxidants and phenolic compounds for human health. These health benefits play an important role in the rapid increase of its production [5]. Many studies reported that strawberry consumption has positive effects on human health, preventing aging, Alzheimer's, obesity, and cancer diseases, especially heart diseases [6–8].

In the strawberry breeding programs, cultivars should be productive throughout the whole season, tolerant to diseases post harvesting, resistant to long transportation, with high fruit firmness and high fruit quality are the main breeding objectives for many years and many studies have been carried out intensively in European countries, especially in the USA, and Asian countries such as China [9, 10]. However, recently, some breeding programs have been carried out in strawberries covering topics such as adaptation to arid conditions or climate change like GOODBERRY supported by the European Union's Horizon 2020 research and innovation program [10]. Breeding programs should use several strawberry genetic resources or different populations and also focus on the development of new genotypes or lines resistant or tolerant to different abiotic stress conditions as breeding criteria. Selection of promising lines in strawberry breeding will be possible not only with traditional breeding methods but also with biotechnological approaches in strawberry breeding. However, identifying major and minor QTL regions associated with such complex traits that are severely affected by environmental conditions and controlled by multiple genes is a very difficult and time-consuming task in strawberries due to its open pollinated and highly heterozygous nature.

Many agriculturally important traits, such as yield and quality, tolerance to environmental stresses, and resistance to certain diseases, are controlled by polygenes, which complicates the breeding process as phenotypic traits only partially reflect the genetic influence of the individuals. These complex traits are defined as quantitative traits (multifactorial traits or polygenic), and specific regions in the genome of genes associated with a trait are described as quantitative trait loci (QTLs). Detection of QTL region or a gene in the plant genome is a complex process. QTL analysis can simply be expressed as the determination of the relationship between phenotype and genotypic data. All the loci are used to divide the population used in mapping into different genotypic groups based on the presence or absence of a particular locus and to determine whether there are significant differences between groups related to that specific trait [11, 12]. Depending on the marker system and population types, if there is a significant difference between the phenotypic means (2 or 3) of the groups, it indicates that the locus has a QTL controlling the relevant trait [13]. QTL mapping also measures the relative effects of alleles on traits. It also provides the basis for marker-assisted selection (MAS), which not only determines the physical location of QTLs on the genome but also shortens breeding time [14]. QTLs determined from several different locations are called more stable QTL (sQTL) and they are preferred for MAS due to their reliability. A genetically diverse and well-segregated population is required for QTL mapping and determining their position in the genome. QTL analysis is performed by determining the correlation between the phenotypic and genetic data with help of different algorithms [15]. The first necessary condition in QTL mapping is intra-species or interspecies populations acquired F2, recombinant inbred lines (RILs), backcross 1 (BC1), double haploid lines (DHLs), closely isogenic lines (NILs), and full-sib F1 (pseudo-testcross) lines [16].

A set of QTLs determined in the same region or confidence interval (CI) region for more than one trait is defined as a set of QTLs or overlapping QTLs (cQTLs) [17]. cQTLs are potentially important QTLs that control multiple traits. The QTL is classified as a major QTL if it accounts for 10% or more of the total phenotypic variation in the population, and as a minor QTL if it accounts for 10% or less. Many mapping definitions such as linkage, association, NIL, F2, or BC can be used to detect QTLs. Classical biparental linkage mapping involves the crossing of parents with the contrasting phenotype for a trait [18]. QTL analysis uses the linkage between two loci and recombination in biparental populations, while association mapping (AM) uses variations generated by landraces and genotypes found in natural populations. The populations used in association mapping are divided into different groups based on recombination, such as biparental populations. In other words, it describes the correlation of traits and markers in more than one biparental population. Thus, an associated locus in the AM means that it has a similar effect on many individuals in the population [19].

Thanks to the advancements in next-generation sequencing technologies and the reduction of sequencing costs, high-density sequencing can be easily performed and big data can be obtained for the detection of more variants. Higher density linkage maps can be constructed by using the obtained big genotypic data and can increase the chance of detecting rare loci. As a result, these advances greatly contributed to the identification of QTL regions in the genome. High-density genetic maps are constructed using genotyping by sequencing (GBS) data and have been used to successfully identify QTLs related to phenological, biochemical fruit traits, and pathogen resistance or tolerance in many different species. In addition, QTLs that control abiotic stress factors such as cold, heat, salinity, frost, and drought can also be more accurately detected with the help of high-density genetic linkage maps. The purpose of this section is to review the current literature on the identification and evaluation of the genomic positions of QTLs that control phenological, morphological and biochemical fruit quality traits, and disease tolerance traits in strawberries.

## 2. Linkage mapping in strawberry

The principle of genetic mapping is that genes segregate on chromosome recombination during meiosis [20]. If two genes do not segregate independently, that means they are closely linked to each other on the same chromosome. The genes that are very close to each other or tightly linked will be transferred together from the parent to the progenies. While the two genes are not close to each other on the same chromosome the chance of them segregating separately from each other is increased. Thus, the genetic distance between genes can be calculated accordingly.

The first step of genetic mapping is the selection of two genetically very distant parents that demonstrate obvious genetic differences in many traits of interest. The parents must be far apart genetically for the exhibition of sufficient polymorphism. However, if two parents are too distant genetically, some undesirable events can accrue, such as sterility in progenies and high segregation distortion during the linkage analysis.

The structure of populations consists of two ways; firstly, inbred lines derived from homozygous parents; the second one is the inbreeding lines or cultivars and individuals derived from the crossing of the allogamy species. Thus, two-way pseudo-testcross, half-sib, and full-sib populations derived from crossings can be used for genetic mapping in outcrossing species. The progenies from the selfing of the F1 generation (F2), backcross (BC), recombinant inbred lines (RILs), double haploids (DHs), and near isogenic lines (NILs) can be preferred for mapping in alloploid species according to the breeding objectives [14, 21, 22]. The selection of population type plays an important role in successfully constructing the linkage maps. Although F2 progenies are obtained from selfing F1 hybrids, BC lines are generated by crossing F1 progenies back with one of the parents. RILs are produced by single-seed selections derived from individuals of the F2 population. The selections should be continued from six to eight generations. Double haploid lines (DHLs) are developed by doubling the gametes of individuals. DHLs regenerate in tissue culture studies after chromosome doubling from pollen grains or haploid embryos derived from crosses.

The first aim of molecular breeding is the construction of linkage groups. The linkage mapping detects genes or markers association with molecular markers and their chromosomal positions and they are valuable tools in order to perform positional cloning of desired/known genes or related regions and to consist of contigs and scaffolds of physical maps at the chromosomal level.

The first linkage mapping report in *Fragaria* was performed using an isozyme marker and a morphological (yellow fruit color) trait, using individuals of a cross between *F. vesca* cultivars Yellow Wonder and Baron Solemacher [23]. Another finding was reported that an isozyme locus (PGI-2) was used in the same population in non-runnering r locus-related morphological traits [24]. The first linkage map in *Fragaria* was constructed similarly in diploid *F. vesca* [25]. The difference between these maps was that they were constructed in the F2 population produced by crossing Baron Solemacher and W6, a non-runnering European cultivar belonging to *Fragaria vesca ssp. vesca*, and *Fragaria vesca ssp. americana* W6, a wild runnering genotype. A total of 75 random amplified polymorphic DNA (RAPD) markers, two isozymes, an STS marker associated with the alcohol dehydrogenase gene, and the runnering locus were used in linkage mapping, and the length of the map in seven linkage groups was calculated as 445 cM.

The genetic linkage map of octoploid strawberries was constructed by using amplified fragment length polymorphism (AFLP) markers [26]. A total of 515 markers and 119 F1 plants of a two-way pseudo-test cross of Capitola (CA75.121–101 × Parker) and CF1116 (Pajaro × (Earliglow × Chandler)) were used in linkage map construction. A total of 28 and 30 linkage groups were constructed in the maternal and paternal linkage maps. The female map covered 1604 cM and the male map length was 1496 cM. The second map of the diploid *Fragaria* population was generated from the F2 population developed from an interspecific hybridization of *F. vesca* 815 and *F. bucharica* 601 [27]. A total of 78 markers were mapped, 68 of them were simple sequence repeats (SSRs), and the map length was computed as 448 cM. This mapping was important because it was the first SSR-based linkage map in strawberry. This map was enriched by different mapping studies using restriction fragment length polymorphism (RFLPs), AFLPs, SSRs, SNPs, expressed sequence tags (ESTs) and gene-specific markers located in seven linkage groups and the total map length was 528.1 cM. This map is considered the reference framework map in the genus *Fragaria* [27–32].

An AFLP-based genetic map was generated by Lerceteau-Köhler et al. [26], and this map was enriched by Rousseau-Gueutin et al. [33]. The researchers explained that a total of 213 genotypes and many AFLP markers, sequence-characterized amplified regions (SCARs) and SSRs were located in linkage groups. In the female map, 367 loci were mapped into 28 linkage groups, and the map length was 2582 cM, while a total of 440 markers were placed into 26 linkage groups in the male map and the length was 2165 cM. The final map is considered as the first comprehensive reference map for octoploid strawberries. This map helps the construction of the homoeologous of the four *Fragaria* x *ananassa* linkage groups using the *F. vesca* reference map. Another linkage map was also constructed in *F. virginiana*. A total of 319 and 331 SSR-based markers were located in 33 and 32 linkage groups which belong to maternal and paternal maps, respectively [34, 35]. One of the most comprehensive linkage maps in octoploid strawberries was constructed with 490 transferrable SSR or gene-specific markers in individuals obtained from Redgauntlet × Hapil. The authors stated that the constructed linkage map can be used as a framework for future genetic studies and to facilitate the marker-assisted selection in octoploid strawberry breeding. This high-density map was constructed which represents 91% of the *Fragaria* x *ananassa* genome having a map length of 2140.3 cM and consisting of 549 markers [36].

A total of 186 SSR markers were mapped into 28 linkage groups in octoploid strawberry, derived from Holiday × Korona crossing, the estimated map length was 2050 cM [37]. Another linkage map in *Fragaria* x *ananassa* was generated using a total of 4474 SSRs (3746 EST-SSRs in *F. vesca*, 603 EST-derived in *Fragaria* x *ananassa,* and 125 transcriptomic-derived from *Fragaria* x *ananassa*) were used for mapping. All markers were first mapped onto a parent-specific linkage map using three different populations such as 02–19 × Sachinoka, Kaorino × Akihime, and 0212921 × 0212921 inbred lines which were assembled onto one linkage map. The constructed map length was calculated as 2364.1 cM [38]. The backcross progenies of a cross between *F. vesca* and *F. viridis* were utilized in linkage mapping and the map length was computed as 241.6 cM [39, 40]. The construction of linkage groups in octoploid strawberries can be used for comparative studies between the diploid and octoploid maps.

The development of NGS technology and a decrease in unit cost in sequencing increased the improvement of the different sequencing platforms such as ddRAD-seq, DarT-seq, and WGRS (whole genome re-sequencing). Construction of the well-saturated linkage maps is very important for the discrimination of the homoeologous belonging to each chromosome in strawberries. Tennessen et al. [41] used a novel approach called Phylogenetics of Linkage-MapAnchored Polyploid Sub-genomes (POLiMAPS) and understanding of the octoploid strawberry sub-genomes in *F. vesca* subsp. *bracteata*, *Fragaria chiloensis*, and *F. virginiana* populations. The researchers stated that one of these sub-genomes is close to diploid *F. vesca* and the diploid *F. iinumae*, others are close to an unknown ancestor of *F. iinumae*. Meanwhile, the IStraw 90 K Axiom® array was developed by the RosBREED project with international cooperation. The IStraw 90 K Axiom® array was mined from the first draft of the *F. vesca* genome (v1.0). This array contributed to the construction of the map for detection of the homoeologous sub-genomes [42]. It was constructed by using 26 *Fragaria* species, 16 *F.* x *ananassa* and 10 individuals belonging to wild strawberry species. Totally, mined 6594 SNP loci were located on the first SNP linkage map generated by Bassil et al. [42]. The digest restriction-associated DNA sequencing and diversity array technology were used for the construction of high-density linkage maps (ddRAD-Seq, [43]; DarT, [44]). In another study, the IStraw90 SNP array and genotyping by sequencing technology (GBS) were utilized by Mahoney et al. [45] in order to construct a highly saturated linkage group in diploid strawberry, *F. iinumae*. Sargent et al. [36] consisted of the highly saturated linkage map using many polymorphic loci (8407 SNPs) and the IStraw 90 K Axiom® and this map contributed to the detection of the *F. ananassa* genome structure. The SNP alleles on the linkage map were compared with the corresponding *F. vesca* and *F. iinumae* alleles, and the haploSNPs were categorized in the current study. HaploSNPs, which are sub-genome specific markers that have saturated common markers, can be used to compare QTL regions in strawberries, revealing the ancestral-associated loci [36]. These linkage maps provided a better understanding of the genome structure of the octoploid strawberry and increased the interest in SNP-based linkage mapping studies for practical breeding applications.

In strawberry breeding, the development of linkage maps is very important in order to reveal the genetic structure of complex traits because this complexity can be only solved using the QTL mapping and/or association mapping approaches [46]. Although a few association mapping studies have been reported in strawberries, many genetic mapping studies have been performed using biparental populations that have been generated from different crosses [26, 32, 33, 47–50].

## 3. QTLs in strawberry

A QTL is defined as the point or part of coding or non-coding chromosomal regions representing a trait quantitatively. QTL mapping is based on biparental or natural populations and aims at the detection of correlation between phenotypic traits and genomic regions. To carry out the QTL mapping, variant detection in genomic data is completed and phenotypic values are evaluated for each individual in the population. Detection of the chromosomal position related to important genes is one of the major aims of QTL studies. In this section, we will review the previous QTL studies about agronomically important traits that have large effects on strawberry breeding.

In the cultivated octoploid strawberry, the detected QTLs were located at homoeologous loci, meaning each locus can be represented up to four times in the genome as homoeologous loci, each presenting two homologous alleles. These QTLs at homoeologous locations can be considered as homoeo-QTLs [32, 51].

## 4. QTLs related to the morphological and biochemical traits in strawberry

The construction of a high-resolution genetic map is significant because the SNP array development has been for the QTL analyses associated with agronomically important traits in strawberry breeding and cultivation. Researchers from different countries have come together to solve the problems in strawberry breeding around the world. They decided to use the strawberry gene resources and aim to develop new strategies depending on the solution to current problems [10]. Therefore, big data that is generated from these projects needs to be mined fast and effectively because QTL analysis of large-scale data is a problem of software that must be merged large amounts of genotypic and phenotypic data into many individuals of a population.

To date, the detected QTLs have been related to fruit quality and reproductive traits. A number of QTLs have also been identified, which are associated with remontancy, flowering time, runner production, fruit shape, fruit firmness, and biochemical fruit traits such as anthocyanin, fatty acids, phenolic compounds, chlorophyll content, gamma-decalactone and mesifurane (strawberry aroma compounds), and titratable acidity [51–53].

In QTL mapping related to flowering time and some vegetative traits in the (*F. vesca*) woodland strawberry population, three QTLs were detected in LG IV, LG VII, and LG VI. These novel QTLs in this study were identified for the first time, and they were located within the previously detected FvTFL1 gene. Additionally, a stable QTL was determined using the phenotypic data obtained from different locations experiments and the flowering time governing by gene(s) located in this region can be associated with these loci. The authors stated that these candidate genes associated with flowering time genes can be used for the detection of the major QTL regions in a closely related octoploid strawberry [54].

Rey-Serra et al. [55] genotyped the F1 population and F2 population crosses in order to identify genetic variability within the populations, using IStraw35k and IStraw90k SNP arrays. A total of 14,595 and 7977 SNPs were mined in F1 and F2 populations, respectively. According to QTL analysis, although a total of 33 QTLs were detected in the F1 FC50 × FD54 population for shape traits in LG III, IV, and VI, eight QTL regions were identified in the F2 population. The detected a major QTL linked by fruit shape in LG III has 25% total phenotypic variance. Another QTL associated with fruit firmness was calculated as 26.9% total phenotypic variance in LG VII. However, the researchers reported that two QTLs were mapped in LGIII and LGIV for the neck without achenes regions. On the other hand, Nagamatsu et al. [56] conducted a study in order to identify QTLs associated with fruit shape using a MAGIC population derived from full-sib families. In genome-wide association analysis, QTL regions were detected within chromosomes 6 and 7 in genome-wide association analysis.

The short-day type is a problem in strawberry cultivation due to limiting the whole season's productivity, since most of the strawberry cultivars have the short-day type. Day-neutrality is a good choice for decreasing the intensity of the harvest in particular seasons. Thus, everbearing or day-neutral cultivars were preferred to control cultural cultivation applications easily. Day-neutrality is the desired situation in a strawberry to spread throughout the harvest season. Weebadde et al. [47] studied flower blooming traits under long-day conditions using a biparental population of Honeoye × Tribute for QTL analysis. The present study was carried out in different locations and eight QTLs linked to the everbearing were detected as major QTLs with one of these loci having nearly 36% of the total phenotypic variation. The researchers reported that day-neutrality in a strawberry was a multigenic trait. A QTL study related to runner production of strawberries was also performed on the same strawberry population [57]. They determined a single nucleotide variant, ChFaM148–184 T, with 32.4% of the total phenotypic variation for runner production. They stated that day-neutrality phenotypic traits might be controlled either by a single gene or tightly linked polygenes. Associated with runner production, flowering time, and repeat were studied by Sooriyapathirana et al. [58] using the SSR markers. A major QTL was detected linked to day-neutrality within the LG IV. Several QTLs were identified associated with flower and runner production in the present study on different linkage groups.

A major QTL was detected on LG IV associated with day-neutrality in *F.* x *ananassa* in a population generated from Capitola and CF1116 [59]. This major QTL was also in the same location in a previous study using the *Fragaria virginiana ssp. glauca* [60]. Verma et al. [61] reported that this QTL was intensely utilized in the strawberry breeding program at UC Davis [62]. The determined QTL had a single dominant allele on one sub-genome governing the transformation from short-day to day-neutrality [59]. Dominant variants like this are quite common in polyploidy species in order to reduce functional day-neutrality [63–65]. Another different major QTL was also detected in the Honeoye and Tribute population on LG IV. This major QTL is very significant to being identified in the same chromosomal region in all of the data from five different locations. The locus ChFaM148–184 T was found highly linked with this QTL and the total phenotypic explained variation was 32.4%. Verma et al. [61] stated that they performed the QTL analysis related to the day-neutrality in the RosBREED strawberry germplasm dataset. The results demonstrated that there is a major QTL in the same linkage group, LG IV [61].

Another comprehensive QTL study was carried out using an F1 population generated from 232 and 1392, selections and a total of 33 QTLs were identified in 17 agronomic and fruit quality traits such as yield or fruit size and fruit quality traits such as soluble solids content (SSC), ascorbic acid, titratable acidity (TA), color and firmness using SSRs, and AFLPs markers [32]. The authors identified a few candidate genes within the linked QTLs. The identified gene, FaGaLUR, related to the d-galacturonic acid pathway in strawberry fruit and is predicted to govern L-ascorbic acid on LG IV. The QTLs related to anthocyanins and acidity were located in LG V. Within only one QTL, FaMYOX gene was identified and joined to myoinositol biosynthesis. Another QTL associated with fruit firmness was located within the same chromosomal region governed by gene FaExp2 in LG VII. The detected QTL expressed an SGR-like gene that was determined to be linked by photosynthetic chlorophyll-protein complexes and this gene was co-located in the same chromosomal position. The authors stated that photosynthetic chlorophyll-protein complexes might be controlled for yield and yield-related traits. Labadie et al. [66] performed QTL analysis using the F1 population of cultivated strawberries (*F.* x *ananassa*). They have phenotyped the individuals in two successive years' fruit quality traits, such as flavonoid, anthocyanin, flavonols, flavan-3-ols, anthocyanin, flavonoids, phenolics, and total antioxidant capacity. A total of 178 QTLs on the female and male linkage maps of 152 flavonoid metabolic and colorimetric traits were detected, and these QTLs overlapped the previously identified QTL regions related to flavonoid and taste-related traits. The colorimetric QTLs were located within the LG III and LG VI by homologs.

| Populations | Analyzed Traits | No. of QTL | Phenotypic Variation | References |
|---|---|---|---|---|
| *F.* x *ananassa* cv. Capitola<br>*F.* x *ananassa* breeding line CF1116 | *Colletotrichum acutatum* and *Phytophthora cactorum* resistance | 10 | 5.8–12.2% | Denoyes-Rothan et al. [67] |
| *F.* x *ananassa* cv. Capitola<br>*F.* x *ananassa* breeding line CF1116 | 34 fruit quality traits | 22 | 6.5–16.0% | Lerceteau-Kohler et al. [68] |
| *F. vesca*, *F. bucharica* | 6 flowering-related traits | 9 | 10.6–30.3% | Sargent et al. [28] |
| *F.* x *ananassa* F1 population derived from Tribute and Honeoye | Flower blooming trait (Day-neutrality) | 8 | 11.5–36% | Weebadde et al. [33] |
| *F.* x *ananassa* line 232<br>*F.* x *ananassa* line 1392 | 17 agronomical and fruit quality traits | 33 | 9.2–30.5% | Zorrilla-Fontanesi et al. [50] |

| Populations | Analyzed Traits | No. of QTL | Phenotypic Variation | References |
|---|---|---|---|---|
| *F.* x *ananassa* cv. Capitola *F.* x *ananassa* breeding line C141116 | 19 different fruit-related traits | 87 | 5–17% | Lerceteau-Kohler et al. [32] |
| *F.* x *ananassa* line 232, *F.* x *ananassa* line 1392 | 48 volatile compounds | 70 | 14.2–92.8% | Zorrilla-Fontanesi et al. [51] |
| *F.* x *ananassa* cv. Tochiotome *F.* x *ananassa* cv. Itigoiyukanbohonnou2gou (Nou2gou) | Strawberry anthracnose resistance | 9 | 7.7–14.4% | limura et al. [69] |
| *F.* x *ananassa* cv. Capitols *F.* x *ananassa* breeding line CF1116 | Perpetual flowering and runnering | 19 | 4.0–59.3% | Gaston et al. [59] |
| *F.* x *ananassa* F2 cross Dover × Camarosa | Fruit-quality traits | — | — | Molina-Hidalgo et al. [70] |
| *F.* x *ananassa* line 232 *F.* x *ananassa* line 1392 | Fruit-quality traits | — | — | Sánchez-Sevilla et al. [53] |
| *F.* x *ananassa* F1 population derived from Tribute and Honeoye | Flower blooming trait (Day-neutrality) Runner production | 2 | 32.4–63.8% | Castro et al. [57] |
| *F.* x *ananassa* cv. Redgauntlet *F.* x *ananassa* cv. Hapil | *Verricillison dahliae* resistance | 11 | — | Antanaviciute et al. [71] |
| *F.* x *ananassa* F1 population derived from Tribute and Honeoye | Flower blooming traits (Day-neutrality) Runner production | major QTLs | 10.0–34.4% | Sooriyapathirana et al. [58] |
| *F. vesca*, *F. bucharica* | Polyphenolic compounds | 76 | — | Urrutia et al. [72] |
| *F.* x *ananassa* University of Florida breeding population sets | *Xanthomonas fragariae* resistance | a major QTL | — | Roach et al. [73] |
| *F.* x *ananassa* F1 cross Delmarvel × Selva | Fruit-quality traits | 27 | 4.8–10.7% | Castro and Lewers [50] |
| *F.* x *ananassa* University of Florida breeding population sets | *Phytophthora cactorum* resistance | 4 | 13.7–25.3% | Mangandi et al. [74] |
| F2 population derived from the cross *F. vesca* *F. semperflorens* cv. Hawaii-4(H4) × *F. vesca* subsp. *vesca* (FV), (denoted H4 × FV) | Flowering time and vegetative growth | 3 | 12–25% | Samad et al. [54] |
| *F.* x *ananassa* University of Florida breeding population sets | *Colletotrichum gloeosporioides* resistance | a major QTL | 17–29.8% | Anciro et al. [75] |
| *F.* x *ananassa* University of Florida breeding population sets | *Colletotrichum acutatum* resistance | a major QTL | 50% | Salinas et al. [76] |
| *F.* x *ananassa* F1 cross Emily × Fenella | *Phytophthora cactorum* resistance | 3 | 10.3–36.5% | Nellist et al. [77] |
| *F.* x *ananassa* cv. Capitola *F.* x *ananassa* breeding line CF1116 | Fruit-quality traits (f flavonoid-related traits) | 178 | 18–44% | Labadie et al. [66] |

| Populations | Analyzed Traits | No. of QTL | Phenotypic Variation | References |
|---|---|---|---|---|
| F1 population cross between 'FC50' and 'FD54' and an F2 population cross between 'Camarosa' and 'Dover', | Fruit shape, firmness, and other quality traits | 33 | 12.6–36.8% | Rey-Serra et al. [55] |
| *F.* x *ananassa* cv. Benihoppe *F.* x *ananassa* 105(14–9) breeding line from cv Sachinoka | Chlorophyll content | 7 | 1.4–26.4% | Siddique et al. [78] |
| RosBREED strawberry germplasm set | Flower blooming trait (Day-neutrality) | — | — | Verma et al. [61] |

Although QTL studies are limited to the last two decades, numerous QTLs related to strawberry fruit quality traits and several reproductive traits have been identified in various strawberry breeding studies around the world in recent years. Lerceteau-Köhler et al. [51] studied the QTLs associated with fruit quality traits such as fruit size, firmness, color, sugars, organic acids, and anthocyanins using an octoploid strawberry F1 population generated from the cross between Capitola and CF116. Totally, 87 QTLs associated with quality traits were identified, and phenotypic variation for all the QTLs ranged from 5% to 17%. The authors reported that QTLs linked to the soluble solids content (SSC) were identified in LG III, V, and VI in the same population. Castro and Lewers [50] identified a total of 27 QTL linked to fruit quality traits and the total phenotypic variation for each QTL was computed, ranging from 4.8% to 10.7%. There are colocations in QTLs associated with anthocyanins and antioxidant capacity or total phenolics while the QTLs were identified in different LGs. These colocations can be very important for indirect selection according to the higher antioxidant capacity of fruits [50]. The authors stated that they identified three QTLs related to SSC that were placed on LG II, LG V, and LG VI and used the van Dijk et al. [37] genetic map to compare the results of two studies; the QTLs for SSC were located on LG VI. Castro and Lewers [50] also performed QTL analysis for SSC/TA rate and titratable acidity (TA), and an association was detected on LG VI. These results demonstrated that these QTLs were associated with each other. The QTL linked by TA was determined in LG IV. The linkage map must be saturated with common markers for comparison of QTL regions obtained from previously described findings, such as the haplo-SNPs developed by Sargent et al. [36]. The highly saturated HK SNP genetic map was used for QTL analysis in fruit quality traits and resistance to pathogens [79]. The detected QTL-linked SSC was located on LG VI close to QTL regions of previous studies. Lerceteau-Köhler et al. [51] and Verma et al. [79] discovered QTLs for sucrose and glucose traits that were very close to the SSC. The detected QTLs in populations with different genetic backgrounds could be located close to each other within the same chromosomal regions because fruit quality traits such as glucose and sucrose have similar missions in metabolism.

Zorrilla-Fontanesi et al. [52] detected a total of 70 QTLs associated with volatile compounds using the 232 × 1392 population and 35 of them were identified in two or three successive years. These QTLs were located on all LGs with the exception of LG II, this distribution can be explained as the pleiotropic effect of the one locus over the volatile organic compounds (VOCs). Because VOCs have a high correlation, the QTLs for different esters and alcohols are also included in a cluster. Therefore, every single locus was determined to be involved in the biosynthesis or regulation of all the related VOCs. The phenotypic variation percentages were computed, ranging from 14.2% to 92.8%. It can be considered that these major QTLs governing the strawberry fruit aroma were controlled by the several set of loci and had a higher effect than the multiple loci because if this trait had been governed by multiple loci, phenotypic variation percentages would not be higher like this. The two major components related to the VOCs, such as mesifurane and γ-decalactone, whose associated QTLs have ranged from 42% to 67.3% of the total phenotypic variation, and they detected one QTL that has above 90% of the total phenotypic variation. Expression analysis was performed using the parental and F1 progenies and the FaOMT gene was detected in the content of mesifurane in strawberries [52]. This expression was found in progenies, and the expression level of the FaOMT gene was calculated to be higher in the ripening periods. γ-decalactone is a type of lactone and the most abundant lactone in red-ripened fruit. It provides the 'peachy' feel in strawberry [80, 81]. Although this lactone was identified as intensively in the line 1392 parent and was not found in the line 232 parent, this lactone transferred to only half of all progenies in the population. The linked loci were placed in the LG III [52].

Sánchez-Sevilla et al. [53] published a study using genome-wide RNA-seq analysis and bulk segregant analysis, and they detected a novel gene associated with the fatty acid desaturase, FaFAD1, governing γ-decalactone content in strawberry. Similarly, Chambers et al. [82] identified the same gene linked to γ-decalactone using another segregated population as a complement to previous research performed by Sánchez-Sevilla et al. [53]. The developed markers related to the genes, FaOMT and FaFAD1, can be used for the evaluation of these population for prediction of the phenotype with 100% accuracy. Cruz-Rus et al. [83] validated these markers using different and wider strawberry germplasm and the validation percentage was found above 91%. Consequently, the obtained results demonstrated that these genes can be used for future strawberry breeding programs.

In strawberries, phenolic compounds such as anthocyanins, flavonoids, and phenylpropanoids accumulate in ripen strawberries [84, 85]. Phenolic compounds have an important role in protection against abiotic and biotic stress conditions. The transcriptomic analysis of different strawberry genotypes was used to reveal genes' associations with phenolic composition. According to differentially expressed analysis, a candidate gene within the ESTs, FaPRX27, was detected in the ripening period. The researchers mapped the FaPRX27 gene in two segregating populations, FaPRX27 and association QTLs were located in the same chromosomal regions (LG III) and they also mapped a QTL linked to fruit color trait [49].

Molina-Hidalgo et al. [70] characterized the rhamnogalacturonate lyase gene (FaRGlyase1) using an oligonucleotide-based microarray platform in strawberries. The gene FaR-Glyase1 was mapped in a linkage map derived from Dover × Camarosa and with a QTL linked to fruit firmness in LG I.

## 5. QTLs related to pathogens resistance in strawberry

One of the most important shortcomings in strawberry breeding is the lack of commercially cultivated strawberry cultivars with high resistance to many destructive pathogens. Developing disease-resistant cultivars or improving existing cultivars in breeding programs is a priority objective for many strawberry breeding programs around the world [86]. Diseases based on pathogens are not only a very big issue limiting strawberry production, but also restricts the breeding programs. Thus, resistance to diseases has become a significant focus point for strawberry breeding due to its large effects. In this regard, many major and minor QTLs linked to resistance to pathogens have been identified in cultivated octoploid strawberries so far.

The first QTL related to disease resistance in a strawberry was performed and several RAPD markers associated with the Rfp1 locus, a gene that confers resistance to *Phytophthora fragariae* var. *Fragariae* have been reported by Haymes et al. [87]. Then, two sequence-characterized amplified region (SCAR) markers closely associated with the Rpf1 gene were found to be linked with resistance to *P. fragariae* [88]. For resistance to *Colletotrichum acutatum* pathogenicity, SCAR markers associated with the Rca2 gene were developed [68]. The fruit quality traits were one of the first QTL analyses among the agronomically important traits in strawberries. The population derived from Capitola and CF1116 had a total of 213 genotypes that had many opposite traits and were used for the creation of the mapping population. QTLs related to 34 fruit quality traits were identified in this population [89]. The phenotypic evaluation was carried out over two consecutive years and a total of 22 QTLs were detected, among them 8 QTLs on the female and 14 on the male map and phenotypic variance (PV) values were calculated ranging from 6.5% to 16% in these traits. Denoyes-Rothan et al. [67] determined QTL regions related to resistances of *Cerastium acutatum* and *Phytophthora cactorum* using the same map developed by Lerceteau-Kohler. Although all progenies were screened for *P. cactorum* reaction levels, selected 185 plants were screened in terms of their resistance and susceptibility to *C. acutatum*, while the entire progeny was screened for reaction to *P. cactorum*. Totally, five QTLs with a PV ranging from 5.8% to 10.2% were identified for the resistance to *C. acutatum*.

A total of 39 full-sib families derived from octoploid strawberries were utilized in order to test for resistance to anthracnose fruit rot (AFR). Additionally, a validation population consisting of 77 advanced selections and 10 cultivars was tested in the second season. The phenotyping of accessions was performed every week, and genotyping data was obtained using the IStraw35 SNP array. A major QTL linked by resistance to *C. acutatum* that is named FaRCa1 QTL was located at 55–56 cM of LG VI and the phenotypic variation of this locus was calculated as 50%. It was identified that this locus has a dominant allelic nature in all trials. In this study, a locus linked by AFR was detected in LG VI, while markers associated with resistance to *C. acutatum* were found in LG VII in the previous finding gene, Rca2 [76].

QTLs linked with resistance to *Verticillium dahlia* were identified in the cultivated octoploid strawberry [71]. Researchers investigated resistance to the crown and root rot diseases caused by *Colletotrichum gloeosporioides*, *P. cactorum*, and leaf spot caused by *Xanthomonas fragariae* in strawberry. The clonal replicates were derived from 139 full-sib families. More than 1100 strawberry plants were monitored for crown and root rot symptoms in two successive seasons [61].

Roach et al. [73] reported that defined QTL associated with resistance to bacterial leaf spot led by *X. fragariae* (FaRXf1) and that this QTL was controlled by a single dominant allele and demonstrated a 1:1 segregation. According to pedigree-based QTL analysis, major and minor QTLs such as Cg1 and Cg2 were determined as linked with resistance to *C. gloeosporioides* and verified in two successive years [74, 75]. Although total phenotypic variation belongs to Cg1, it explained 26% of the total PV, while Cg2 was identified as minor and explained 5% of the total PV. The associated QTLs that resistance to pathogens such as *P. cactorum* or phytophthora, FaRPc2, were identified and tested [74, 79]. It was noticed that this locus represented 35% of total the phenotypic variation in *P. cactorum*. Three major QTLs for resistance to *P. cactorum*, a water-borne disease, were detected using the segregation population. These QTLs account for 37% phenotypic variation in the population. On the other hand, the researchers reported that several loci associated with crown rot disease caused by *P. cactorum* using a genome-wide association study (GWAS) of 114 individuals were identified in the present study. The loci linked by crown rot were mapped within the LG VI and VII, while the manhattan plot obtained from GWAS demonstrated that associated loci were located on chromosomes 5 and 7 [77].

Gray mold is a pathogenic disease that causes postharvest decay in many different species. This disease, caused by *Botrytis cinerea*, infects different plant tissues and is more destructive, especially in mature fruits and this increases post-harvest economic losses. Petrasch et al. [90] detected the QTLs related to gray mold using the 50 K Axiom SNP array [91] and five full-sib families. A total of nine QTLs were found most significant according to LOD score and these QTLs were located on chromosomes 3, 4, 5, and 7 different sub-genomes. Since the identified QTLs were minor QTLs and distributed among the different sub-genomes of chromosomes, the genes responsible for gray mold can be controlled by polygenes.

## 6. Marker-assisted selection (MAS) in strawberry

Marker-assisted selection is one of the most important breeding techniques that shorten the breeding period in modern molecular breeding studies. To date, different marker technics such as SSRs and SNPs were used in octoploid strawberry breeding programs for MAS associated with many agronomical traits [44, 52, 92].

Simple sequence repeat markers are preferred in molecular studies due to their high level of polymorphism, abundance in the whole genome, co-dominant nature, and reliability. However, this system is not cost-effective for screening of large breeding populations. Another reason is that it produces more than a pair of alleles due to having sub-genomes of octoploid strawberries. Thus, SSRs are not practical because of their multi-allelic nature for large breeding populations and for cloning of associated genes. Although rapid DNA isolation protocols and next-generation genotyping platforms were recently developed by Noh et al. [93] for strawberry breeding, these applications are very expensive and laborious because of the need to obtain high-quality DNA in species with a high content of phenolic compounds, such as strawberry.

In the last two decades, high-resolution melting (HRM) and TaqMan-based markers have been utilized for MAS in strawberry breeding effectively. On the other hand, probe-based methods such as Kompetitive Allele-Specific PCR (KASP) markers can be designed for any species because they are flexible. However, these markers are not feasible for multiplex genotyping while HRM is a post-PCR analysis and does not require sequence validation of the target region. In this regard, SNP-based DNA markers such as endpoint genotyping and KASP should be continued for use in marker-assisted breeding in strawberries.

There is a desirable strategy in MAS is that detected QTL is governed by one or a few major loci or genes. Especially if loci on major QTLs were located in a single sub-genome and this is a positive situation for effective MAS in the cultivated octoploid strawberry. The validated DNA loci associated with traits have been widely used in numerous breeding programs, and the detected linked DNA loci have recently been verified and will be used in breeding programs [61].

The c-decalactone is a volatile compound and is responsible for 'peach-like' aroma in strawberries. Chambers et al. [82] and Sánchez-Sevilla et al. [53] identified the fatty acid desaturase gene (FaFAD1), which is responsible for the c-decalactone biosynthesis process. Sánchez-Sevilla et al. [53] discovered a desaturase gene (FaFAD1) on LG III and they tested all cultivars. Although the gene responsible for this volatility was amplified in less than half of the cultivars, it did not amplify in the rest of the cultivars. The researchers developed a functional SCAR marker in the upstream region of the strawberry genome. The gene created a 500 bp PCR product if the strawberry cultivars have this gene, if not, it does not generate any PCR product. An SSR marker responsible for this volatile compound was developed on an 11 kb upstream genic region related to FaFAD1. It generated a PCR product with 205 bp.

One of the HRM primer pairs developed in the FaFAD1 genic region, GDHRM5 has been used in the UF breeding program. Developing a codominant marker is desired to distinguish individuals with homozygous and heterozygous loci. Recently, a candidate gene associated with mesifurane, the FaOMT gene, was replaced on LG VII [52]. The authors developed the marker based on agarose gel and they identified the functional PCR amplification in 248 bp and the non-functional locus in 217 bp (after 30 bp deletion). Thus, the scoring of the products was performed according to the absence or presence of amplification of the mesifurane gene.

Remontancy is defined as the unresponsiveness of the day-light and it is an important trait to provide throughout the whole season of production in strawberry cultivation. Gaston et al. [59] detected the QTL linked to the day-neutrality, transition from short-day to remontant blooming, governed by the FaPFRU locus identified in the UC Davis breeding program. Another genetic mapping discovery revealed that an SSR locus detected was mostly associated with FaPFRU [94]. This SSR marker was validated in RosBREED germplasm and about 90% of the accessions that were not producing any bands on 129 bp did not flower under the long day conditions.

The screening of populations using the DNA marker for disease resistance in strawberry become more popular, particularly in the last decades. The resistance to *C. acutatum* pathogenicity test is one of these tests, and the Rca2 locus was found to be associated with this trait [68]. It has a dominant allele that is responsible for resistance, and these AFLP markers were converted into two SCAR markers, which amplified a 240 bp PCR product. Thus, these SCAR markers have been used easily in separation by agarose gel-based electrophoresis. The resistance to root and crown rot caused by *P. cactorum* locus was found as the FaRPc2 on the linkage group VII [74]. The FaRPc2 is detected in the major QTL region, which is responsible for about 35–40% of the phenotypic variation [61]. Van de Weg [95] identified the gene resistance to red stele in strawberries. The Rpf1 gene has been used for MAS as resistance factor R1 to *P. fragariae* var. *fragariae*. Recently, one SSR marker associated with R1 was developed and amplification was performed in 99% of the strawberry samples [61]. The previous study took place using a total of 49 *F.* x *ananassa* individuals to verify the Rfp1 gene that is associated with disease resistance. The results demonstrated that 17 accessions created the PCR fragment while 32 individuals were evaluated as susceptible [96]. The angular leaf spot disease caused by *X. fragariae* is the main problem in strawberries. Recently, a single dominant locus located in linkage group VI was associated with this bacterial disease resistance in octoploid strawberries [73].

The reason for the high interest and appreciation for strawberries is their sweetness. And with the method developed by Schwieterman et al. [97], the sweetness can be predicted in strawberry individuals. The QTL region associated with the soluble solids content (SSC) is an indicator of sweetness in strawberries [98], EMFv006 was located on LG VI using the 'Capitola' CF1116 population [51]. The QTL analysis was performed using the 609 strawberry accessions, which consisted of worldwide promising lines or selections and cultivars in 2011 and 2012 [96, 99] The detected QTL was determined very closely near to the EMFv006 SSR marker on LG VI [61]. Verma et al. [61] investigated the association of this marker with the SSC trait in different germplasm sets, but there was no association between the trait and marker EMFv006. That means using this marker in MAS studies may not be a good idea.

In this section, we summarized QTLs associated with important agronomical traits and different disease resistance traits, which are very important in strawberry breeding. And these QTLs can be converted into DNA markers that are closely linked to the desired traits. These markers can be used in MAS in strawberry breeding due to their easy scoring, not being complicated, and giving more accurate results. These markers were developed for MAS in many genetic resources as a common idea for solving the problems encountered in strawberry breeding.

## 7. Future prospects in strawberry breeding

Thanks to the increase of genomic studies in plant breeding, it consists of many areas, such as the development of new approaches and strategies in order to shorten processes, such as the development of promising new cultivars or genotypes with desired characteristics. In this way, advances in strawberry breeding depending on genomic developments are very important processes for octoploid cultural strawberries. Detection of the four sub-genomes of the strawberry genome and different next-generation sequencing methods will provide the opportunity to make a more accurate genetic assembly. The genomic selection in the early period can take place with a more accurate and easier explanation of the functions of the SNPs associated with all traits of the strawberry. The identification and characterization of candidate gene regions linked by economically important traits will facilitate advances in genomic selection. Achieving the targeted objectives in strawberry breeding will enable high-density genetic and physical maps to be obtained and effective marker-assisted selection with the reduction of sequencing unit costs.

In breeding studies, geneticists, breeders, and even farmers want to monitor the process of identifying the underlying biological problems of a phenotypic trait and those associated with genomics. Therefore, many studies have been conducted on the detection of QTLs associated with many traits. However, not only the determination of QTL regions but also the detection of structural and copy number variations will allow the related trait to be examined from a wider perspective. As a result, researching the functions of candidate genes associated with traits and post-transcriptional translation processes, including epigenetic variations, supplies the necessary infrastructure in genomic sequencing. In the future, it will be possible to replacc undesirable loci associated with desired traits with methods such as CRISPR-cas9.

Therefore, new loci associated with traits will be discovered in many breeding programs with new technological and genomic approaches in future strawberry breeding.

## 8. Conclusion

In order to detect more stable QTL regions in plants, it is basically needed to have a high-quality genome assembly. By increasing the quality of the octoploid strawberry genome, it can be ensured that the four sub-genomes are brought together or differentiated with a good assembly. In addition, better coverage of individuals' genomes will be ensured with next-generation SNP sequences. In this way, it will allow a more accurate description of the genome functions of the relevant QTL regions. It will also allow the discovery of additional chromosomal regions that control economically important fruit traits and the development of functional markers targeted by candidate genes. Their high-resolution mapping along with the reduction of unit cost in sequencing will allow the discovery of rare markers associated with economically important traits.

Recently, a quality genome assembly and the mapping of QTLs associated with economically important traits and their use in MAS have greatly facilitated breeding programs. To date, most of the QTLs detected are environment specific. The number of experiments, the year, and the application and cultivation conditions are very important for QTL studies. In particular, the selection of populations such as F2, F3, and BC to be used for mapping is very important, as are other conditions. The distance of associated QTLs, the impact of major QTLs on the population, environmental factors, population size used in mapping, and based on individual errors affect the accuracy of QTLs. However, increasing the number of individuals in the population allows the estimation and determination of the phenotypic variance value of minor QTLs. On the other hand, the techniques used for the validation of QTLs are very important for the use of linked markers in MAS. Some important loci may be eliminated due to the selection of inadequate verification methods of linked markers. Therefore, the discovery of QTLs associated with economically important traits in strawberries for selection at an early stage is very significant. In this process, this summary can be useful for QTLs linked to strawberry traits.

## Author details

Harun Karci, Habibullah Tevfik, Nesibe Ebru Kafkas* and Salih Kafkas
Department of Horticulture, Faculty of Agriculture, Çukurova University, Adana, Turkey

*Address all correspondence to: ebruyasakafkas@gmail.com

## References

[1] Amil-Ruiz F, Blanco-Portales R, Muñoz-Blanco J, Caballero JL. The strawberry plant defense mechanism: A molecular review. Plant & Cell Physiology. 2011;**52**(11):1873-1903

[2] Shulaev V, Korban SS, Sosinski B, Abbott AG, Aldwinckle HS, Folta KM, et al. Multiple models for Rosaceae genomics. Plant Physiology. 2008;**147**:985-1003

[3] Kole C, editor. Genome mapping and molecular breeding in plants. In: Fruits and Nuts. Strawberry. 2007. Vol IV. pp. 189-205

[4] Edger PP, Poorten TJ, VanBuren R, Hardigan MA, Colle M, McKain MR, et al. Origin and evolution of the octoploid strawberry genome. Nature Genetics. 2019;**51**(3):541-547

[5] Ertürk YE, Geçer MK, Karadaş K. Türkiye'de Çilek Üretimi ve Pazarlaması. 2017;**46**(1):13-20

[6] Maas JL, Galletta GJ, Stoner GD. Ellagic acid, an anticarcinogen in fruits, especially in strawberries: A review. HortScience. 1991;**26**(1):10-14

[7] Zhang Y, Seeram NP, Lee R, Feng L, Heber D. Isolation and identification of strawberry phenolics with antioxidant and human cancer cell antiproliferative properties. Journal of Agricultural and Food Chemistry. 2008;**56**(3):670-675

[8] da Silva PM, de Carvalho JE, Lajolo FM, Genovese MI, Shetty K. Evaluation of antiproliferative, anti-type 2 diabetes, and antihypertension potentials of ellagitannins from strawberries (*Fragaria × ananassa* Duch.) using In vitro models. Journal of Medicinal Food. 2010;**13**(5):1027-1035

[9] Diamanti J, Capocasa F, Balducci F, Battino M, Hancock J, Mezzetti B. Increasing strawberry fruit sensorial and nutritional quality using wild and cultivated germplasm. PLoS One. 2012;**7**(10):e46470

[10] Mezzetti B, Giampieri F, Zhang YT, Zhong CF. Status of strawberry breeding programs and cultivation systems in Europe and the rest of the world. Journal of Berry Research. 2018;**8**(3):205-221

[11] Tanksley SD. Mapping polygenes. Annual Review of Genetics. 1993;**27**(1):205-233

[12] Young ND. QTL mapping and quantitative disease resistance in plants. Annual Review of Phytopathology. 1996;**34**(1):479-501

[13] Sehgal D, Singh R, Rajpal VR. Quantitative trait loci mapping in plants: Concepts and approaches. Molecular Breeding for Sustainable Crop Improvement. 2016;**11**(2):31-59

[14] Doerge RW. Mapping and analysis of quantitative trait loci in experimental populations. Nature Reviews Genetics. 2002;**3**(1):43-52

[15] Miles C, Wayne M. Quantitative trait locus (QTL) analysis. Nature Education. 2008;**1**(1):208

[16] Schneider K. Mapping populations and principles of genetic mapping. The Handbook of Plant Genome Mapping: Genetic and Physical Mapping. 2005:1-21

[17] Zhou NN, Tang KX, Jeauffre J, Thouroude T, Arias DCL, Foucher F, et al. Genetic determinism of prickles in rose. Theoretical and Applied Genetics. 2020;**133**(11):3017-3035

[18] Portis E, Cericola F, Barchi L, Toppino L, Acciarri N, Pulcini L, et al.

Association mapping for fruit, plant and leaf morphology traits in eggplant. PLoS One. 2015;**10**(8):e0135200

[19] Paudel B, Zhuang Y, Galla A, Dahal S, Qiu Y, Ma A, et al. WFhb1-1 plays an important role in resistance against Fusarium head blight in wheat. Scientific Reports. 2020;**10**(1):1-15

[20] Paterson AD, Kennedy JL, Petronis A. Evidence for genetic anticipation in non-mendelian diseases. American Journal of Human Genetics. 1996;**59**(1):264

[21] Burr B, Burr FA, Thompson KH, Albertson MC, Stuber CW. Gene mapping with recombinant inbreds in maize. Genetics. 1988;**118**(3):519-526

[22] He P, Li JZ, Zheng XW, Shen LS, Lu CF, Chen Y, et al. Comparison of molecular linkage maps and agronomic trait loci between DH and RIL populations derived from the same rice cross. Crop Science. 2001;**41**(4):1240-1246

[23] Williamson SC, Yu H, Davis TM. Shikimate dehydrogenase allozymes: Inheritance and close linkage to fruit color in the diploid strawberry. The Journal of Heredity. 1995;**86**(1):74-76

[24] Hongrun Y, Davis TM. Genetic linkage between runnering and phosphoglucoisomerase allozymes, and systematic distortion of monogenic segregation ratios in diploid strawberry. Journal of the American Society for Horticultural Science. 1995;**120**(4):687-690

[25] Davis TM, Yu H. A linkage map of the diploid strawberry, *Fragaria vesca*. Journal of Heredity. 1997;**88**(3):215-221

[26] Lerceteau-Köhler E, Guérin G, Laigret F, Denoyes-Rothan B. Characterization of mixed disomic and polysomic inheritance in the octoploid strawberry (*Fragaria* x *ananassa*) using AFLP mapping. Theoretical and Applied Genetics. 2003;**107**(4):619-628

[27] Sargent DJ, Davis TM, Tobutt KR, Wilkinson MJ, Battey NH, Simpson DW. A genetic linkage map of microsatellite, gene-specific and morphological markers in diploid *Fragaria*. Theoretical and Applied Genetics. 2004;**109**(7):1385-1391

[28] Sargent DJ, Clarke J, Simpson DW, Tobutt KR, Arús P, Monfort A, et al. An enhanced microsatellite map of diploid *Fragaria*. Theoretical and Applied Genetics. 2006;**112**(7):1349-1359

[29] Sargent DJ, Rys A, Nier S, Simpson DW, Tobutt KR. The development and mapping of functional markers in *Fragaria* and their transferability and potential for mapping in other genera. Theoretical and Applied Genetics. 2007;**114**(2):373-384

[30] Sargent DJ, Cipriani G, Vilanova S, Gil-Ariza D, Arús P, Simpson DW, et al. The development of a bin mapping population and the selective mapping of 103 markers in the diploid *Fragaria* reference map. Genome. 2008;**51**(2):120-127

[31] Sargent DJ, Kuchta P, Girona EL, Zhang H, Davis TM, Celton J-M, et al. Simple sequence repeat marker development and mapping targeted to previously unmapped regions of the strawberry genome sequence. Plant Genome. 2011;**4**(3):165-177

[32] Zorrilla-Fontanesi Y, Cabeza A, Torres AM, Botella MA, Valpuesta V, Monfort A, et al. Development and bin mapping of strawberry genic-SSRs in diploid *Fragaria* and their transferability across the Rosoideae subfamily. Molecular Breeding. 2011;**27**(2):137-156

[33] Rousseau-Gueutin M, Lerceteau-Köhler E, Barrot L, Sargent DJ, Monfort A, Simpson D, et al. Comparative genetic mapping between octoploid and diploid *Fragaria* species reveals a high level of colinearity between their genomes and the essentially disomic behavior of the cultivated octoploid strawberry. Genetics. 2008;**179**(4):2045-2060

[34] Spigler RB, Lewers KS, Main DS, Ashman TL. Genetic mapping of sex determination in a wild strawberry, *Fragaria virginiana*, reveals earliest form of sex chromosome. Heredity (Edinb). 2008;**101**(6):507-517

[35] Goldberg MT, Spigler RB, Ashman TL. Comparative genetic mapping points to different sex chromosomes in sibling species of wild strawberry (*Fragaria*). Genetics. 2010;**186**(4):1425-1433

[36] Sargent DJ, Yang Y, Šurbanovski N, Bianco L, Buti M, Velasco R, et al. HaploSNP affinities and linkage map positions illuminate subgenome composition in the octoploid, cultivated strawberry (*Fragaria × ananassa*). Plant Science. 2015;**242**:140-150

[37] van Dijk T, Pagliarani G, Pikunova A, Noordijk Y, Yilmaz-Temel H, Meulenbroek B, et al. Genomic rearrangements and signatures of breeding in the allo-octoploid strawberry as revealed through an allele dose based SSR linkage map. BMC Plant Biology. 2014;**14**(1):1-16

[38] Isobe SN, Hirakawa H, Sato S, Maeda F, Ishikawa M, Mori T, et al. Construction of an integrated high density simple sequence repeat linkage map in cultivated strawberry (*Fragaria × ananassa*) and its applicability. DNA Research. 2013;**20**(1):79-92

[39] Nier S, Simpson DW, Tobutt KR, Sargent DJ. A genetic linkage map of an inter-specific diploid *Fragaria* BC1 mapping population and its comparison with the *Fragaria* reference map (FV x FN). The Journal of Horticultural Science and Biotechnology. 2006;**81**(4):645-650

[40] Cipriani G, Pinosa F, Bonoli M, Faedi W. A new set of microsatellite markers for *Fragaria* species and their application in linkage analysis. The Journal of Horticultural Science and Biotechnology. 2006;**81**(4):668-675

[41] Tennessen JA, Govindarajulu R, Ashman TL, Liston A. Evolutionary origins and dynamics of octoploid strawberry subgenomes revealed by dense targeted capture linkage maps. Genome Biology and Evolution. 2014;**6**(12):3295-3313

[42] Bassil NV, Davis TM, Zhang H, Ficklin S, Mittmann M, Webster T, et al. Development and preliminary evaluation of a 90 K axiom® SNP array for the allo-octoploid cultivated strawberry *Fragaria × ananassa*. BMC Genomics. 2015;**16**(1):1-30

[43] Davik J, Sargent DJ, Brurberg MB, Lien S, Kent M, Alsheikh M. A ddRAD based linkage map of the cultivated strawberry, *Fragaria* x *ananassa*. PLoS One. 2015;**10**(9):e0137746

[44] Sánchez-Sevilla JF, Horvath A, Botella MA, Gaston A, Folta K, Kilian A, et al. Diversity arrays technology (DArT) marker platforms for diversity analysis and linkage mapping in a complex crop, the octoploid cultivated strawberry (*Fragaria × ananassa*). PLoS One. 2015;**10**(12):e0144960

[45] Mahoney LL, Sargent DJ, Abebe-Akele F, Wood DJ, Ward JA, Bassil NV, et al. A high-density linkage map of the ancestral diploid strawberry, *Fragaria iinumae*, constructed with single nucleotide polymorphism markers

from the IStraw90 array and genotyping by sequencing. The Plant Genome. 2016;**9**(2):1-14

[46] Collard BCY, Mackill DJ. Marker-assisted selection: An approach for precision plant breeding in the twenty-first century. Philosophical Transactions of the Royal Society B: Biological Sciences. 2008;**363**(1491):557-572

[47] Weebadde CK, Wang D, Finn CE, Lewers KS, Luby JJ, Bushakra J, et al. Using a linkage mapping approach to identify QTL for day-neutrality in the octoploid strawberry. Plant Breeding. 2008;**127**(1):94-101

[48] Sargent DJ, Fernandéz-Fernandéz F, Ruiz-Roja JJ, Sutherland BG, Passey A, Whitehouse AB, et al. A genetic linkage map of the cultivated strawberry (*Fragaria* × *ananassa*) and its comparison to the diploid *Fragaria* reference map. Molecular Breeding. 2009;**24**(3):293-303

[49] Ring L, Yeh SY, Hücherig S, Hoffmann T, Blanco-Portales R, Fouche M, et al. Metabolic interaction between anthocyanin and lignin biosynthesis is associated with peroxidase FaPRX27 in strawberry fruit. Plant Physiology. 2013;**163**(1):43-60

[50] Castro P, Lewers KS. Identification of quantitative trait loci (QTL) for fruit-quality traits and number of weeks of flowering in the cultivated strawberry. Molecular Breeding. 2016;**36**(10):1-19

[51] Lerceteau-Köhler E, Moing A, Guérin G, Renaud C, Petit A, Rothan C, et al. Genetic dissection of fruit quality traits in the octoploid cultivated strawberry highlights the role of homoeo-QTL in their control. Theoretical and Applied Genetics. 2012;**124**(6):1059-1077

[52] Zorrilla-Fontanesi Y, Rambla JL, Cabeza A, Medina JJ, Sánchez-Sevilla JF, Valpuesta V, et al. Genetic analysis of strawberry fruit aroma and identification of O-methyltransferase FaOMT as the locus controlling natural variation in mesifurane content. Plant Physiology. 2012;**159**(2):851-870

[53] Sánchez-Sevilla JF, Cruz-Rus E, Valpuesta V, Botella MA, Amaya I. Deciphering gamma-decalactone biosynthesis in strawberry fruit using a combination of genetic mapping, RNA-Seq and eQTL analyses. BMC Genomics. 2014;**15**(1):1-15

[54] Samad S, Kurokura T, Koskela E, Toivainen T, Patel V, Mouhu K, et al. Additive QTLs on three chromosomes control flowering time in woodland strawberry (*Fragaria vesca* L.). Horticulture Research. 2017;**4**:17020

[55] Rey-Serra P, Mnejja M, Monfort A. Shape, firmness and fruit quality QTLs shared in two non-related strawberry populations. Plant Science. 2021;**311**:111010

[56] Nagamatsu S, Tsubone M, Wada T, Oku K, Mori M, Hirata C, et al. Strawberry fruit shape: Quantification by image analysis and qtl detection by genome-wide association analysis. Breeding Science. 2021;**71**(2):167-175

[57] Castro P, Bushakra JM, Stewart P, Weebadde CK, Wang D, Hancock JF, et al. Genetic mapping of day-neutrality in cultivated strawberry. Molecular Breeding. 2015;**35**(2):1-16

[58] Sooriyapathirana SS, Mookerjee S, Weebadde CK, Finn CE, Lewers K, Bushakra JM, et al. Identification of QTL associated with flower and runner production in octoploid strawberry (*Fragaria* × *ananassa*). Journal of Berry Research. 2015;**5**(3):107-116

[59] Gaston A, Perrotte J, Lerceteau-Köhler E, Rousseau-Gueutin M, Petit A, Hernould M, et al. PFRU, a single dominant locus regulates the balance between sexual and asexual plant reproduction in cultivated strawberry. Journal of Experimental Botany. 2013;**64**(7):1837-1848

[60] Powers L. Inheritance of period of blooming in progenies of strawberries. The American Society for Horticultural Science. 1954;**64**:293-298

[61] Verma S, Osorio LF, Lee S, Bassil NV, Whitaker VM. Genome-Assisted Breeding in the Octoploid Strawberry. Cham: Springer; 2018. pp. 161-184

[62] Hancock JF, Lavín A, Retamales JB. Our southern strawberry heritage: *Fragaria chiloensis* of Chile. HortScience. 1999;**34**(5):814-816

[63] Albani MC, Battey NH, Wilkinson MJ. The development of ISSR-derived SCAR markers around the seasonal flowering locus (SFL) in *Fragaria vesca*. Theoretical and Applied Genetics. 2004;**109**(3):571-579

[64] Doyle JJ, Flagel LE, Paterson AH, Rapp RA, Soltis DE, Soltis PS, et al. Evolutionary genetics of genome merger and doubling in plants. Annual Review of Genetics. 2008;**42**:443-461

[65] Feldman M, Levy AA, Fahima T, Korol A. Genomic asymmetry in allopolyploid plants: Wheat as a model. Journal of Experimental Botany. 2012;**63**(14):5045-5059

[66] Labadie M, Vallin G, Petit A, Ring L, Hoffmann T, Gaston A, et al. Metabolite quantitative trait loci for flavonoids provide new insights into the genetic architecture of strawberry (*Fragaria × ananassa*) fruit quality. Journal of Agricultural and Food Chemistry. 2020;**68**(25):6927-6939

[67] Denoyes-Rothan B, Lerceteau-Köhler E, Guérin G, Bosseur S, Bariac J, Martin E, et al. QTL analysis for resistances to *Colletotrichum acutatum* and *Phytophthora cactorum* in octoploid strawberry (*Fragaria* x *ananassa*). In: Acta Horticulturae, Eucarpia Symposium, on Fruit Breeding and Genetics, ISHS in Belgium; 2004

[68] Lerceteau-Köhler E, Guérin G, Denoyes-Rothan B. Identification of SCAR markers linked to Rca2 anthracnose resistance gene and their assessment in strawberry germplasm. Theoretical and Applied Genetics. 2005;**111**(5):862-870

[69] Iimura K, Tasaki K, Nakazawa Y, Amagai M. QTL analysis of strawberry anthracnose resistance. Breeding Research. 2013;**15**(3):90-97

[70] Molina-Hidalgo FJ, Franco AR, Villatoro C, Medina-Puche L, Mercado JA, Hidalgo MA, et al. The strawberry (*Fragaria × ananassa*) fruit-specific rhamnogalacturonate lyase 1 (FaRGLyase1) gene encodes an enzyme involved in the degradation of cell-wall middle lamellae. Journal of Experimental Botany. 2013;**64**(6):1471-1483

[71] Antanaviciute L, Surbanovski N, Harrison N, McLeary KJ, Simpson DW, Wilson F, et al. Mapping QTL associated with verticillium dahliae resistance in the cultivated strawberry (*Fragaria* x *ananassa*). Horticulture Research. 2015;**2**:15009

[72] Urrutia M, Rambla JL, Alexiou KG, Granell A, Monfort A. Genetic analysis of the wild strawberry (*Fragaria vesca*) volatile composition. Plant Physiology and Biochemistry. 2017;**121**:99-117

[73] Roach JA, Verma S, Peres NA, Jamieson AR, van de Weg WE, Bink MCAM, et al. FaRXf1: A locus

conferring resistance to angular leaf spot caused by *Xanthomonas fragariae* in octoploid strawberry. Theoretical and Applied Genetics. 2016;**129**(6):1191-1201

[74] Mangandi J, Verma S, Osorio L, Peres NA, van de Weg E, Whitaker VM. Pedigree-based analysis in a multiparental population of octoploid strawberry reveals QTL alleles conferring resistance to *Phytophthora cactorum*. G3 Genes, Genomes, Genet. 2017;7(6):1707-1719

[75] Anciro A, Mangandi J, Verma S, Peres N, Whitaker VM, Lee S. FaRCg1: A quantitative trait locus conferring resistance to Colletotrichum crown rot caused by *Colletotrichum gloeosporioides* in octoploid strawberry. Theoretical and Applied Genetics. 2018;**131**(10):2167-2177

[76] Salinas N, Verma S, Peres N, Whitaker VM. FaRCa1: A major subgenome-specific locus conferring resistance to *Colletotrichum acutatum* in strawberry. Theoretical and Applied Genetics. 2019;**132**(4):1109-1120

[77] Nellist CF, Vickerstaff RJ, SobczykMK,Marina-MontesC,WilsonFM, Simpson DW, et al. Quantitative trait loci controlling *Phytophthora cactorum* resistance in the cultivated octoploid strawberry (*Fragaria* × *ananassa*). Horticulture Research. 2019;**6**:60

[78] Siddique MI, Han K, Lee J, Lee ES, Lee YR, Lee HE, et al. QTL analysis for chlorophyll content in strawberry (*Fargaria* × *ananassa* Duch.) leaves. Agriculture. 2021;**11**(11):1163

[79] Verma S, Zurn JD, Salinas N, Mathey MM, Denoyes B, Hancock JF, et al. Clarifying sub-genomic positions of QTLs for flowering habit and fruit quality in U.S. strawberry (*Fragaria ananassa*) breeding populations using pedigree-based QTL analysis. Horticulture Research. 2017;**4**:17062

[80] Douillard C, Guichard E. Comparison by multidimensional analysis of concentrations of volatile compounds in fourteen frozen strawberry varieties [aroma, furaneol, mesifurane]. Sciences des Aliments. 1989;**9**(1):53-73

[81] Ménager I, Jost M, Aubert C. Changes in physicochemical characteristics and volatile constituents of strawberry (cv. Cigaline) during maturation. Journal of Agricultural and Food Chemistry. 2004;**52**(5):1248-1254

[82] Chambers AH, Pillet J, Plotto A, Bai J, Whitaker VM, Folta KM. Identification of a strawberry flavor gene candidate using an integrated genetic-genomic-analytical chemistry approach. BMC Genomics. 2014;**15**(1):1-15

[83] Cruz-Rus E, Sesmero R, Ángel-Pérez JA, Sánchez-Sevilla JF, Ulrich D, Amaya I. Validation of a PCR test to predict the presence of flavor volatiles mesifurane and γ-decalactone in fruits of cultivated strawberry (*Fragaria* × *ananassa*). Molecular Breeding. 2017;**37**(10):1-15

[84] Fait A, Hanhineva K, Beleggia R, Dai N, Rogachev I, Nikiforova VJ, et al. Reconfiguration of the achene and receptacle metabolic networks during strawberry fruit development. Plant Physiology. 2008;**148**(2):730-750

[85] Tulipani S, Mezzetti B, Capocasa F, Bompadre S, Beekwilder J, De Vos CHR, et al. Antioxidants, phenolic compounds, and nutritional quality of different strawberry genotypes. Journal of Agricultural and Food Chemistry. 2008;**56**(3):696-704

[86] Faedi W, Baruzzi G, Lucchi P. Outstanding strawberry selections from

italian breeding activity. Acta Horticulturae. 2003;**626**:125-132

[87] Haymes KM, Henken B, Davis TM, Van De Weg WE. Identification of RAPD markers linked to a *Phytophthora fragariae* resistance gene (Rpf1) in the cultivated strawberry. Theoretical and Applied Genetics. 1997;**94**(8):1097-1101

[88] Haymes KM, Van De Weg WE, Arens P, Maas JL, Vosman B, Den Nijs APM. Development of SCAR markers linked to a *Phytophthora fragariae* resistance gene and their assessment in European and North American strawberry genotypes. Journal of the American Society for Horticultural Science. 2000;**125**(3):330-339

[89] Lerceteau-Köhler E, Moing A, Guérin G, Renaud C, Courlit S, Camy D, et al. QTL analysis for fruit quality traits in octoploid strawberry (*Fragaria* x *ananassa*). In: Acta Horticulturae, Eucarpia Symposium on Fruit Breeding and Genetics (No. 663). Vol. 11. ISHS in Belgium; 2004

[90] Petrasch S, Mesquida-Pesci SD, Pincot DDA, Feldmann MJ, López CM, Famula R, et al. Genomic prediction of strawberry resistance to postharvest fruit decay caused by the fungal pathogen Botrytis cinerea. G3 Genes|Genomes|Genetics. 2022;**12**(1):1-14

[91] Hardigan MA, Feldmann MJ, Lorant A, Bird KA, Famula R, Acharya C, et al. Genome synteny has been conserved among the octoploid progenitors of cultivated strawberry over millions of years of evolution. Frontiers in Plant Science. 2020;**10**:1789

[92] Darwish O, Shahan R, Liu Z, Slovin JP, Alkharouf NW. Re-annotation of the woodland strawberry (*Fragaria vesca*) genome. BMC Genomics. 2015;**16**(1):1-9

[93] Lee S, Noh YH, Roach JA, Mangandi J, Verma S, Whitaker VM, et al. A high-throughput genotyping system combining rapid DNA extraction and high-resolution melting analysis in allo-octoploid strawberry. In: Acta Horticulturae, VIII International Strawberry Symposium. Vol. 1156. ISHS, Belgium; 2017. pp. 89-94

[94] Perrotte J, Gaston A, Potier A, Petit A, Rothan C, Denoyes B. Narrowing down the single homoeologous FaPFRU locus controlling flowering in cultivated octoploid strawberry using a selective mapping strategy. Plant Biotechnology Journal. 2016;**14**(11):2176-2189

[95] Van De Weg WE. Resistance to *Phytophthora fragariae* var. *Fragariae* in strawberry: The Rpf2 gene. Theoretical and Applied Genetics. 1997;**94**(8):1092-1096

[96] Mathey M. Phenotyping diverse strawberry (*Fragaria* spp.) germplasm for aid in marker-assisted breeding, and marker-trait association for red stele (*Phytophthora fragariae*) resistance marker Rpf1. [Masters thesis] 2013.

[97] Schwieterman ML, Colquhoun TA, Jaworski EA, Bartoshuk LM, Gilbert JL, Tieman DM, et al. Strawberry flavor: Diverse chemical compositions, a seasonal influence, and effects on sensory perception. PLoS One. 2014;**9**(2):e88446

[98] Perkins-Veazie P. Growth and ripening of strawberry fruit. In: Horticultural Reviews. Oxford, UK: John Wiley & Sons, Inc.; 1995. pp. 267-297

[99] Salinas N. Validation of molecular markers associated with perpetual flowering (PF) and soluble solids content (SSC) in strawberry (*Fragaria x ananassa* Duch. ex Rozier). [Masters thesis]. 2015.

Section 4

# Biochemical Content of Strawberries

Chapter 9

# Volatile Components of Strawberries

*Iryna Zamorska*

## Abstract

Strawberries of such cultivars as "Ducat", "Honey" and "Polka" were studied to define the content of aromatic volatiles using the methods of highly efficient liquid chromatography. About 49 components were identified in the composition of volatiles of ripe strawberries, namely esters, aldehydes, ketones, furanone, organic acids, aroma compounds, lactones, terpenic compounds, and alkanes. Their shares are the following: esters—12.8–41.8%, aldehydes—5.9–15.9%, ketones—8.7–35.6%, furanone—22.7–24.4%, and organic acids—2.47–21.85%. Depending on a cultivar, typical volatile components of strawberries are ethyl butanoate (10.1–30.65%), trans-2-hexenal (5.31–15.55%), acetoin (8.20–35.67), 2.5–dimethyl-4-methoxy-3(2H)-furanone (mesifurane) (19.08–19.92%) and 2.5-dimethyl-4-hydroxy-3(2H)-furanone (3.43–4.40%). A peculiar feature of volatile compounds of Polka strawberries is the highest ester share—41.77% of total content of volatiles, for Ducat strawberries it would be the share of ketones (35.88%), and for Honey strawberries—γ-decalaktone (12.41%). A high aroma activity of ripe strawberries is recorded on 2.5-dimethyl-4-methoxy-3(2H)-furanone (mesifurane) and 2.5-dimethyl-4-hydroxy-3(2H)-furanone (furaneol). Sweet, caramel flavors are typical for strawberries of the studied cultivars. Strawberries of Polka cultivar have a pronounced aroma.

**Keywords:** strawberries, volatiles, compounds, activity of volatile components

## 1. Introduction

Strawberries are valued for their harmonious taste and attractive, pronounced aroma, which is formed under the effect of volatiles. Strawberry aroma is known to be very specific and more than 360 compounds constitute its components, such as esters, aldehydes, ketones, alcohols, lactones, terpenic compounds, furanone [1–4]. Esters are considered to be the major components (over 130 different esters were identified); according to various data, they constitute 25–90% of the total amount of volatiles of ripe strawberries and add some flower or fruit flavors to strawberries [1, 2]. About 49 components were identified in volatile compound composition in strawberry varieties "Polka", "Ducat" and "Honey": esters, aldehydes, ketones, furanone, organic acids, aroma compounds, lactones, terpenic compounds, and alkanes. The most meaningful ones are: esters—12.8–41.8%, aldehydes—5.9–15.9%, ketones—8.7–35.6%, furanone—22.7–24.4% and organic acids—2.47–21.85% [5].

 IntechOpen

Aldehydes and furanone represent a large amount of volatile compounds of strawberries, their share being 50%. The latter adds a grass, sweet, or caramel flavor to the strawberry aroma [6–8]. Instead, according to other data, furanone is the main source of the strawberry aroma. Their level in the most aromatic cultivars exceeds a corresponding indicator of other cultivars by 20 times [4]. A small amount of aroma volatiles can be referred to as terpenic and sulfur compounds which can also have a considerable impact on strawberry aroma [9]. The most important chemical compounds which form strawberry aroma methyl butanoat, ethylbutanoate, 2-methylbutanoat, ethyl hexanoat, methylhexanoat, methyl 2-methylpropanoat [10], 4-hydroxy-2,5-dimethyl-3(2H)-furanone and 4-methoxy-2,5-dimethyl-3(2H)-furanone [6, 11, 12], (Z)-3-hexenal; butan-2,3-dion and linalool [13].

The most common ethers are: ethyl butanoate, butylacetate, ethylcrotonate, ethylcapronate, ethyl 3-hydroxycapronate, which give them fresh grass tone. Among aldehydes, those will be trans-2-hexenal and pentanal. Palmitic acid, carvone, and acetoin were found in all studied strawberry cultivars. Terpenic compounds are represented by nerolidol and linalool which give spicy notes to a strawberry flavor [14, 15]. Tetradecane was defined in the alkane class. Furanone were represented by 2,4-dioxy-2,5-dimethyl-3(2H)-furan-3-one, 2,5-dimethyl-4-methoxy-3(2H)-furanone (mesifurane) and 2,5-dimethyl-4-hydroxy-3(2H)-furanone (furaneol) [5].

Genetic peculiar features of a cultivar, a maturity degree, and storage conditions have an effect on the composition and concentration of volatiles [9]. The aroma of strawberries can change during storage [3, 16] and processing them into canned products [17]. For instance, the availability of methyl ester is a characteristic feature for "Hokowase", "Kent", "SengaGigana" and "Annapolis" strawberries, the share of which is equal to 70% of the total content of volatiles, whereas the availability of ethyl3-methylbutonoat and 3-methylacetate is more typical for cultivars "Kent" and "Micmac", and that of hexylacetate—for "Honeoye". Linalool was found in "SengaSengana" and "Annelie" strawberries [1].

Ethylbutanoate (10.1–30.65%), trans-2-hexenal (5.31–15.55%), acetoin (8.20–35.67), 2,5-dimethyl-4-methoxy-3(2H)-furanone (mesifurane) (19.08–19.92%) and 2,5-dimethyl-4-hydroxy-3(2H)-furanone (3.43–4.40%) were typical compounds for the flavor of the strawberry varieties studied. In addition, strawberries of Honey variety showed high content of Hexanoic (caproic) acid (9.54%) and Hexadecanoic acid (5.08%) [5].

The aroma of strawberries is formed in the process of ripening under the effect of enzymes. It is a known fact that when the strawberry color changes from white to a total red one the content of volatiles in them increases by 14 times. In green, unripe strawberries aldehydes and alcohols predominate and they add grass, green aroma, and in ripe strawberries—esters and furanone [10]. According to [18], EA and aldehydes predominate in green strawberries, in white berries—ketones and alkanes, and in red strawberries—esters, acids, furanes, and alcohols [18]. I has been experimentally proved that in the process of ripening the concentration of grass aroma components, such as hexanal, trans-2-hexanol and cys-3-hexenylacetat, decreases gradually [19]. Instead, the total ether content, including esters, increases, the content of furanone and lactones grows rapidly, the former were not found in green strawberries. Besides, during ripening the concentration of general aldehydes, aromatic compounds and alcohols undergo change [4].

Aldehydes and esters result from the enzymatic oxidation of lipids and a fermentative bio-synthesis of alcohols and acids, the latter occurs with the participation of alcohol-dehydrahenaza [3]. Alcoholacetyltransferaza plays a decisive role in a taste bio-synthesis during strawberry ripening through the formation of esters which are the main components of strawberry volatiles [20], this takes place together with volatile free fatty acids in mesocarp cells and strawberry protoplasts [21]. The role of Fragaria 3 ananassaxinon-oxydoreductasa in bio-synthesis of 2,5-dimethyl-4-hydroxy-3(2H)-furanone has been proved [22], and the formation of 2,5-dimethyl-4-hydroxy-3 (2H)-furan-3-one catalyzes β-glucozidaza [23, 24].

The environmental conditions, in particular lighting, are the major factor in the formation of a strawberry aroma. It has been proved that shading by 47% leads to a considerable decrease in the concentration of hexanal, hexanal, ethyl-methyl butyrate, and methylbutyrat in strawberries, as compared with well-lighted plants [21], which is due to the deceleration of a photosynthesis process in the plants, and, as a result, the decrease of the number of primary metabolism products; the latter are the raw materials for the synthesis of volatiles [25, 26].

## 2. The change of strawberry volatiles during the storage in a refrigerated state

When strawberries are harvested, their aroma can enhance during the first days of the storage which results from the fruit aging and the increased synthesis as well as the accumulation of volatile esters in fruit tissues and/or a resistance decrease to the diffusion of these compounds from the fruits because of the tissue aging [23]. It has been proved that during four days of the storage at a temperature of 15°C the concentration of volatiles in the strawberries increases by 7 times; besides, according to the data of some authors the content of esters and furanone increases [9].

The refrigerator storage of strawberries can cause changes in their aroma: its losses or gains of some outside, this decreases quality considerably [3]. Any storage method for strawberries facilitates the losses of aromatic compounds [25]. Both temperature and storage duration have a serious effect on fruit aroma [27].

According to the data of Pelayo-Zaldívar et al. [28], the loss of aroma of strawberries during the storage can be due to the accumulation of acetaldehyde and ethanol in them, the additional synthesis of methyl ether, and the formation of a new profile of aromatic compounds. Besides, the losses can be caused by the changes of non-volatile components of aroma such as sugars and citric acid, the changes of phenol compounds, and the increase of their resistance under the effect of low temperatures. El Hadi et al. [27] consider the temperature to be the main factor that influences fruit taste and aroma.

Pérez with the co-authors [29] states that during the storage strawberry aroma changes and it occurs due to the slowdown of the formation of some volatile esters and the sharp reduction of furaneol amount. Ozcan and Barringer [30] present the data about a low activity of the enzyme of spirit, acetyltransferase in strawberries during the storage at 1°C.

According to Larsen and Watkins [31], the storage of strawberries at high content of carbon dioxide in the atmosphere enhances the resistance to physiological decay, but it slows down the development of aroma.

It was proved that as a result of the refrigerator storage of strawberries, cultivar Polka, at temperature 0 ± 1°C and air relative humidity 90–95% during 9 days the content of volatile compounds decreased by 36.3%, that of esters—by 66.0%, which was caused by lowering the temperature and weakening the activity of the enzyme alcoholdehydrohenasa in the conditions of the lowered temperature [32].

During the storage, the content of furanes also decreased—by 45.3%. Similar data was received when the sum of acids, ketones, and terpenes was determined, their share decreased by 9.7, 32.4, and 64.9%, respectively, as compared with the initial content [5]. A serious decrease was typical for some compounds, namely, ethyl butanoate—by 6.8 times, ethyl crotonate—by 11.3 times, ethyl caproate—by 6.1 times. The content of furanes decreased by 1.6 (mesifurane) and by 3 times (furaneol).

Along with a number of biochemical transformations in strawberries after the storage, their organoleptic properties also change. For instance, when strawberries are stored in a modified atmosphere not only the change in aroma, consistency, and a degree of sweetness are recorded; some undesirable features appear such as rancidity, fermented smell, etc. Also, strawberries that were stored in the atmosphere with a high content of carbon dioxide showed undesirable features more often [33].

## 2.1 Change of volatile compounds of frozen strawberries

The freezing process causes changes in the aromatic profile of strawberries. For instance, the storage of strawberries within one week with a follow-up defrosting at the ambient temperature facilitates the enhanced level of acetaldehyde, cis-3-hexenal, hexenal, ethylacetate, and methyl acetate [34]. The scientists [35] advise about the worsening of strawberry aroma during the freezing and continuous storage. It has been proved that ethers, which are characteristic components of fresh strawberry aroma, in fact, are not recorded in frozen berries, whereas the content of carbon compounds remains at the same level.

The appearance of an unpleasant aroma in frozen berries is associated with the formation of $H_2S$ which releases due to pH decrease of cell sap resulted from the cell damage during the freezing process [36]. According to the data presented by Schreier [37], the freezing process leads to the reduction of the concentration of the majority of aromatic compounds, but the content of 2,5-dimethyl-4-methoxy-3(2H)-furanone in them increases. Douillard and Guichard [38] point out the increase of the concentration of nerolidol and the decrease of the ether share in frozen berries.

## 2.2 Aroma of canned strawberries

The processing of fruits and berries has a serious impact on the taste and aroma of the finished output, the reduction of its effect favors quality preservation. The aroma of the processed strawberry output depends, to a great extent, on fresh berries.

The aroma of strawberry compotes (stewed fruit) is formed under the effect of furanone (15.5–23.5%) and aromatic acids (48.4–76.1%) which add sweet caramel and sour-sweet scents [5].

The aroma of strawberry compotes (stewed fruit) consists of a mixture of compounds: esters, aldehydes, aromatic spirits, aromatic acids, lactones, ketones, furanone, and terpenes. Among them, a large share belongs to acids—48.4–76.1 from the total content of volatiles and to furanes—15.5–23.5%. A considerable amount of aromatic spirits—21.1% and esters—4.2% were recorded in compotes made of Polka strawberries.

Typical volatile compounds of the compotes made of strawberry cultivars Polka, Ducat, and Honey are hexanoic (caproic) acid (7.6–23.3% from the total content of volatiles), 2-ethylhexanoic (capronic) acid (6.9–8.6%), trans-cinnamic acid (22.5–30.2%) which add sour-sweet aroma to them.

Large amounts of 2,4-dioxy-2,5-dimethyl-3(2H)-furan-3-one (1.1–2.3%), 2,5-dimethyl-4-methoxy-3(2H)-furanone (mesifurane) (7.6–16.0%) and 2.5-dimethyl-4-oxy-3(2H)-furanone (6.4–9.8%) are recorded in compotes (stewed fruit); they all add sweet caramel scents [5].

The presence of furfural (0.3%) and 5-hydroxymethylfurfural (0.2%) in compotes confirms a non-fermentative darkening during heat treatment [39, 40].

In addition to the above-mentioned compounds, ethyl butanoate (3.2%), which adds fresh grassy scents, and hydrocinnamicalcohol (15.9%) make their serious contribution to the aroma of the compotes made of cultivar Polka. Vanillin was recorded in the compotes made of cultivars Ducat and Honey, adding typical vanilla scents; its concentrations were 1.4 and 0.8%, respectively. There was γ-decalactone (3.2%) in the composition of volatile compounds of the compotes made of cultivar Honey which added fruit, a sweet scent to aroma [15].

The compotes had 2H-pyran-2,6(3H)-dion (0.7–2.0%), 3,5-hydroxy-2-dimethyl-4H-pyran-4-on (0.2%) and 3.4-dihydropyran (0.2–0.4%), which were the products of Mayar's reaction, resulted from the reaction of glucose with glutamine acid, glycine, butalamine, lisyn,hydroxyproline and/or fenilalanine [39].

In strawberry compotes terpene compounds include linalool (0.5–0.6% depending on a cultivar), α-Terpineol (0.2–0.7%), which were present in fresh strawberries [16], they add spicy scents to berry aroma, and also oxydebisabolol A (0.1–0.8%), trans-linalool oxide (0.3–0.5%), which were not found in fresh strawberries.

There were ethers, aldehydes, acids, lactones, furane derivatives, and terpenes in the composition of volatile compounds in strawberry juices. Acids—60.1% and furanone—28.9% from the total amount of juice volatiles had a large share among aromatic juice compounds. Among the total amount of volatile compounds, a large share belonged to 2,5-dimethyl-4-methoxy-3(2H)-furanone (mesifurane) (21% of the total sum of juice volatiles), linoleic acid (16.6%), trans-cinnamic acid (16.0%), palmitoleic acid (9.5%), 2.5-dimethyl-4-oxy-3(2H)-furanone (6.7%), hexanoic (caproic) acid (6.2%), 2-ethyl hexanoic (capronic) acid (4.8%). Furfural was found among volatile compounds of natural unclarified strawberry juice; it was identified earlier in other processed strawberry products and this confirms the features of non-fermentative darkening of the product.

Ethyl butanoate in the amount of 1.2% from the total content of volatile compounds and isoamyl butanoate—1.4% were identified among ethers in a strawberry juice. The share of each other ether found in a juice is at the level of 0.1–0.6%. The aldehyde content does not exceed 0.5%. Also, 2H-pyran-2,6(3H)-dion, the share of which was 1.7% of the total sum of volatile compounds, was identified, which confirms Mayar's reaction.

Terpene compounds of a natural strawberry unclarified juice include linalool (0.4% from the total sum of volatiles), α-terpineol (0.3%) that add spicy scents to juice aroma, and oxydebisabolol A (0.9%), oxydebisabolol B (0.5%), nerolidol (0.8%) which add sweet flower scents [5].

The aroma of strawberry jams is formed under the effect of acids, alcohols, and ethers [39, 40], which can be both of natural origin and the result of heat treatment. Acids add a lot to the aroma of jams: 2-methylbutyric acid, hexanoic (caproic) acid, octadecanoic acid, dodecanoic acid, tetradecanoic acid, palmitoleic acid, trans-cinnamic acid; alcohols: 1-hexanol, 3-methyl-3-buten-2-ol, linalool, linalool oxide, α-terpineol, trans-nerolidol, benzyl alcohol [39].

As a result of high-temperature treatment, sugar caramelization, and Mayar's reaction, strawberry products obtain boiled, burnt, and caramel tastes [29, 41–43].

Contrary to this, green fruit scents which are typical for fresh berries are less pronounced [2, 30]. High concentrations of furaneols add characteristic caramel and sweet scents to the aroma of strawberry jams [29, 43].

According to Lambert et al. [17], the sterilization of strawberry puree with juice at 120°C for 20 min causes serious losses of flower aromas along with the formation of geraniol and vanillin. In addition, the concentration of butylacetate increases by 1.7 times, 2-Hexen-1-al—by 3.2, butyric acid—by 2, 2-methylbutanoic acid—by 1.6, hexanoic acid—by 1.8, furaneol—by 3.0, nerolidol—1.7, octanoic acid—by 1.7 and γ-decalactone—by 1.5 times, as compared with fresh strawberries.

The main volatile compounds of strawberry jams belong to the classes of acids, alcohols, and ethers [39, 40] and they are both of natural origin and can be the result of heat treatment. The most common among acids are 2-Methylbutyric acid, hexanoic acid, octanoic acid, dodecanoic acid, tetradecanoic acid, palmitoleic acid, trans-cinnamic acid. The most common alcohols are 1-hexanol, 3-methyl-3-buten-2-ol, linalool, cis-epoxy-linalool, cis-Linalool oxide, α-Terpineol, trans-nerolidol, benzyl alcohol. Active compounds include ethyl butanoate, methyl and ethylcapronates, 3-Hydroxybutanoic acid methyl ester, and 3-Hydroxy hexanoic acid methylester [43].

Thirty-eight components, including esters, aldehydes, ketones, furanone, acids, aromatic compounds, lactones, and terpene compounds, were identified in the composition of volatiles of strawberry jams made of cultivars Polka, Ducat, and Honey. The most meaningful shares are: acids—65.6–76.8%, furanes—8.3–14.6% and aldehydes—3.4–10.8%. The share of esters in jams exceeds 0.7–3.1% of the total volatile content. It is important to mention that in strawberry jams made of Polka cultivar the share of esters and aldehydes is much higher: 3.1 and 10.8%, that of furanes and acids, on the contrary, is the lowest—8.3 and 65.6% which proves strong expression of scents typical for fresh strawberries.

Characteristic compounds for strawberry jam flavor made of the studied cultivars are hexanoic (caproic) acid (0.84–6.89 mg/kg), which is 6.9–22.9% of the total volatile amount depending on their quantity for each cultivar, hexadecanoic acid (2.5–12.4%), 2-ethyl hexanoic (capronic) acid (3.1–10.7%), trans-cinnamic acid (17.5–25.3%), linoleic acid (0.3–7.2%), furil hydroxy methylketone (3.1–6.0%), 2,5-dimethyl-4-methoxy-3(2H)-furanone (mesifurane) (7.4–13.7%), furfural (0.8–3.1%), 5-hydroxymethylfurfural (0.8–5.2%), vanillin (0.2–0.8%) [5].

The availability of furfural (0.8–3.1%), 5-hydroxymethylfurfural (0.8–5.2%) 5-methylfurfural (0.7%) in strawberry jams indicates non-fermentative darkening during thermal treatment [39, 40].

Small amounts of 2H-pyran-2,6(3H)-dion and 3,5-hydroxy-2-dimethyl-4H-pyran-4—(0.09–0.66 mg/kg) which, depending on the cultivar, is 0.5–0.9% of the total volatile content in jams, were found; and according to [39] they are the products of Maiyar reaction resulted from the reaction of glucose with glutamic acid, glycine, butylamine, lysine, hydroxyproline and/or phenylalanine (amino acids) [5].

Terpenic compounds of strawberry jams are presented by small amounts of limonene (0.1 mg/kg, which is 0.8%) and α-terpineol (0.2–1.3%), they were found in fresh berries [16] these compounds add aromatic scent to fresh berries [15, 16]; oxyde bisabolol A (0.1–0.3%), trans-linalool oxide (0.1–0.3%), cis-linalool oxide (0.3%), however no data concerning their presence in fresh strawberries is available.

## 3. Conclusions

Thus, the aroma of fresh strawberries is developed during ripening and it depends on the environmental conditions, a strawberry cultivar, a degree of maturity, and post-harvest conditions. The aroma of strawberries consists of a complex mixture of compounds including ethers, aldehydes, alcohols, ketones, lactones, furanone, and terpene compounds. Refrigerating, freezing, and heat treatment of berries have an impact on the change of aroma. The understanding of the nature of these changes will make it possible to predict the quality of the refrigerated berries and the canned output.

## Author details

Iryna Zamorska
Uman National University of Horticulture, Uman, Ukraine

*Address all correspondence to: zil197608@gmail.com

## References


[1] Larsen M, Poll L. Odour thresholds of some important aroma compounds in strawberries. Zeitschrift für Lebensmittel-Untersuchung und Forschung. 1992;**195**(2):120-123. DOI: 10.1007/BF01201770

[2] Larsen M, Poll L, Olsen CE. Evaluation of the aroma composition of some strawberry (Fragaria ananassa Duch) cultivars by use of odour threshold values. Zeitschrift für Lebensmittel-Untersuchung und Forschung. 1992;**195**(6):536-539. DOI: 10.1007/BF01204558

[3] Forney CF, Kalt W, Jordan MA. The composition of strawberry aroma is influenced by cultivar, maturity and storage. HortScience. 2000;**35**(6):1022-1026. DOI: 10.21273/HORTSCI.35.6.1022

[4] Kafkas E, Kafkas S, Koch-Dean M, et al. Comparison of methodologies for the identification of aroma compounds in strawberry. Turkish Journal of Agriculture and Forestry. 2005;**29**(5):383

[5] Zamorska IL. Theoretical explanation and development of the storage and canning technologies of strawberries. Qualifying scientific work as manuscript.

[6] Perez AG, Sanz C, Olias R, et al. Aroma quality evaluation of strawberry cultivars in southern Spain. In: Proceedings of the III International Strawberry Symposium Acta Hort. 439. 1996. pp. 337-340. DOI: 10.17660/ActaHortic.1997.439.53

[7] Jetti RR. Fruit Quality Evaluation of Strawberries (Fragaria ananassa). In: Grown in California and Oregon. Oregon State University; 2006

[8] Kim YH, Kim KH, Szulejko JE, et al. Quantitative analysis of fragrance and odorants released from fresh and decaying strawberries. Sensors. 2013;**13**(6):7939-7978. DOI: 10.3390/s130607939

[9] Forney CF. Horticultural and other factors affecting aroma volatile composition of small fruit. HortTechnology. 2001;**11**(4):529-538. DOI: 10.21273/HORTTECH.11.4.529

[10] Vandendriessche T, Vermeir S, Mayayo MC, et al. High-throughput flavor evaluation of strawberry cultivars: Focus on aroma development during ripening. Acta Horticulturae. 2012;**945**:227-232. DOI: 10.17660/ActaHortic.2012.945.30

[11] Orruño E, Apenten RO, Zabetakis I. The role of β-glucosidase in the biosynthesis of 2, 5-dimethyl-4-hydroxy-3 (2H)-furanone in strawberry (Fragaria× ananassa cv. Elsanta). Flavour and Fragrance Journal. 2001;**16**(2):81-84. DOI: 10.1002/1099-1026(200103/04)16:23.0.CO;2-H

[12] Siegmund B, Bagdonaite K, Leitner E. Furaneol and mesifuran in strawberries—an analytical challenge. In: Expression of Multidisciplinary Flavour Science: Proceedings of the 12th Weurman Symposium. Zürich; 2010:537-540

[13] Hui YH, Pérez AG, Sanz C. Strawberry Flavor. In: Hui YH, Chen F, Nollet LML, editors. Handbook of Fruit and Vegetable Flavors. Hoboken, New Jersey: John Wiley and Sons, Inc.; 2010. pp. 431-449

[14] Zabetakis I, Holden MA. Strawberry flavour: Analysis and biosynthesis. Journal of the Science of Food and Agriculture. 1997;**74**(4):421-434. DOI: 10.1002/(SICI)1097-0010(199708)74:4<421:AID-JSFA817>3.0.CO;2-6

[15] Ulrich D, Komes D, Olbricht K, Hoberg E. Diversity of aroma patterns in wild and cultivated Fragaria accessions. Genetic Resources and Crop Evolution. 2007;**54**(6):1185-1196. DOI: 10.1007/s10722-006-9009-4

[16] Bianchi G, Lovazzano A, Lanubile A, Marocco A. Aroma quality of fruits of wild and cultivated strawberry (Fragaria spp.) in relation to the flavour-related gene expression. Journal of Horticultural Research. 2014;**22**(1):77-84. DOI: 10.2478/johr-2014-0009

[17] Lambert Y, Demazeau G, Largeteau A, Bouvier JM. Changes in aromatic volatile composition of strawberry after high pressure treatment. Food Chemistry. 1999;**67**(1):7-16. DOI: 10.1016/S0308-8146(99)00084-9

[18] Zhao J, Liu J, Wang F, Wang S, Feng H, Xie X, et al. Volatile constituents and ellagic acid formation in strawberry fruits of selected cultivars. Food Research International. 2020;**138**:109767. DOI: 10.1016/j.foodres.2020.109767

[19] Azodanlou R, Darbellay C, Luisier JL, Villettaz JC, Amadò R. Changes in flavour and texture during the ripening of strawberries. European Food Research and Technology. 2004;**218**(2):167-172. DOI: 10.1007/s00217-003-0822-0

[20] Aharoni A, Keizer LC, Bouwmeester HJ, et al. Identification of the SAAT gene involved in strawberry flavor biogenesis by use of DNA microarrays. The Plant Cell. 2000;**12**(5):647-661. DOI: 10.1105/tpc.12.5.647

[21] Yamashita I, Iino K, Nemoto Y, et al. Studies on flavor development in strawberries. 4. Biosynthesis of volatile alcohol and esters from aldehyde during ripening. Journal of Agricultural and Food Chemistry. 1977;**25**(5):1165-1168. DOI: 10.1021/jf60213a027

[22] Raab T, López-Ráez JA, Klein D, et al. FaQR, required for the biosynthesis of the strawberry flavor compound 4-hydroxy-2, 5-dimethyl-3 (2H)-furanone, encodes an enone oxidoreductase. The Plant Cell. 2006;**18**(4):1023-1037. DOI: 10.1105/tpc.105.039784

[23] Zabetakis I, Gramshaw JW, RobinsonDS.2,5-Dimethyl-4-hydroxy-2H-furan-3-one and its derivatives:analysis, synthesis and biosynthesis—A review. Food Chemistry. 1999;**65**(2):139-151

[24] Miszczak A, Forney CF, Prange RK. Development of aroma volatiles and color during postharvest ripening ofKent'strawberries. Journal of the American Society for Horticultural Science. 1995;**120**(4):650-655

[25] Watson R, Wright CJ, McBurney T, Taylor AJ, Linforth RST. Influence of harvest date and light integral on the development of strawberry flavour compounds. Journal of Experimental Botany. 2002;**53**(377):2121-2129. DOI: 10.1093/jxb/erf088

[26] Sanz CC. Furaneol flavor compounds in ripening strawberry (Fragaria x ananassa Duch.) [doctoral dissertation]. 1993

[27] El Hadi MAM, Zhang FJ, Wu FF, et al. Advances in fruit aroma volatile research. Molecules. 2013;**18**(7):8200-8229. DOI: 10.3390/molecules18078200

[28] Pelayo-Zaldívar C, Abda JB, Ebeler SE, et al. Quality and chemical changes associated with flavor of 'Camarosa'strawberries in response to a $CO_2$-enriched atmosphere. HortScience. 2007;**42**(2):299-303

[29] Pérez AG, Olías R, Sanz C, et al. Furanones in strawberries: Evolution during ripening and postharvest shelf life. Journal of Agricultural and Food

Chemistry. 1996;**44**(11):3620-3624. DOI: 10.1021/jf960099m

[30] Ozcan G, Barringer S. Effect of enzymes on strawberry volatiles during storage, at different ripeness level, in different cultivars, and during eating. Journal of Food Science. 2011;**76**(2). DOI: 10.1111/j.1750-3841.2010.01999.x

[31] Larsen M, Watkins CB. Firmness and aroma composition of strawberries following short-term high carbon dioxide treatments. HortScience. 1995;**30**(2):303-305. DOI: 10.21273/HORTSCI.30.2.303

[32] de León-Sánchez FD, Pelayo-Zaldívar C, Rivera-Cabrera F, et al. Effect of refrigerated storage on aroma and alcohol dehydrogenase activity in tomato fruit. Postharvest Biology and Technology. 2009;**54**(2):93-100. DOI: 10.1016/j.postharvbio.2009.07.003

[33] Shamaila M, Powrie WD, Skura BJ. Sensory evaluation of strawberry fruit stored under modified atmosphere packaging (MAP) by quantitative descriptive analysis. Journal of Food Science. 1992;**57**(5):1168-1184. DOI: 10.1111/j.1365-2621.1992.tb11290.x

[34] Modise DM. Does freezing and thawing affect the volatile profile of strawberry fruit (Fragaria×ananassa Duch.)? Postharvest Biology and Technology. 2008;**50**:25-30. DOI: 10.1016/j.postharvbio.2008.03.009

[35] Ueda Y, Iwata T. Off-odor of strawberry by freezing. Journal of the Japanese Society for Horticultural Science. 1982;**51**(2):219-223. DOI: 10.2503/jjshs.51.219

[36] Hong D, Ueda Y, Chachin K, et al. Off-flavor production in frozen strawberries. Postharvest Biology and Technology. 1996;**1**(9):31-39

[37] Schreier P. Quantitative composition of volatile constituents in cultivated strawberries, Fragaria ananassa cv. Senga Sengana, Senga Litessa and Senga Gourmella. Journal of the Science of Food and Agriculture. 1980;**31**(5):487-494

[38] Douillard C, Guichard E. The aroma of strawberry (Fragaria ananassa): Characterisation of some cultivars and influence of freezing. Journal of the Science of Food and Agriculture. 1990;**50**(4):517-531. DOI: 10.1002/jsfa.2740500410

[39] Barren D, Etiévant PX. The volatile constituents of strawberry jam. Zeitschrift für Lebensmittel-Untersuchung und Forschung. 1990;**191**(4-5):279-285. DOI: 10.1007/BF01202426

[40] Kimura K, Ida M, Yosida Y, et al. Comparison of keeping quality between pressure-processed jam and heat-processed jam: Changes in flavor components, hue, and nutrients during storage. Bioscience, biotechnology, and biochemistry. 1994;**58**(8):1386-1391. DOI: 10.1271/bbb.58.1386

[41] Sloan JL, Bills DD, Libbey LM. Heat-induced compounds in strawberries. Journal of Agricultural and Food Chemistry. 1969;**17**(6):1370-1372. DOI: 10.1021/jf60166a020

[42] Avasoo M, Johansson L. Evaluation of thermal processing technologies for strawberry jam. 2011. Available from: http://ukkpfhd.livsmedelsakademin.se/sites/default/files/media/evaluation_of_thermal_processing_technologies_for_strawberry_jam.final_.pdf.

[43] Lesschaeve I, Langlois D, Etiévant P. Volatile compounds in strawberry jam: Influence of cooking on volatiles. Journal of Food Science. 1991;**56**(5):1393-1398

Chapter 10

# Phytochemical Constituents and Pharmacological Activities of Strawberry

*Rushendran Rapuru, Sivakumar Bathula and Ilango Kaliappan*


## Abstract

Strawberry is a widely grown hybrid species of single ovary fruit that is indehiscent and the scientific name of strawberry is *Fragaria ananassa* belongs to the family Rosaceae which is native to America, Europe, Asia and cultivated worldwide for its fruits. The fruits are rich in vitamin C. The phytochemical constituents such as gallotannin, ellagitannin, ellagic acid anthocyanins, proanthocyanins, catechin, cyanidin, agrimonium, coumaroyl glycosides, pedunculagin, pelargonidin, flavonols, chlorogenic acid, salicylic acid, caffeic acid, and chlorogenic acid have various medicinal benefits like regulating heart function, antidiarrheal, antioxidant, diuretic, decline blood pressure, elevate good cholesterol, esophageal cancer, etc. This chapter deals with nutritional value, phytoconstituents, and its structure along with medicinal uses of Strawberry.


**Keywords:** Strawberry, Fragaria X ananassa, Phytochemical constituents, Medicinal properties, Nutritional facts

## 1. Introduction

The strawberry is the common familiar fruit that emanated 250 years ago [1] and is consumed worldwide either fresh or processed [2]. The scientific name of strawberry is *Fragaria* x *ananassa* and falls under the family Rosaceae which has great medicinal value. The letter **"x"** in between the botanical name indicates that strawberry is produced due to the hybridization of two species such as *Fragaria virginiana* and *Fragaria chiloensis* [3]. The native of *Fragaria virginiana* is North America and *Fragaria chiloensis* is Chile, especially due to the shape of a heart and red color it is used as a symbol of venus and Goddess of love [4]. In the 18th century, the fruit was originated in Europe later developed by most of the countries in the 19th century which is suitable in all aspects such as altitude, climate, day length, and type of production. It was produced commercially for immediate consumption as well as a processed form like canned, frozen, juice, and preserved berries [5]. The taxonomical classification of *Fragaria x ananassa* was listed in **Table 1**.

The taxonomical classification of *Fragaria x ananassa* is listed in **Table 1** [6]. The various *Fragaria ananassa* Duch species are available in the universe such as diploids

| | |
|---|---|
| Domain | Eukayota |
| Kingdom | Plantae |
| Phylum | Spermatophyta |
| Subphylum | Angiospermae |
| Class | Dicotyledone |
| Order | Rosales |
| Family | Rosaceae |
| Genus | *Fragaria* |
| Species | *ananassa* |

**Table 1.**
*Taxonomical classification of* Fragaria x ananassa.

like *Fragaria viridis*, *Fragaria daltoniana*, *Fragaria nubicola*, *Fragaria nipponica*, *Fragaria vesca, Fragaria nilgerrensis*, *Fragaria iinumae*; tetraploids like *Fragaria orientalis, Fragaria moupinensis;* hexaploids like *Fragaria moschata*; octaploids like *Fragaria* X *ananassa, Fragaria virginiana, Fragaria ovalis,* and their characteristics and with their location are enlisted in **Table 2** [7]. This chapter focuses on the phytochemical constituents and pharmacological activities of *Fragaraia* x *ananassa*. This chapter will give the information regarding all the phytochemical constituents present in the strawberry, chemical structures of phytoconstituents along with their medicinal uses which may assist researchers for the development of new drugs and also to the common people regarding the medicinal uses such as antioxidants, anticancer, anti-obesity, anti-diabetic, antihypertension, anti-inflammatory, antiosteoporotic, antimicrobial, anthelmintic, immunostimulatory, neuroprotective, anti-platelet, etc. so that they can habituate intake of strawberry in their routine life as a prophylaxis treatment naturally.

| **Name of the species** | **Native location** | **Characteristics of fruit** |
|---|---|---|
| **Diploids (2n = 14)** | | |
| *Fragaria viridis* Duch. | Europe to Central Asia | Pink-red in color, aromatic, small, firm, pit seeds. |
| *Fragaria daltoniana* J.Gay | Himalayas (10000–15000 ft) | Bright red in a color, tasteless, elongated oval in shape. |
| *Fragaria nubicola* Lindl ex. Lacaita | Himalayas (5000–13000 ft) | Bright red in a color, soft, aromatic, raised seeds, long oval in shape. |
| *Fragaria nipponica* Makino | Japan (Honshu island) | — |
| *Fragaria vesca L.* | Circumpolar | Aromatics, soft, red in color, long oval shape, raised seeds. |
| *Fragaria nilgerrensis* Schlect | Southeast Asia | Pink in color, round, small, unpleasant taste, sunken seeds. |
| *Fragaria iinumae* Makino | Mountains of Japan | Long oval in shape, sunken seeds, tasteless, bright red. |
| **Tetraploids (2n = 18)** | | |
| *Fragaria orientalis* Losink | Korea | Round or conical in shape, soft, sunken seeds, aromatic. |

| Name of the species | Native location | Characteristics of fruit |
|---|---|---|
| *Fragaria moupinensis* (French.) Card | China | Round, small, sunken seeds, unpleasant taste, pink in color. |
| **Hexaploids (2n = 42)** | | |
| *Fragaria moschata* Duch. | Europe to Korea | Red in color, soft, irregular to ovoid, aromatic, raised achenes, musky or vinous. |
| **Octaploids (2n = 56)** | | |
| *Fragaria* X *ananassa* Duch. | America, Asia, Europe, and cultivated worldwide. | Red in color, large and variable in all traits. |
| *Fragaria virginiana* Duch. | North America | Soft, deep red in color, aromatic, sunken seeds, ovoid shape. |
| *Fragaria ovalis* (Lehn.) | Rocky Mountains | Pink in color, round, flavorful, small. |
| *Fragaria chiloensis* (L.) Duch. | Argentina, Hawaii, Cile, North America | White flesh, round to oblate in shape, small, red-brown in color, firm. |
| *Fragaria inturupensis* Staudt | Japan | — |
| *Fragaria mandschurica* Staudt | Manchuria | — |

**Table 2.**
*Different* Fragaria ananassa *Duch species available worldwide.*

## 2. Nutritional facts of strawberry

Strawberry is familiarly utilized in the industry for making jellies, ice cream, marmalades, yogurts, jam, and other dessert products. It is also used in the making of deodorants and cosmetics for odor and as a flavoring agent. It is abundant in carbohydrates and contains fats (0.3 g) and proteins (0.67 g). It is having a magnificent source of vitamins and contains pantothenic acid, thiamine, riboflavin, vitamin B6, Niacin, vitamin C, Folate, vitamin K and vitamin E. It also contains elements such as potassium, sodium, manganese, iron, zinc, and calcium [8]. Strawberry is selected as the best healthy food choice based on the strawberry nutrient profile **Table 3**. The contents of dietary fibers and fructose may regulate blood sugar levels. The strawberry seed oil is abundant in unsaturated fatty acids. It contains high content of vitamin C when it aggregates with folate plays a critical role in accentuating the micronutrient content [9]. Folate or folic acid is a highly available supplement in strawberries that helps in the development of a baby during pregnancy [10]. It contains calcium and vitamin D are the supplements that mainly assist in the prevention and treatment of osteoporosis [11]. Iron can treat anemia, production of hemoglobin and prevent disproportionality affects in pregnant women and young children [12, 13]. Vitamin K plays a key role in most of the functions such as an anticancer, insulin-sensitizing molecule, bone-forming, and anti calcification, etc. [14–16]. Generally, vitamin C is involved in many biological roles such as antioxidant, antithrombotic, antiviral, sepsis treatment, and boost-up immune system. In the present pandemic scenario, especially this supplement is playing a crucial role in the treatment of COVID-19 [17, 18]. The elements like choline, riboflavin, and fat-soluble vitamins like vitamin A, E, and K are having antioxidant properties [19]. The other

| Proximates | Minerals | Vitamins |
|---|---|---|
| Water (90.95 g) | Calcium (16 mg) | Vitamin K<br>Phylloquinone (2.2 μg) |
| Energy (32 kcal) | Selenium (0.4 mg) | Thiamin (0.024 mg)<br>Vitamin A (0.024 mg, 1 μg) |
| Total lipid (0.30 g) | Manganese<br>(0.386 mg) | Vitamin C (58.8 mg)<br>Vitamin E (0.29 mg) |
| Ash (0.40 g) | Zinc (0.14 mg) | Niacin (0.386 mg)<br>α,β,γ,δ-tocopherol (0.29 mg, 0.01 mg, 0.08 mg, 0.01 mg) |
| Fructose (2.44 g) | Copper (0.048 mg) | Riboflavin (0.022 mg) |
| Glucose (1.99 g) | Potassium (153 mg) | Vitamin B6 (0.047 mg) |
| Sucrose (0.47 g) | Sodium (1 mg) | Pantothenic acid (0.125 mg) |
| Dietry fibers (2 g) | Iron (0.41 mg) | Choline (5.7 mg) |
| Carbohydrates (7.68 g) | Magnesium (13 mg) | Folate (24 μg) |
| Protein (0.67 g) | Phosphorus (24 mg) | Vitamin B12 (0 μg) |
| Sugar (4.89 g) | | Betaine (0.2 mg) |

**Table 3.**
*Nutritional composition of strawberry per 100 g.*

supplements like pantothenic acid, zinc, manganese, copper, selenium, sodium may assist in immunomodulatory effect and tocopherol having antioxidant activity [20, 21]. The nutrient niacin also plays an important role in controlling hypophosphatemia, cardioprotective and may improve penile erection by elevating blood flow [22–24].

## 3. Phytochemical constituents

Ni Shi et al. [25] have quantified different active biomolecules by utilizing HPLC under 260, 355, and 500 nm range. The molecules such as carotenoids [26, 27], vitamin C [28–30], ellagitannin, coumaroyl hexoside, procyanidin dimer, catechin. Phenolic compounds are characterized into three categories.

1. Anthocyanins like cyanidin glucoside, pelargonidin glucoside [31], pelargonidin rutinoside and pelargonidin malonyl glucoside [32, 33].

2. Ellagic acid derivatives such as ellagitannin, ellagic acid rhamnoside, agrimoniin and lambertianin.

3. Flavonols such as quercetin hexuronide, kaempferol glucoside, kaempferol hexuronide, kaempferol malonyl hexoside [34].

Anthocyanins are the most important polyphenolic compounds [35, 36] have been determined by more than different anthocyanin pigments from different varieties. Pelargonidin-3-glucoside is one of the main compounds of anthocyanin which is genetically independent [37–40]. Conjugation of arabinose, rutinose, and rhamnose are the most common substituting sugar in anthocyanins of strawberries [41, 42]. Ellagitannins are the combination of ellagic acid and hexahydroxydiphenic acid but the aggregation of gallotannins is known as hydrolyzable tannins. Phane Quideau., 2009 has been recognized specifically ellagitannins are the active principle in medicinal plants [43]. Koponen et al., screened and analyzed ellagitannins from food consumed by an individual in Finland. The specific compounds such as Ellagitannin, p-Coumaroyl hexose, HHDP-galloyl-glucose, Bis-HHDP-glucose, Galloyl-HHDP-glucose, Dimer of galloyl-bis-HHDP, Sanguiin H-6, Galloyl-bis-HHDP-glucose, Methyl-EA-pentose conjugates, Ellagic acid, Ellagic acid pentoside are detected.. It also contains other phenolic compounds like flavonols which are recognized as kaempferol and quercetin derivatives [44]. The total polyphenol molecules of *Fragaria* x *ananassa* are enlisted in **Table 4** [45–47].

| Class | Group | Compound | Ref. |
|---|---|---|---|
| Phenolic acids<br>Hydrolyzable tannins | Hydroxycinnamic acids<br>Ellagitannins | Ellagitannin, p-Coumaroyl hexose<br>HHDP-galloyl-glucose<br>Bis-HHDP-glucose<br>Galloyl-HHDP-glucose<br>Dimer of galloyl-bis-HHDP<br>Sanguiin H-6<br>Galloyl-bis-HHDP-glucose<br>Methyl-EA-pentose conjugates<br>Ellagic acid, Ellagic acid pentoside | [46, 47] |
| Flavonoids | Flavonols | Quercetin-rutinoside<br>Quercetin-3-malonyglucoside<br>Quercetin-3-glucuronide<br>Quercetin-glucuronide<br>Quercetin-glucoside<br>Kaempferol-glucunoride<br>Kaempferol-3-malonyglucoside<br>Kaempferol-3-glucoside<br>Kaempferol-coumaroyl-glucoside | |
| | Flavanols | Proanthocyanidin trimer (EC-4,8-EC-4,8-C)<br>Proanthocyanidin B1 (EC-4,8-C)<br>Proanthocyanidin B3 (C-4,8-C)<br>(þ)-Catechin | |
| | Anthocyanins | Cyanidin-3-glucoside<br>Cyanidin-3-rutinoside<br>Cyanidin-3-malonylglucoside<br>Cyanidin-3-malonylglucosyl-5-glucoside<br>Pelargonidin-3-galactoside<br>Pelargonidin-3-glucoside<br>Pelargonidin-3-rutinoside<br>Pelargonidin-3-arabinoside<br>Pelargonidin-3,5-diglucoside<br>Pelargonidin-3-malylglucoside<br>Pelargonidin-3-malonylglucoside<br>Pelargonidin-3-acetylglucoside<br>Pelargonidin-disacharide (hexose þ pentose) acylated with acetic acid<br>5-Pyranopelargonidin-3-glucoside | |

**Table 4.**
*List of polyphenol molecules present in* Fragaria x ananassa.

## 4. Pharmacological activity

Basu et al., Youdim et al., and Mazz were reported that the strawberry is having a special role in cardiovascular diseases, when this fruit included in the regular diet it will inhibit platelet aggregation and inflammation, improve endothelial function, plasma lipid profile, elevate low-density lipid profile which is oxidation resistance and also modulate the metabolism of eicosanoid [46, 48–50]. Most of the researchers evaluated that due to the presence of vitamin C and phenolic compounds in the strawberry having **antioxidant** capacity [51–54]. It has been concluded that

**Phenolic acids and Hydrolyzable tannins**

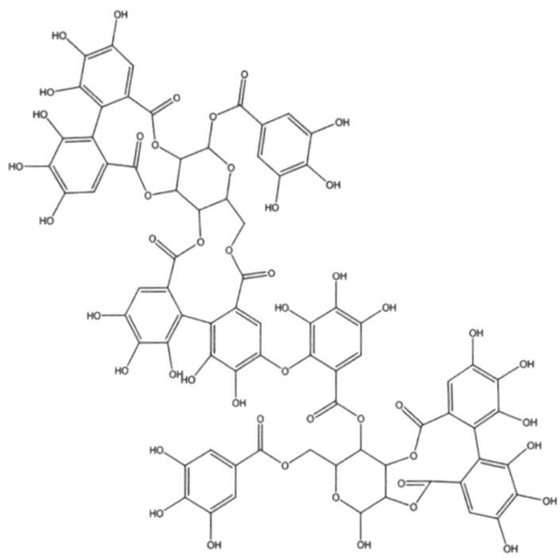

**Ellagitannin**

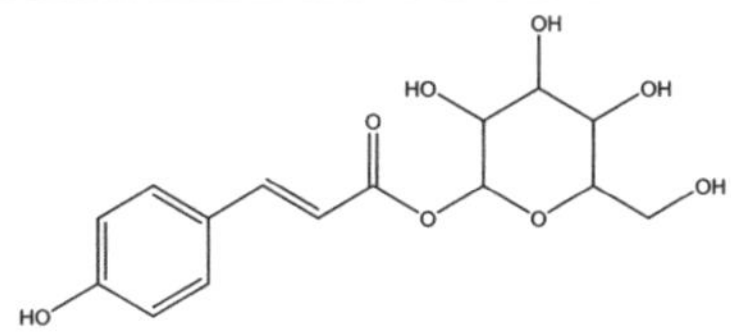

**p-Coumaroyl hexose**

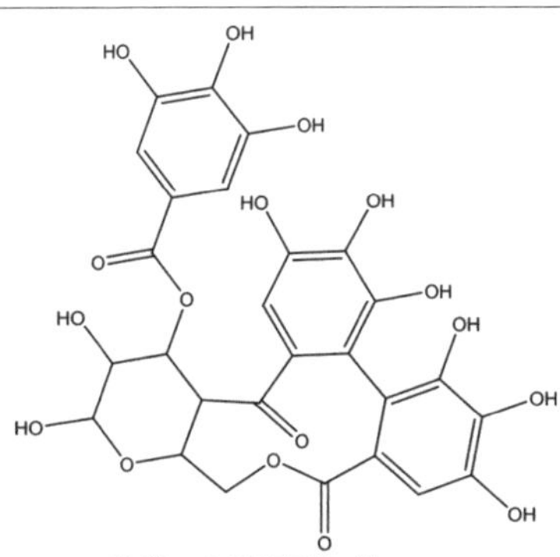

**Galloyl-HHDP-glucose**

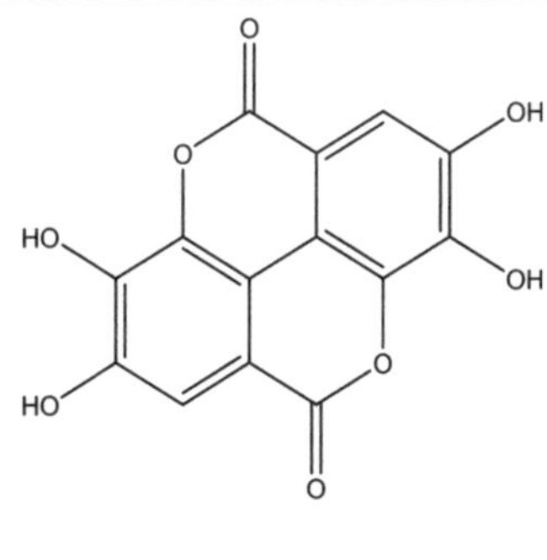

**Ellagic acid**

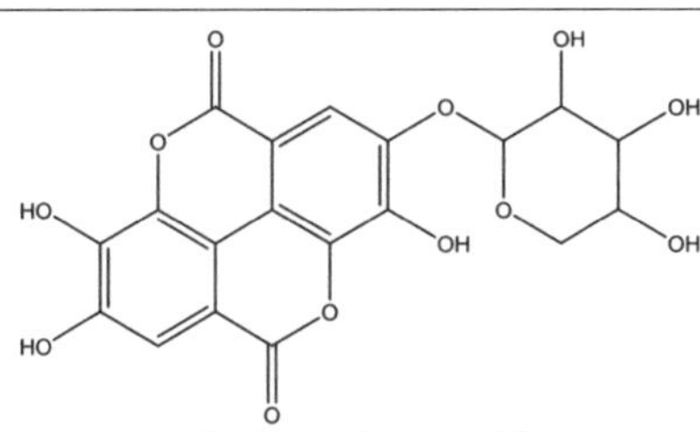

**Ellagic acid pentoside**

| **Flavanols and Flavonols** | |
|---|---|
| 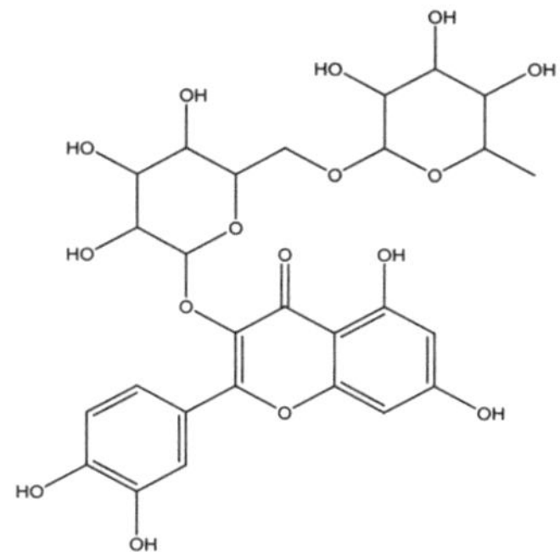 **Quercetin-rutinoside** | 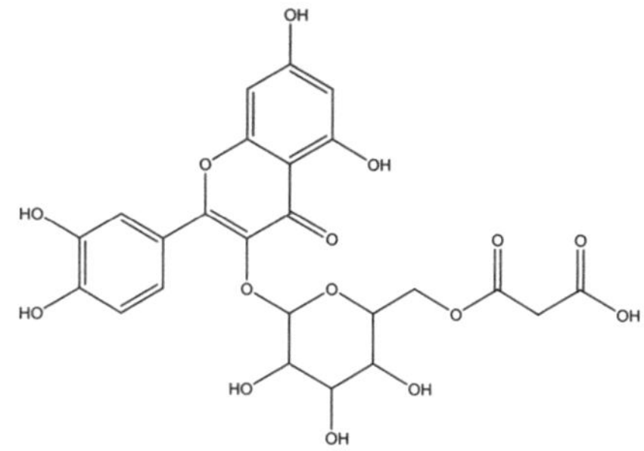 **Quercetin-3-O-malonyglucoside** |
| 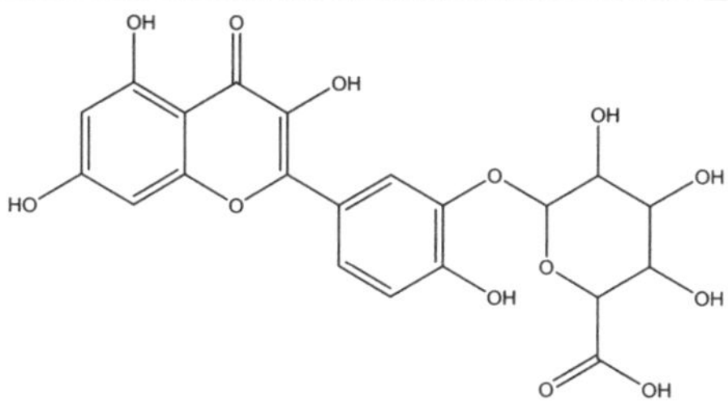 **Quercetin-3-glucuronide** | 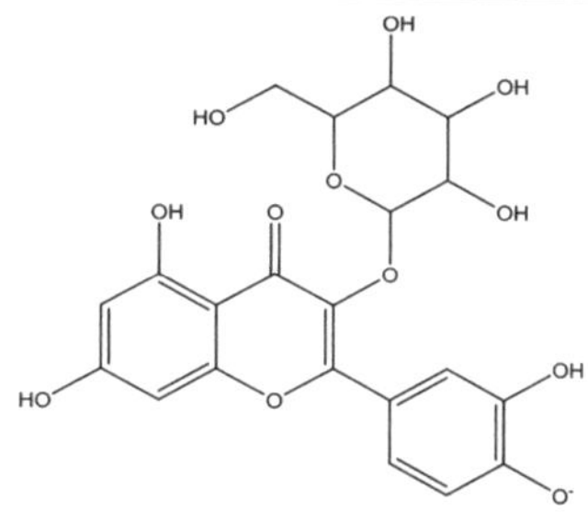 **Quercetin-3-glucoside** |
| 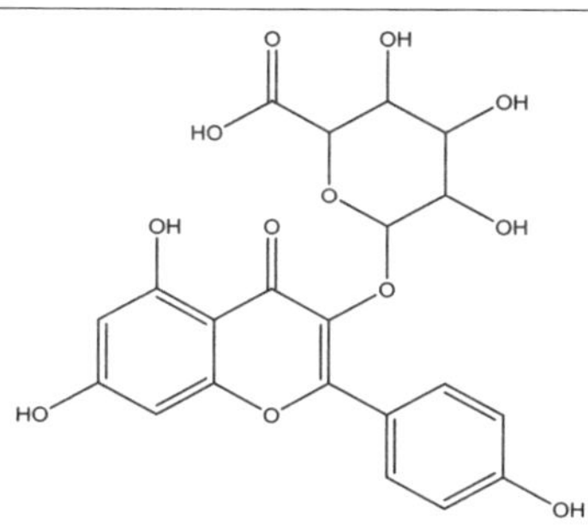 **Kaempferol-3-glucunoride** | 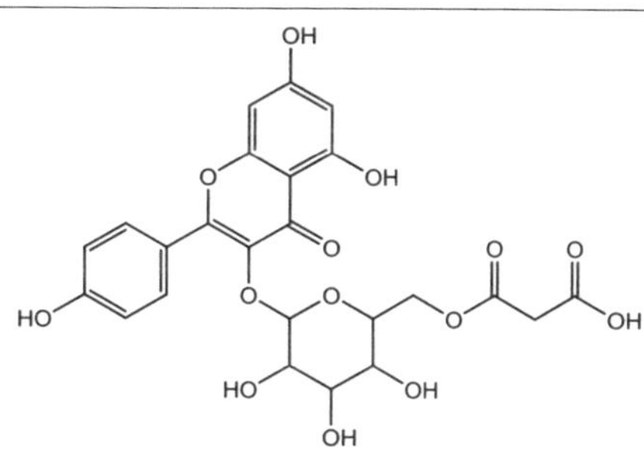 **Kaempferol 3-(6''-malonylglucoside)** |
| 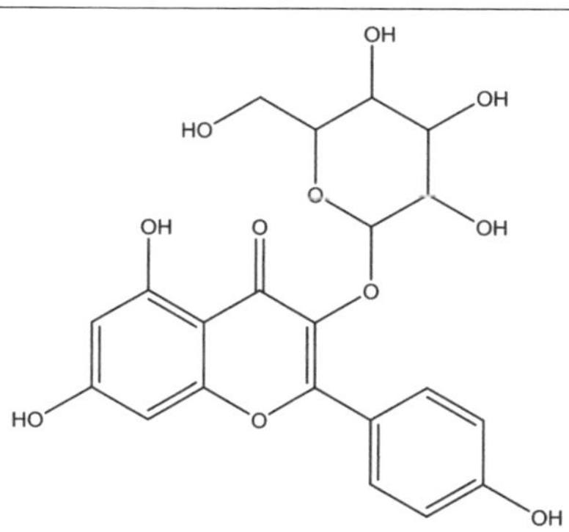 **Kaempferol-3-glucoside** | 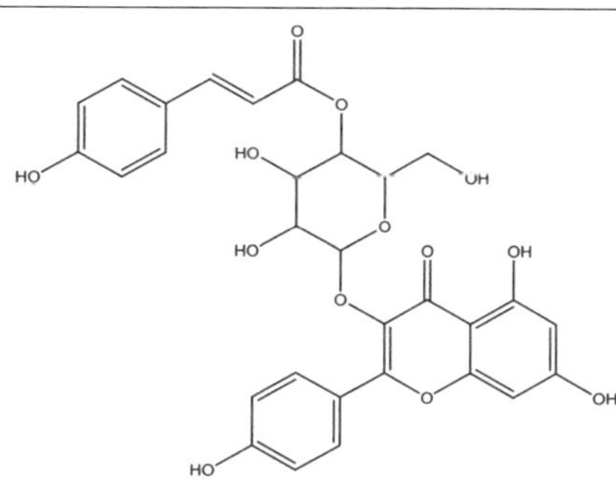 **Kaempferol 3-(4''-p-coumaroylglucoside)** |

| | |
|---|---|
| 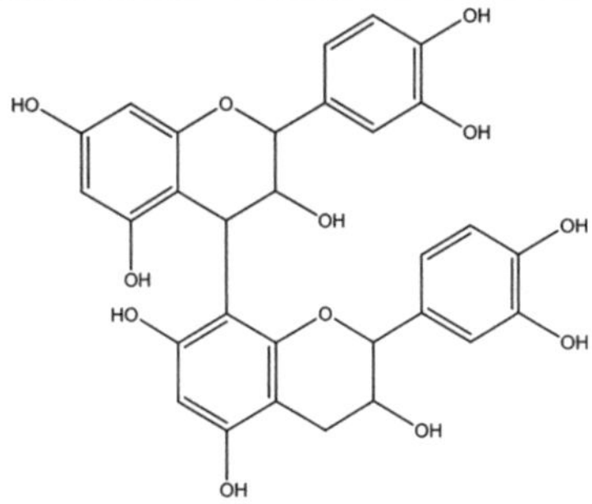 **Proanthocyanidin B1** | 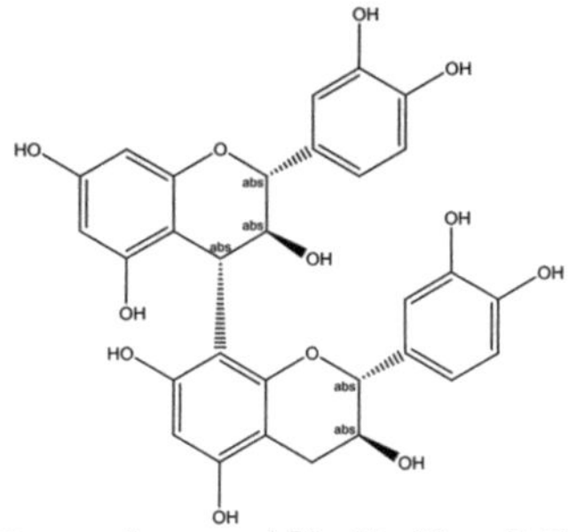 **Proanthocyanidin B3(C-4,8-C) (þ)-Catechin** |
| **Anthocyanins** | |
| 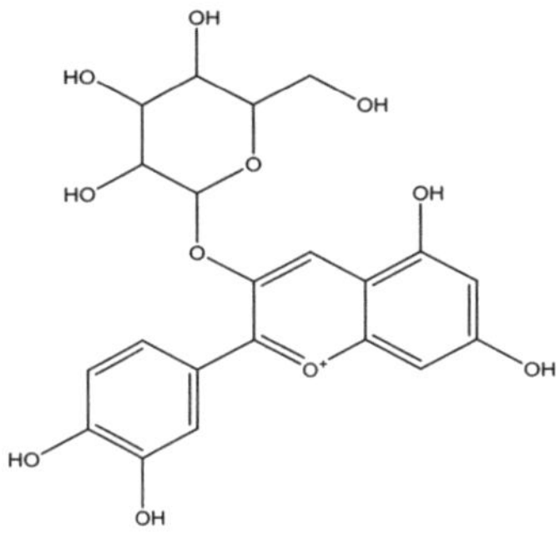 **Cyanidin-3-glucoside** | 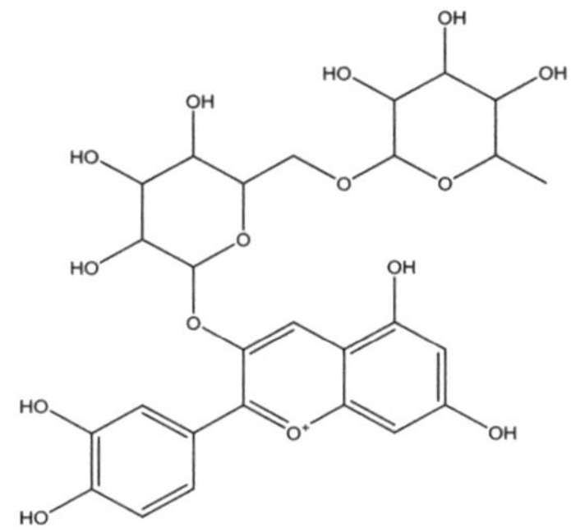 **Cyanidin-3-rutinoside** |
| 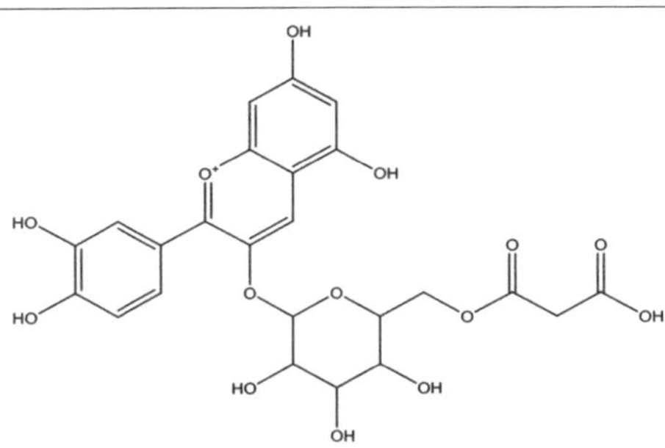 **Cyanidin-3-malonylglucoside** | 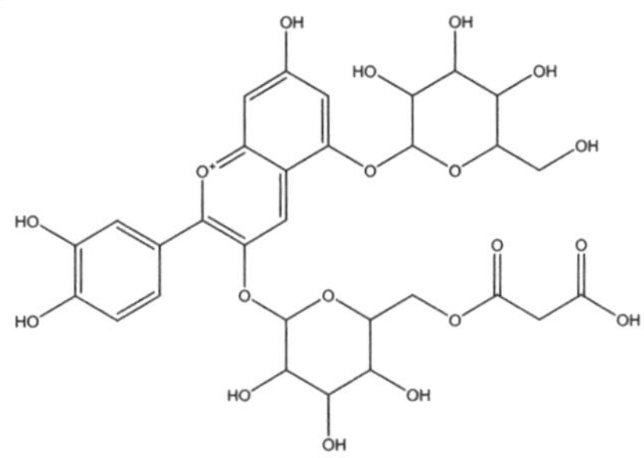 **Cyanidin-3-malonylglucosyl-5-glucoside** |
| 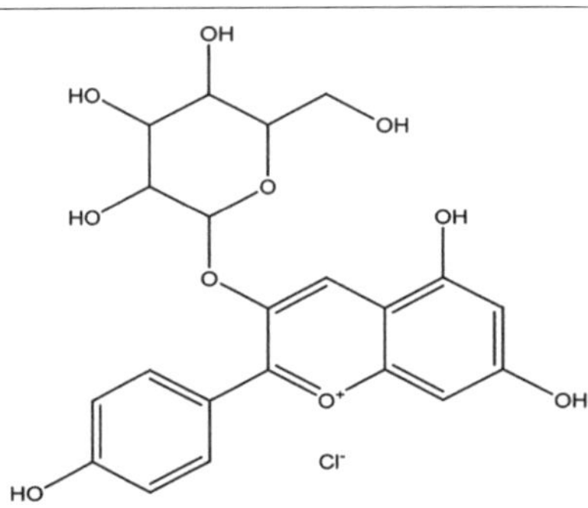 **Pelargonidin-3-galactoside** | 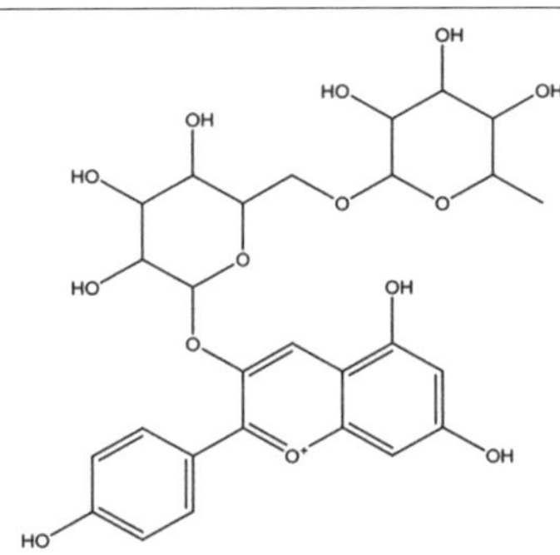 **Pelargonidin-3-rutinoside** |

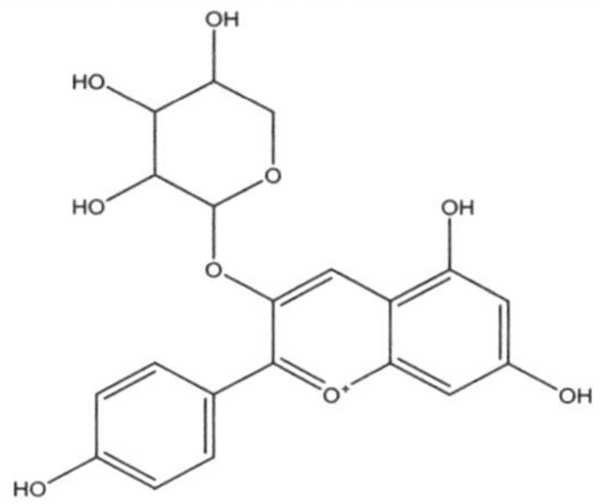

**Pelargonidin-3-arabinoside**

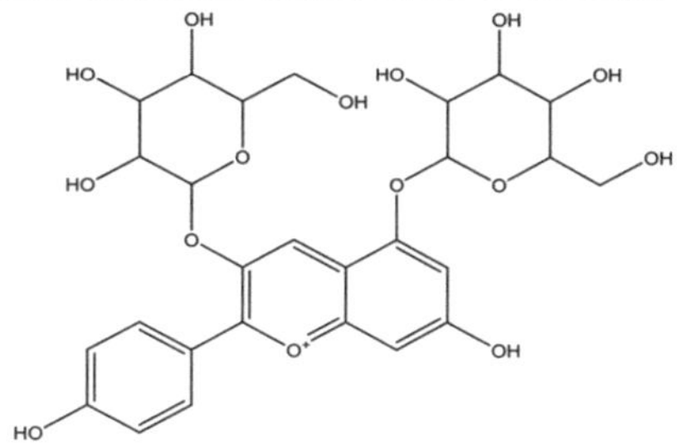

**Pelargonidin-3,5 diglucoside**

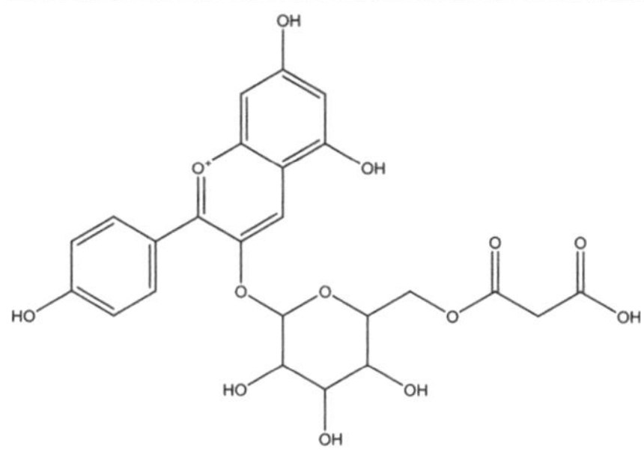

**Pelargonidin-3-malonylglucoside**

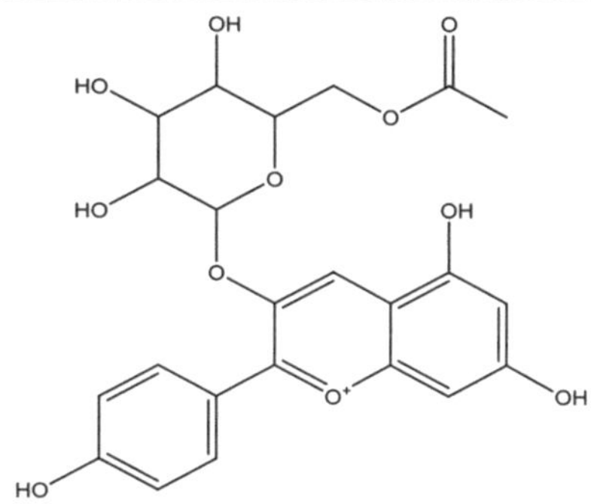

**Pelargonidin-3-acetylglucoside**

strawberry is 40 times more efficient than cereals, 10 times more efficient than vegetables, and 4 times more efficient than other fruits [55]. The polyphenols like anthocyanin, 2,5-dimethyl-4-hydroxy-3-[$^2$H] furanone, and ellagic acid shown the antioxidant property (*in vitro* model) and have high bioavailability (*in vivo* model) [56, 57]. The specific phytoconstituents like carotenoids, phytosterols, vitamins (A, C, and folic acid), vitamin precursors like calcium and selenium, polyphenols, and triterpene esters are reported **chemoprotective** activity in carcinogenesis (multi-stage) [58, 59]. Depending on the strawberry species anthocyanin [60–64], ellagitannins, and proanthocyanidins having **anti-tumorigenic** properties against multiple human cancer cells of *in vitro* and *in vivo* animal models [65, 66]. Tamara et al., 2019 evaluated that the methanolic extract of strawberry significantly declines 3 T3-L1 adipocyte differentiation, CCAAT adipogenic transcription factors, C/REB-α enhancer-binding protein, peroxisome proliferation-activated receptor (PPAR-γ), resistin, angiotensinogen and it stimulates the AMP-activated protein kinase (AMPKα), Sirtuin 1 (Sirt1) and peroxisome proliferator-activated receptor-gamma coactivator 1-alpha (PGC 1α) suggested that mitochondrial biogenesis elevation. Their evidence concluded that the strawberry methanolic extract has a capability of **antiobesity** activity [67]. Mahendri et al., 2020 their team effort was reported that anthocyanin is one of the photo molecules present in the strawberry having an **anti-diabetic** effect by declining blood sugar levels, improving endothelial dysfunction, and blood vessel inflammation [68, 69]. Susan (2010), Pinto (2008), and Apostolidis (2006) et al., evaluated that the strawberry is having the potency to treat hypertension by inhibiting α-glucosidase, α-amylase, and angiotensin-converting enzyme activities which are linked to type-2 diabetes. They finally concluded that the strawberry is having **antihypertension** properties [70–72]. Gasapirrini et al., 2017 they screened and proposed that the strawberry is having an **anti-inflammatory** effect by activation of the Nrf2-AMPK pathway and NF-kB pathway which also elevates the antioxidant defense [73]. The list of pharmacological properties are incorporated in **Figure 1**.

Al-Sanea et al., 2021 they made tremendous work by selecting the combination of ginger and strawberry methanolic extract were used for silver nanoparticle synthesis (AgNPs) and evaluated by utilizing TEM analysis, Fourier-transform infrared, spectroscopy ultraviolet–visible spectroscopy, zeta potential, dynamic light scattering technique, and docking studies that the combination is having antiviral activity against SARS-CoV-2 by targeting proteins like Mpro, ADP ribose phosphatase, NSP14, NSP16, PLpro, AAK1 and Cathepsin L [74]. **Fisetin** is one of the phytomolecules found in strawberries and other fruits as well as having different biological activities such as antioxidant, anti-inflammatory, **anti-osteoporotic**, anticancer, **antimicrobial**, and antidiabetic activities [75]. Especially Kaempferol glycosides like kaempferitrin (kaempferol 3,7-dirhamnoside) is one of the biomolecules of flavonoids possessing different pharmacological activities such as anti-apoptotic, pro-apoptotic, **anthelmintic**, ant-convulsant, antidepressant, **immunostimulatory**, cardioprotective, and natriuretic [76]. Afrin. S et al., their immense work has proved that the combination of strawberry tree honey and 5-fluorouracil is effective in adenocarcinoma and **metastatic colon cancer**. The combined treatment declined cell viability, oxidative stress, elevated MAPK, ATF-6, XBP-1 markers, enhanced cell cycle arrest, apoptotic genes, and modulates regulatory genes in cell line studies [77]. Specifically, anthocyanin and ellagic acid phytomolecules have a **neuroprotective** property from oxidative damage [78, 79]. Alarcon et al., 2015 their team effort finally reported that the strawberry possesses **anti-platelet** activity by declining specific signaling molecules such as sP-selectin, sCD40L, RANTES, and IL-1b levels [80]. The natural therapy of strawberries gives relief to those who are suffering from **constipation** and **hemorrhoids** [81].

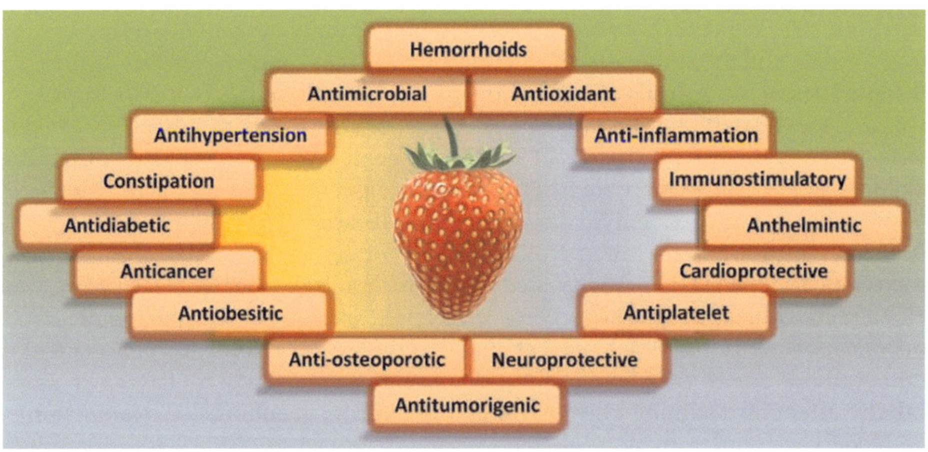


**Figure 1.**
*Pharmacological properties of strawberry.*

## 5. Conclusion

Strawberry is the only fruit available in all the seasons, mildly acid to bracingly sour taste to eat depending on the ripeness and individual variety. It contains many micronutrients, minerals, and most fat and water-soluble vitamins with rich quantities due to the presence of their phytochemical constituents. It is capable of treating different diseases such as inflammation, hypertension, obesity, diabetes, cancer, scavenge free radicals, constipation, hemorrhoids, cardiac, nerve disorders, etc., and phytomolecules like flavonoids and phenolic compounds are playing a crucial role in it.

## Author details

Rushendran Rapuru[1], Sivakumar Bathula[2] and Ilango Kaliappan[2*]

1 Department of Pharmacology, SRM College of Pharmacy, Chengalpattu, India

2 Department of Pharmaceutical Chemistry, SRM College of Pharmacy, Chengalpattu, India

*Address all correspondence to: ilangok1@srmist.edu.in

## References

[1] Wilhelm S, Sagen JA. A History of the Strawberry. Berkeley, California: University of California, Agr. Public; 1972. p. 4031

[2] Terefe NS, Kleintschek T, Gamage T, Fanning KJ, Netzel G, Versteeg C, et al. Comparative effects of thermal and high pressure processing on phenolic phytochemicals in different strawberry cultivars. Innovative Food Science & Emerging Technologies. 2013;**19**:57-65

[3] Panico AM, Garufi F, Nitto S, Di Mauro R, Longhitano RC, Magrì G, et al. Antioxidant activity and phenolic content of strawberry genotypes from Fragaria x ananassa. Pharmaceutical Biology. 2009;**47**(3):203-208

[4] David T. Strawberry: A Brief History. 2012. Available from: https://ipm.missouri.edu/meg/2012/5/Strawberry-A-Brief-History/

[5] https://www.britannica.com/print/article/568585

[6] https://www.cabi.org/isc/datasheet/24406.22. 2019

[7] Gunduz K. Strawberry: Phytochemical composition of strawberry (Fragaria× ananassa). In: Nutritional composition of fruit cultivars. Turkey: Academic Press; 2016. pp. 733-752

[8] Nutrition Facts and Analysis for Strawberries, raw, 100 g; USDA Nutrient Database, SR-21. Nutritiondata.com. Conde Nast, 2014. US Department of Agriculture, Agriculture Research Service. USDA National Nutrient for Standard References, Release 23. United States: Fruits and fruit juices; 2010. pp. 785–787. Available from: http://www.ars.usda.gov/Services/docs.htm?docid=8964. Accessed: January 10, 2011

[9] Tulipani S, Romandini S, Suarez JMA, Capocasa F, Mezzetti B, Battino M, et al. Folate content in different strawberry genotypes and folate status in healthy subjects after strawberry consumption. Biofactors. 2008;**34**(1):47-55

[10] Verhoef H, Veenemans J, Mwangi MN, Prentice AM. Safety and benefits of interventions to increase folate status in malaria-endemic areas. British Journal of Haematology. 2017;**177**(6):905-918

[11] Bendich A, Leader S, Muhuri P. Supplemental calcium for the prevention of hip fracture: Potential health-economic benefits. Clinical Therapeutics. 1999;**21**(6):1058-1072

[12] Georgieff MK, Krebs NF, Cusick SE. The benefits and risks of iron supplementation in pregnancy and childhood. Annual Review of Nutrition. 2019;**39**:121-146

[13] Weiss G, Gordeuk VR. Benefits and risks of iron therapy for chronic anaemias. European Journal of Clinical Investigation. 2005;**35**:36-45

[14] Khalil Z, Alam B, Akbari AR, Sharma H. The medical benefits of vitamin K2 on calcium-related disorders. Nutrients. 2021;**13**(2):691

[15] DiNicolantonio JJ, Bhutani J, O'Keefe JH. The health benefits of vitamin K. Open Heart. 2015;**2**(1):e000300

[16] Gröber U, Reichrath J, Holick MF, Kisters K. Vitamin K: An old vitamin in a new perspective. Dermato-Endocrinology. 2014;**6**(1):e968490

[17] Toro AP, Rodriguez JR, Miranda-Massari JR, Borges RM, Marcial V,

Olalde J, et al. Vitamin C and COVID-19: An orthomolecular perspective on physiological mechanisms. J Orthomol Med. 2021;**36**(3):1-9

[18] Li R, Guo C, Li Y, Qin Z, Huang W. Therapeutic targets and signaling mechanisms of vitamin C activity against sepsis: A bioinformatics study. Briefings in Bioinformatics. 2021;**22**(3): bbaa079

[19] Alasalvar C, Chang SK, Bolling B, Oh WY, Shahidi F. Specialty seeds: Nutrients, bioactives, bioavailability, and health benefits: A comprehensive review. Comprehensive Reviews in Food Science and Food Safety. 2021;**20**(3):2382-2427

[20] Neamat-Allah AN, Mahsoub YH, Mahmoud EA. The potential benefits of dietary β-glucan against growth retardation, immunosuppression, oxidative stress and expression of related genes and susceptibility to Aeromonas hydrophila challenge in Oreochromis niloticus induced by herbicide pendimethalin. Aquaculture Research. 2021;**52**(2):518-528

[21] Laskoś K, Pisulewska E, Waligórski P, Janowiak F, Janeczko A, Sadura I, et al. Herbal additives substantially modify antioxidant properties and tocopherol content of cold-pressed oils. Antioxidants. 2021; **10**(5):781

[22] Martin M, Richa C. Korean red ginseng benefits for erectile dysfunction. Men's health. 2021. Available from: https://ro.co/health-guide/ginseng-ed/

[23] Edema O, Okoye O, Unuigbe E, Odonmeta B, Obodo J. POS-255 niacin lowers serum phosphate levels in nigerian chronic kidney disease patients. A randomized control trial. Kidney International Reports. 2021;**6**(4):S107-S108

[24] Li H, Xu X, Lu L, Sun R, Guo Q, Chen Q, et al. The comparative impact among different intensive statins and combination therapies with niacin/ ezetimibe on carotid intima-media thickness: A systematic review, traditional meta-analysis, and network meta-analysis of randomized controlled trials. European Journal of Clinical Pharmacology. 2021;**77**:1-13

[25] Shi N, Clinton SK, Liu Z, Wang Y, Riedl KM, Schwartz SJ, et al. Strawberry phytochemicals inhibit azoxymethane/ dextran sodium sulfate-induced colorectal carcinogenesis in Crj: CD-1 mice. Nutrients. 2015;**7**(3): 1696-1715

[26] Sharpless KE, Arce-Osuna M, Thomas JB, Gill LM. Value assignment of retinol, retinyl palmitate, tocopherol, and carotenoid concentrations in standard reference material 2383 (baby food composite). Journal of AOAC International. 1999;**82**(2):288-296

[27] Maiani G, Pappalardo G, Ferro-Luzzi A, Raguzzini A, Azzini E, Guadalaxara A, et al. Accumulation of β-carotene in normal colorectal mucosa and colonic neoplastic lesions in humans. Nutrition and Cancer. 1995;**24**:23-31

[28] Cordenunsi BR, Genovese MI, do Nascimento JRO, Hassimotto NMA, dos Santos RJ, Lajolo FM. Effects of temperature on the chemical composition and antioxidant activity of three strawberry cultivars. Food Chemistry. 2005;**91**(1):113-121

[29] Tulipani S, Mezzetti B, Capocasa F, Bompadre S, Beekwilder J, De Vos CR, et al. Antioxidants, phenolic compounds, and nutritional quality of different strawberry genotypes. Journal of Agricultural and Food Chemistry. 2008; **56**(3):696-704

[30] Tulipani S, Romandini S, Busco F, Bompadre S, Mezzetti B, Battino M. Ascorbate, not urate, modulates the plasma antioxidant capacity after strawberry intake. Food Chemistry. 2009;**117**(1):181-188

[31] Buend'ıa B, Gil MI, Tudela JA, et al. HPLC-MS Analysis of proanthocyanidin oligomers and other phenolics in 15 strawberry cultivars (dagger). Journal of Agricultural and Food Chemistry. 2010; **58**:3916-3926

[32] Kay CD, Mazza G, Holub BJ, Wang J. Anthocyanin metabolites in human urine and serum. British Journal of Nutrition. 2004;**91**(6):933-942

[33] Vitaglione P, Donnarumma G, Napolitano A, Galvano F, Gallo A, Scalfi L, et al. Protocatechuic acid is the major human metabolite of cyanidin-glucosides. The Journal of nutrition. 2007;**137**(9):2043-2048

[34] Hertog MG, Hollman PC, Venema DP. Optimization of a quantitative HPLC determination of potentially anticarcinogenic flavonoids in vegetables and fruits. Journal of Agricultural and Food Chemistry. 1992; **40**(9):1591-1598

[35] Clifford MN. Anthocyaninsdnature, occurrence and dietary burden. Journal of the Science of Food and Agriculture. 2000;**80**:1063-1072

[36] Lopes-da-Silva F, de Pascual-Teresa S, Rivas-Gonzalo J, Santos-Buelga C. Identification of anthocyanin pigments in strawberry (cv Camarosa) by LC using DAD and ESI-MS detection. European Food Research and Technology. 2002;**214**(3):248-253

[37] Da Silva FL, Escribano-Bailón MT, Alonso JJP, Rivas-Gonzalo JC, Santos-Buelga C. Anthocyanin pigments in strawberry. LWT-Food Science and Technology. 2007;**40**(2): 374-382

[38] Bridle P, García-Viguera C. Analysis of anthocyanins in strawberries and elderberries. A comparison of capillary zone electrophoresis and HPLC. Food Chemistry. 1997;**59**(2):299-304

[39] Hong V, Wrolstad RE. Characterization of anthocyanin-containing colorants and fruit juices by HPLC/photodiode array detection. Journal of Agricultural and Food Chemistry. 1990;**38**(3):698-708

[40] Lukton A, Chichester CO, Mackinney G. Characterization of a second pigment in strawberries. Nature. 1955;**176**(4486):790-790

[41] Bakker J, Bridle P, Bellworthy SJ. Strawberry juice colour: A study of the quantitative and qualitative pigment composition of juices from 39 genotypes. Journal of the Science of Food and Agriculture. 1994;**64**(1):31-37

[42] Goiffon JP, Mouly PP, Gaydou EM. Anthocyanic pigment determination in red fruit juices, concentrated juices and syrups using liquid chromatography. Analytica Chimica Acta. 1999;**382**(1–2): 39-50

[43] Phane Quideau S. Chemistry and Biology of Ellagitannins: An Underestimated Class of Bioactive Plant Polyphenols. France: World Scientific; 2009

[44] Koponen JM, Happonen AM, Mattila PH, Törrönen AR. Contents of anthocyanins and ellagitannins in selected foods consumed in Finland. Journal of Agricultural and Food Chemistry. 2007;**55**(4): 1612-1619

[45] Wang SY, Millner P. Effect of different cultural systems on antioxidant capacity, phenolic content, and fruit quality of strawberries (fragaria × aranassa duch.). Journal of Agricultural and Food Chemistry. 2009;**57**(20): 9651-9657

[46] Basu A, Rhone M, Lyons TJ. Berries: Emerging impact on cardiovascular health. Nutrition Reviews. 2010;**68**(3): 168-177

[47] Aaby K, Ekeberg D, Skrede G. Characterization of phenolic compounds in strawberry (Fragaria× ananassa) fruits by different HPLC detectors and contribution of individual compounds to total antioxidant capacity. Journal of Agricultural and Food Chemistry. 2007; **55**(11):4395-4406

[48] Youdim KA, Martin A, Joseph JA. Incorporation of the elderberry anthocyanins by endothelial cells increases protection against oxidative stress. Free Radical Biology and Medicine. 2000;**29**(1):51-60

[49] Mazza G. Anthocyanins and heart health. Annali-Istituto Superiore Di Sanita. 2007;**43**(4):369

[50] Giampieri F, Tulipani S, Alvarez-Suarez JM, Quiles JL, Mezzetti B, Battino M. The strawberry: Composition, nutritional quality, and impact on human health. Nutrition. 2012;**28**(1): 9-19

[51] Halvorsen BL, Carlsen MH, Phillips KM, Bøhn SK, Holte K, Jacobs DR Jr, et al. Content of redox-active compounds (ie, antioxidants) in foods consumed in the United States. The American Journal of Clinical Nutrition. 2006;**84**(1):95-135

[52] Proteggente AR, Pannala AS, Paganga G, Buren LV, Wagner E, Wiseman S, et al. The antioxidant activity of regularly consumed fruit and vegetables reflects their phenolic and vitamin C composition. Free Radical Research. 2002;**36**(2): 217-233

[53] Pellegrini N, Serafini M, Colombi B, Del Rio D, Salvatore S, Bianchi M, et al. Total antioxidant capacity of plant foods, beverages and oils consumed in Italy assessed by three different in vitro assays. The Journal of Nutrition. 2003; **133**(9):2812-2819

[54] Ahouagi VB, Mequelino DB, Tavano OL, Garcia JAD, Nachtigall AM, Boas BMV. Physicochemical characteristics, antioxidant activity, and acceptability of strawberry-enriched ketchup sauces. Food Chemistry. 2021; **340**:127925

[55] Halvorsen BL, Holte K, Myhrstad MC, Barikmo I, Hvattum E, Remberg SF, et al. A systematic screening of total antioxidants in dietary plants. The Journal of Nutrition. 2002; **132**(3):461-471

[56] Henning SM, Seeram NP, Zhang Y, Li L, Gao K, Lee RP, et al. Strawberry consumption is associated with increased antioxidant capacity in serum. Journal of Medicinal Food. 2010;**13**(1):116-122

[57] Hannum SM. Potential impact of strawberries on human health: A review of the science. Critical Reviews in Food Science and Nutrition. 2004;**44**(1): 1-17

[58] Han C, Ding H, Casto B, Stoner GD, D'Ambrosio SM. Inhibition of the growth of premalignant and malignant human oral cell lines by extracts and components of black raspberries. Nutrition and Cancer. 2005;**51**(2): 207-217

[59] Duthie SJ. Berry phytochemicals, genomic stability and cancer: Evidence for chemoprotection at several stages in the carcinogenic process. Molecular Nutrition & Food Research. 2007;**51**(6): 665-674

[60] Zhang Y, Seeram NP, Lee R, Feng L, Heber D. Isolation and identification of strawberry phenolics with antioxidant and human cancer cell antiproliferative properties. Journal of Agricultural and Food Chemistry. 2008;**56**(3):670-675

[61] Bagchi D, Sen CK, Bagchi M, Atalay M. Anti-angiogenic, antioxidant, and anti-carcinogenic properties of a novel anthocyanin-rich berry extract formula. Biochemistry (Moscow). 2004; **69**(1):75-80

[62] Wang LS, Hecht SS, Carmella SG, Yu N, Larue B, Henry C, et al. Anthocyanins in black raspberries prevent esophageal tumors in rats. Cancer Prevention Research. 2009;**2**(1): 84-93

[63] Wang LS, Stoner GD. Anthocyanins and their role in cancer prevention. Cancer Letters. 2008;**269**(2):281-290

[64] Seeram NP, Adams LS, Henning SM, Niu Y, Zhang Y, Nair MG, et al. In vitro antiproliferative, apoptotic and antioxidant activities of punicalagin, ellagic acid and a total pomegranate tannin extract are enhanced in combination with other polyphenols as found in pomegranate juice. The Journal of Nutritional Biochemistry. 2005;**16**(6): 360-367

[65] McDougall GJ, Ross HA, Ikeji M, Stewart D. Berry extracts exert different antiproliferative effects against cervical and colon cancer cells grown in vitro. Journal of Agricultural and Food Chemistry. 2008;**56**(9):3016-3023

[66] Schmidt BM, Erdman JW Jr, Lila MA. Differential effects of blueberry proanthocyanidins on androgen sensitive and insensitive human prostate cancer cell lines. Cancer Letters. 2006; **231**(2):240-246

[67] Forbes-Hernández TY, Cianciosi D, Ansary J, Mezzetti B, Bompadre S, Quiles JL, et al. Strawberry (Fragaria× ananassa cv. Romina) methanolic extract promotes browning in 3T3-L1 cells. Food & Function. 2020;**11**(1):297-304

[68] Deayu Putri M, Wiboworini B, Dirgahayu P. The effect of strawberry on type 2 diabetes mellitus: A review. International Journal of Nutrition Sciences. 2020;**5**(1):1-6

[69] Blahova J, Martiniakova M, Babikova M, Kovacova V, Mondockova V, Omelka R. Pharmaceutical drugs and natural therapeutic products for the treatment of type 2 diabetes mellitus. Pharmaceuticals. 2021;**14**(8):806

[70] Cheplick S, Kwon YI, Bhowmik P, Shetty K. Phenolic-linked variation in strawberry cultivars for potential dietary management of hyperglycemia and related complications of hypertension. Bioresource Technology. 2010;**101**(1): 404-413

[71] Pinto MDS, Kwon YI, Apostolidis E, Lajolo FM, Genovese MI, Shetty K. Functionality of bioactive compounds in Brazilian strawberry (Fragaria x ananassa Duch.) cultivars: Evaluation of hyperglycemia and hypertension potential using in vitro models. Journal of Agricultural and Food Chemistry. 2008;**56**(12):4386-4392

[72] Apostolidis E, Kwon YI, Shetty K. Potential of select yogurts for diabetes and hypertension management. Journal

of Food Biochemistry. 2006;**30**(6): 699-717

[73] Gasparrini M, Forbes-Hernandez TY, Giampieri F, Afrin S, Alvarez-Suarez JM, Mazzoni L, et al. Anti-inflammatory effect of strawberry extract against LPS-induced stress in RAW 264.7 macrophages. Food and Chemical Toxicology. 2017;**102**:1-10

[74] Al-Sanea MM, Abelyan N, Abdelgawad MA, Musa A, Ghoneim MM, Al-Warhi T, et al. Strawberry and ginger silver nanoparticles as potential inhibitors for SARS-CoV-2 assisted by in silico modeling and metabolic profiling. Antibiotics. 2021;**10**(7):824

[75] Antika LD, Dewi RM. Pharmacological aspects of fisetin. Asian Pacific Journal of Tropical Biomedicine. 2021;**11**(1):1

[76] Patel DK. Pharmacological activities and Therapeutic potential of kaempferitrin in medicine for the treatment of human disorders: A review of medicinal importance and health benefits. Cardiovascular & Hematological Disorders Drug Targets. 2021;**21**(2):104-114

[77] Afrin S, Giampieri F, Cianciosi D, Alvarez-Suarez JM, Bullon B, Amici A, et al. Strawberry tree honey in combination with 5-fluorouracil enhances chemosensitivity in human colon adenocarcinoma cells. Food and Chemical Toxicology. 2021;**156**:112484

[78] Silva RF, Pogacnik L. Polyphenols from food and natural products: Neuroprotection and safety. Antioxidants. 2020;**9**(1):61

[79] Gupta A, Singh AK, Kumar R, Jamieson S, Pandey AK, Bishayee A. Neuroprotective potential of ellagic acid: A critical review. Advances in Nutrition. 2021;**12**(4):1211-1238

[80] Alarcón M, Fuentes E, Olate N, Navarrete S, Carrasco G, Palomo I. Strawberry extract presents antiplatelet activity by inhibition of inflammatory mediator of atherosclerosis (sP-selectin, sCD40L, RANTES, and IL-1β) and thrombus formation. Platelets. 2015; **26**(3):224-229

[81] Sarfaraz S, Sarwar G, Baig L, Khalid A, Fatima W. Screening of strawberry juice for role in fecal evacuation. The Pharma Innovation. 2015;**4**(6):7-9

Section 5

# Pests and Diseases of Strawberries

Chapter 11

# Pests, Diseases, Nematodes, and Weeds Management on Strawberries

*Refik Bozbuga, Selman Uluisik, Pınar Aridici Kara, Semiha Yuceer, Hale Gunacti, Pakize Gok Guler, Elen Ince, Hatice Nilufer Yildiz and Ozcan Tetik*

## Abstract

Strawberry is an important crop for many features, including being rich in vitamins and minerals. In addition to fresh consumption, it has been appealing to a wide range of consumers in recent years. Its cultivation is in flat areas, slopes, and areas where other crops are limited. Many pests and diseases that are the main biotic stress factors cause significant crop losses in strawberry cultivation. The aim of this chapter is to reveal biotic stress factors and their management. Several plant-parasitic nematodes, fungal diseases, weeds, pests, virus diseases, and bacterial diseases are the main biotic stress factors in plant growing and fruit ripening. The preparation of this book chapter is based on previously published sources and researches and manuscripts. In this section, it is aimed to provide readers with new perspectives in terms of collecting data on nematodes, diseases, pests, weeds, and fruit ripening of strawberry plants. The effect and mechanism of those biotic stress factors on strawberry growing are discussed and revealed in this chapter.

**Keywords:** strawberry, biotic stress, plant-parasitic nematodes, fungal diseases, weeds, pests, virus diseases, bacterial diseases, fruit ripening

## 1. Introduction

Strawberry can grow in diverse ecological conditions, and it is an important crop for many countries. Many diseases and pests cause damage to not only strawberry foliar and roots but also directly to fruits. During the ripening, pests and diseases are also precisely essential. Therefore, in the beginning, it would be better to focus on the ripening process.

Fleshy fruit ripening is explained by a series of biochemical and physiological changes involving complex changes in taste, aroma, color, texture, and sugar, coordinated by plant hormones. It has a noteworthy influence on fruit quality, postharvest shelf-life, and consumer acceptance. Changes in color, taste, aroma, texture, and nutritional value during the ripening period make the fleshy fruit attractive and delicious for consumers. These changes that occur with ripening are governed by external (i.e., light, temperature, and irrigation) and internal factors (i.e., genetic regulation and hormonal control) that allow fruit characteristics to develop [1]. As it is known primarily in the model plant tomato, the climacteric fruit ripening process is coordinated by ethylene perception and signal transmission [2]. Although ethylene was not thought to have played a role in the ripening of strawberry fruit, recent studies have shown that the ripening process is much more complex in non-climacteric fruits and controlled by ethylene [3], abscisic acid [4], auxin (indole-3-acetic acid [5], gibberellic acid [6], jasmonate [7] and brassinosteroids [8].

Following the ripening, pest, disease, and weeds of strawberry plants and interaction with yield losses are given attention in this chapter.

## 2. Hormones involved in strawberry ripening

Ethylene controls almost all ripening processes in climacteric fruits; however, many hormones are actively involved in the ripening process of non-climacteric fruits, which makes these fruits attractive for research. In this context, strawberry (*Fragaria x ananassa*) fruit has become a model of non-climacteric fruit. A complete hormonal profile of woodland strawberry *Fragaria vesca* fruit was reported that auxin is produced mainly in achenes (seeds), whilst abscisic acid (ABA), ethylene, gibberellins, and bioactive free base cytokinins are chiefly produced in receptacles [9]. The report also indicated that ABA promotes ripening while auxin delays it. Moreover, endogenous auxin GA levels are greatly reduced in the late stages of strawberry ripening when the abscisic acid (ABA) level increases dramatically [10]. Indole-3-acetic acid (IAA) has a significant role in cell expansion, determination of fruit size, and ripening of strawberry fruits. A recent RNA-seq study describes the expression profile of auxin biosynthesis and signaling during the development and maturation of F × ananassa [11]. Based on this study, the auxin content drops by 50% in the receptacle but remains constant during ripening, supporting the idea that auxins may involve in strawberry fruit ripening in later stages.

Auxin has been shown to delay ripening by altering the expression of many genes associated with ripening [5]. Expression of FaPL and FaEGase, which are the most important enzymes responsible for softening, increased at the beginning of strawberry ripening and decreased with exogenous auxin application [12]. However, in another study, the expression of two genes encoding Xyloglucan endotransglycosylase/hydrolases (XTH), FaXTH1, and FaXTH2, significantly up-regulated by auxins treatment [13]. In the same study, gibberellins and abscisic acid up-regulated both gene expressions. Increases of FaAux/IAA1 and FaAux/IAA2 transcripts increased by the influence of naphthalene acetic acid (NAA) at the stage of large green and white fruit [14]. During strawberry fruit ripening, the ABA content increasingly grows from the green stage to the red stage (and the commencement of this rise overlaps with declines in IAA levels [15]. In other words, the ripening of strawberries is controlled by ABA in an ethylene-independent manner [16]. Exogenous application of ABA to strawberry fruits has fluctuating results during fruit ripening. IAA has been shown to play an important role in inducing cell division and expansion, which is related to the early stages of strawberry fruit development. In the later stages of ripening, ABA and sucrose are the main molecules that play a role in controlling gene expression. Expression of ABA and sucrose signaling genes and ripening-related genes such as endo-β-(1,4)-glucanases 2 (CEL2), 9-cis-epoxycarotenoid dioxygenase 2 (NCED2),

MYB5, Sucrase synthase (SuSy), as endo-β-(1,4)-glucanases 1 (CEL1), Sucrose non-fermenting 1 (SNF1)—related protein kinase 2(SnRK2.2), and9-cis-epoxycarotenoid dioxygenase 1 (NCED1) were all considerably up-regulated by ABA or sucrose treatment alone, and especially with ABA + sucrose treatment [17]. However, postharvest ripening of strawberry fruits varied from the fruits attached to the plant, proposing ripening is related to the signal activated by ABA, as the application of ABA caused the modified amassing of numerous compounds including sugars, ABA, anthocyanins, and ABA-GE [18]. Treatment of ABA positively regulated the expression of FaRGL [19]. ABA is a crucial signal molecule in the advancement of strawberry ripening which was proved by the study [20]. Downregulation of FaNCED1 (9-cis-epoxycarotenoi dioxygenase, a key gene in ABA biosynthesis) inhibits ripening. A recent study showed that ABA biosynthesis is firmly connected by response and forward loops to limit ABA contents for fruit growing and to rapidly raise ABA contents for the commencement of fruit ripening [21]. To summarize the role of ABA in strawberry ripening, the expression of ABA-related genes significantly increases, and the hormone regulates many ripening-related metabolic pathways.

Gibberellic acid (GA) plays in the regulation of the growth of non-climacteric fruits, especially strawberries [6]. The expansion of receptacle cells during fruit development is coordinated by endogenous GAs. Among the bioactive GAS (GA1, GA3, and GA4), GA1 and GA4 are the most abundant in the early stages of strawberry fruit development and drop to lower levels as the fruit ripens. Exogenous treatment of GA3 retarded red color development and the loss of fruit firmness throughout the ripening period was significantly reduced in strawberry cultivars [22]. In a recent study, the application of GA3 affected the fruits quality of strawberries by changing organic acid and individual phenolic compound composition [23].

Although strawberry is known as climacteric fruit, ethylene could play a role at early stages of fruit ripening in strawberries [24]. Initially, an increase in FaPG1 gene expression was found in response to ethylene [25], and in later studies, several other genes, such as FaPG1, FaGal1, and FaGal2, were involved in cell wall modification were found to be modified by ethylene application [26]. Exogenous ethephon treatment increased the expression of biosynthesis and signaling genes, FaERF2 and FaACO1, and influenced the phytochemical profile of phenolic compounds, vitamin C contents, anthocyanins, and sugars [27]. Studies have shown that ethylene elicitors or inhibitors affect some significant feature qualities in strawberry fruits, including firmness [28]. Different hormonal treatments differentially affected hemicellulose metabolism during strawberry fruit ripening and under postharvest conditions. For example, postharvest 1-methylcyclopropene treatment up-regulated FaXynA and FaXynC expressions [29]. The physiological consequences of ethylene on strawberry fruit have been shown to depend on the developing phase of the fruit [30]. Up-regulation of ethylene-responsive transcription factor, ERF105 likc gene is significantly induced under cold stress, showing that ethylene could also play an important role in abiotic stress resistance in strawberries. Although ethylene appears to be involved in a secondary role compared to abscisic acid in non-climacteric strawberry ripening, this does not disregard that ethylene may adjust some certain occurrences linked to the ripening progression.

Polyamines (PAs), ubiquitous aliphatic amines and biogenic regulators present in all living organisms, are involved in many developmental and physiological processes involving plant aging, stress, and plant growth [31]. The content of spermine (Spm) rises strongly following the commencement of fruit coloring to red, and Spm is the dominant component of the ripe strawberry fruits. The predominance of spm in ripe fruit over other PAs is due to abundant expression of the strawberry S-adenosyl-L-Met decarboxylase gene (FaSAMDC), which encodes an enzyme that produces a residue required for PA biosynthesis [32]. Polyamine oxidase 5, FaPAO5, negatively adjusts strawberry fruit ripening, as down-regulation of FaPAO5 stimulated Spd, Spm, and ABA amassing, which ultimately enhanced ripening. The opposite results were shown in FaPAO5-overexpressing in the same study [33]. The results showed that FaPAO5 plays a role in the terminal catabolism of Spd and Spm. Application of putrescine (PUS) reduced the adverse effects of osmotic stress of the nutrient solution and increased plant resistance against salt stress, showing that PAs are important regulators against abiotic stress conditions [34].

## 3. Plant-parasitic nematode problems on strawberry plants

Several pathogens and pests cause damage to strawberry plants. Several plant-parasitic nematodes damage strawberry plants that some feed on roots, others in foliar parts. Parasitic nematodes cause yield losses, crop size, and quality [35]. Many kinds of research have been conducted on Strawberry nematodes.

More than 4000 plant-parasitic nematodes are found on the earth [36]. In strawberry (*Fragaria x ananassa* Duch.) from soil and foliage, plant-parasitic nematodes are present in 10 genera and 15 species belonging to the order of Dorylaimida and Tylenchida [37]. *Xiphinema pachtaicum*, *Meloidogyne hapla*, *Aphelenchoides ritzemabosi*, *Criconema nutabile*, *Pratylenchus microdorus*, *Ditylenchus dipsaci*, *Longidorus caespiticola*, *Meloidogyne arenaria*, *Longidorus elongates*, *Pratylenchus penetrans*, *Helicotylenchus dihystera*, *Paratylenchus pseudoparietinus*, *Aphelenchoides besseyi*, *Tylenchorhynchus claytoni*, *Aphelenchoides fragariae*, are plant-parasitic nematodes in strawberry fields in the soil in Bulgaria [37]. *P. penetrans* is cause reddish-brown lesions on roots and increase fungus infections in the roots of strawberry [38]. *Aphelencoides fragariae*, *Aphelencoides ritzemabosi M. hapla*, *M. arenaria*, and *P. penetrans* species cause severe damage in strawberry fields and need to be controlled [37]. The virus vector nematodes: *Longidorus Criconemoides*, *Helicotylenchus Tylenchorhynchus* are frequently found in strawberry fields [37]. The root-knot nematodes, genus *Meloidogyne* Goeldi, are the most challenging in strawberry cultivation [39]. A sting nematode, *Belonolaimus longicaudatus,* is also one of the most damaging nematodes in strawberry plants [39–41]. *Aphelenchoides fragariae*, *M. hapla*, *Aphelenchoides besseyi*, *Aphelenchoides bicaudatus*, *Aphelenchoides ritzemabosi*, *Ditylenchus dipsaci*, *Longidorus elongatus*, *Meloidogyne javanica*, *Criconemella onoensis*, *Meloidogyne incognita*, *Helicotylenchus dihystera*, *Pratylenchus penetrans*, *Pratylenchus brachyurus*, *Xiphinema*, *Pratylenchus vulnus* and *Pratylenchus zeae* in Brazil [42–45].

Incidence and population density are *Aphelenchoides fragariae* (34–98), *Aphelenchoides ritzemabosi* (23–70), *Aphelenchoides besseyi* (8–11), and *Ditylenchus dipsaci* (8–16) in strawberry 15 g leaves of plant tissues [37]. Some major species of Strawberry in the USA are *M. hapla, A. besseyi, B. longicaudatus, A. fragariae,* and *P. penetrans* [39–41, 46]. Nematode existence incidence and density are vastly variable that *M. hapla* 55–125 nematodes/100 $cm^3$ soil) and for *M. arenaria* (14-61nematodes/100 $cm^3$) in Bulgaria [37].

*Meloidogyne* spp. is the utmost major species of plant-parasitic nematodes on strawberry plants in Egypt, and *Aphelenchoides* sp. may exceedingly decrease strawberry yields [35]. The second stage of *Meloidogyne* species is the infective stage of nematodes and J2s penetrate plant roots, modify cell development, and cause root gall formation. During the nematode infection, nematode feeds on those cells termed giant cells, and females of root-knot nematodes develop within the galls [47–49]. Root-knot nematodes cause root galls in plants [47].

*Pratylenchus* species also damage strawberry plants. They are migratory nematodes that cause root lesions when they enter and migrate completely throughout the roots [49]. Nematodes cause damage to reducing roots, and nematodes absorb water and nutrients [49] and therefore decrease plant growth, shorten the crop cycle, decrease production, and cause leaf drop [39]. *Pratylenchus penetrans* is also a noteworthy nematode that is related to the occurrence of a disease that causes strawberry root rot [50, 51].

Some nematodes cause damage to others in the foliar part of strawberry, and they may be found in different densities in many countries. Plant-parasitic nematodes in the soil of strawberry plant may found as: *Helicotylenchus* (421.3 nematodes/100 $cm^3$ of soil), *Scutellonema* (1.0 nematodes/100 $cm^3$ of soil), *Meloidogyne* (3.9 nematodes/100 cm3 soil), Hemicycliophora (5.3 nematodes/100 $cm^3$ soil), *Ditylenchus* (0.3 nematodes/100 $cm^3$ soil), *Pratylenchus* (1.4 nematodes/100 $cm^3$ soil), *Xiphinema* (0.4 nematodes/100 $cm^3$ soil), *Trichodorus* (0.2 nematodes/100 $cm^3$ soil) and *Mesocriconema* (0.2 nematodes/100 $cm^3$ soil) in Brazil [52]. Root-knot nematodes: *M. javanica*, *M. arenaria*, *M. incognita, M. hapla,* and other nematodes: *A. fragariae, P. penetrans, Hemicycliophora* spp. and *D. dipsaci* are found in strawberry fields in Spain [53, 54]. *A. fragariae, D. dipsaci, Criconemoides morgensis, Hirschmanniella imamure, Meloidogyne arenaria, P. penetrans, H. dihystera, Tylenchorhynchus claytoni, Psilenchus hilarulus,* and *M. incognita* are found, but the root nematodes: *M. arenaria* and *M. incognita* are utmost common species among them in strawberry fields in Korea [55]. Similarly, the species of *Pratylenchus, Xiphinema, Helicotylenchus, Rotylenchus,* and *Ditylenchus* are associated with strawberries [56].

Some strawberry cultivars can be resistant or susceptible to nematode species. The cultivars: San Andreas, Monterey, Camino Real, Oso Grande, Aromas, and Albion) are resistant to *M. incognita, M. javanica, Pratylenchus zea,* and P. *brachyurus*. However, some strawberry cultivars, such as Camarosa are susceptible to *M. hapla* and *M. arenaria* [57].

## 4. Fungal disease on strawberry plants

*Colletotrichum acutatum* and *Botrytis cinerea* are the most common pathogens in the strawberry field. Among the fungal pathogens, *B. cinerea* causes significant economic losses in the strawberry industry. In wet conditions, more than 80% of strawberry flowers and fruit can be lost if plants are not sprayed with fungicide [58, 59]. Strawberry quarantine agents include *Colletotrichum acutatum, Botrytis cinerea,* and *Phytophthora* spp. It is included in the EPPO A2 list, but as of 2015, only *Phytophthora fragariae, Verticillium dahliae, Verticillium albo-atrum,* and *Fusarium oxysporum* are included in the final A2 list [60].

Common Leaf spot (*Mycosphaerella fragariae* [Tul.] Lindau): The leaf spot pathogen, *Mycosphaerella fragariae, also* recognized as Mycosphaerella leaf spot, "rust or white spot. This disease starts on leaves as purplish spots that look like leaf scorch. Plant vigor, fruit quality, and yield are reduced by leaf spot disease [61]. Typical symptoms are on the leaves small and circular leaf spots. Leaf lesions start as small, deep purple, irregular-shaped necrotic spots on the upper leaf surface. *M. fragariae* also causes spots on fruit, petioles and cause black seeds. Plants are mostly susceptible early in the growing season. Spores of the fungus form on the spots and are spread by rain, by farm implements, or on hands when plants are wet [61, 62]. The use of resistant varieties is the utmost practical and efficient method to control leaf spot disease. Timely applications of protective fungicides and non-infected nursery plants are suggested (**Figure 1**) [63].

Antracnose (*Cercospora fragariae*): Anthracnose leaf spot disease of strawberry which caused by *Colletotrichum acutatum, C. gloeosporioides,* and *C. fragariae* species. Lesions are circular, irregular purplish/reddish, with very light-colored centers seeming merely on the upper surface and small. The spots look like usual leaf spots (caused by *M. fragariae*) but are tinier with lighter centers and further uneven shapes [64]. Lesions are dark brown, almost black, and slightly sunken on petioles and fruit. Spore production appears on the leaf, in the white centers, which develop dotted with tiny dark stroma or knots of fungal cells. Infested buds and flowers may become dry and weakened. Spread of pathogen inoculums by rain and by irrigation as well as by movement of farmer beings and animals [65]. The control of leaf spots is very difficult; fungicides may not be very effective against anthracnose of strawberries. Anthracnose can be lessened using disease-free nursery plants and elimination of infected fruit and plants from fruiting fields followed by prompt fungicide applications subsequently each rain period. In addition, resistant cultivars should be used (**Figure 2**).

Powdery Mildew (*Sphaerotheca macularis f.sp. fragariae*): Powdery mildew, caused by the fungus *Sphaerotheca macularis f.sp. fragariae* results in purplish or reddish blotches on leaves and sometimes a powdery growth. This fungus is an obligate pathogen and survives only in the living tissues of its host. This pathogen affects all aboveground parts of (leaves, flowers, etc.) the strawberry plant. In susceptible cultivars, white Powderly mycelium mass develops on the lower leaf surface. Flowers and fruit in all stages of development are susceptible to attack. Infected mature fruit remains soft and pulpy. The major effect of this pathogen is that it weakens plants and reduces yields [61, 66, 67]. The application of protective fungicides is an effective method

**Figure 1.**
*Symptoms of common leaf spot (*Mycosphaerella fragariae *[Tul.] Lindau) in strawberry leaves.*

**Figure 2.**
*Symptoms of Antracnose (*Cercospora fragariae*) in strawberry plants.*

of control. In addition to controlling powdery mildew, to use of resistant cultivars recommends. Farmers should prefer plantings with disease-free plants and mowing as suggested to remove infected plants (**Figure 3**) [68].

Gray Mold (*Botrytis cinerea*): Gray mold caused Botrytis cinerea by is a significant disease in strawberry production, which utterly affects the yield and quality of strawberries. *B. cineria* affects fruit in the field, storage, transport, and market. This pathogen is also known as 'Botrytis rot fruit' which causes huge losses in the field (more than 80% loss) during rainy and cloudy periods, just before or during harvest and storage [69]. *B. cinerea* can live in soils such as sclerotia or mycelium. Infection usually begins in fruits that come into contact with the soil. Gray mold affects flowers and green or mature fruit. Infection begins in the flower and may enter the calyx or stem, later causing fruit rot [61]. Control measures that can be taken to diminish losses due to Botrytis fruit rot contain preventing excessive vegetative growth by regulating plant density, removing diseased fruits from the field, timely fertilization, harvesting the fruit before it is fully ripe to prevent injuries, and rapid transfer of the harvested fruits to the cold storage. Protective fungicide application is recommended during the flowering period (**Figure 4**) [70].

Fungal disease also damages the roots of strawberries. *Rhizoctonia solani, Fusarium oxysporum, Pythium* sp. cause root rot disease in strawberries. In addition to the difference in the disease factor in a region, factors such as root rot formation, accumulation of soil water, oxygen deficiency in the soil, and temperature were also found to be efficient in the growth of the disease [71]. Roots die as a result of prolonged water accumulation in the soil. Instead of these roots, short thick new rootlets are formed [71]. Root rot disease agents (*Rhizoctonia solani, Fusarium oxysporum, Pythium sp.*)

**Figure 3.**
*Symptoms of powdery mildew (*Sphaerotheca macularis f.sp. fragariae*) in strawberry plants.*

**Figure 4.**
*Symptoms of Gray Mold (*Botrytis cinerea*) in strawberry fruits.*

cause stagnation in development, shrinkage of leaf surfaces, shortening of leaf stalks are in the form of drying of leaves and wilting of plants as the disease progresses. In the sub-soil part, due to the disease, the hairy roots quickly turn black and rot. Easily peeling of the bark is one of the most typical features of the disease [71]. When the roots are damaged, there is a pause in plant growth, shortening, and shrinking of the leaves. As the disease progresses, the main roots turn black and rot. With the intense death of the hairy roots, the plant loses its vitality and efficiency and dies suddenly [71]. Macrophomina crown rot (*Macrophomina phaseolina*) is also an important strawberry disease that causes plant stunting, drying of older leaves, wilting of leaves, and discoloration, which are some of the symptoms in strawberries plants [71].

Phytophthora Crown Rot of Strawberry (*Phytophthora cactorum*) causes the collapse of plants, and dark red discoloration of the crown is seen. Plants are stunted, or young leaves are wilted as initial symptoms, the disease progresses, widespread necrosis appears that is homogeneously brown in tissues [71].

## 5. Bacterial disease on strawberry plants

*Xanthomonas fragariae* and *X. arboricola* pv *fragariae* are important diseases of strawberries.

### 5.1 *Xanthomonas fragariae* (*Xf*) Kennedy and King, 1962, Bacterial angular leaf spot

Angular leaf spot is a potentially threatening disease of strawberry. The causative agent was identified as *Xanthomonas fragariae* and was firstly reported in the USA [72]. *Xanthomonas fragariae* is regulated as a quarantine organism in most EU countries [73]. The disease starts with small water-soaked blotches on young leaves. These symptoms form angular spots. These spots are usually bordered by small veins. Observing the lesions appear dark green under light, but using transmitted light, they become translucent. In high humidity and high-temperature conditions (over 20°C), a sticky bacterial ooze forms on the leaves. Disease lesions may coalesce as they grow and then appear as irregular stains on the upper side of the leaf. Reddish-brown lesions then become necrotic. Vascular tissue in the trunk may also be infected [74]. Bacteria cause latent infections by moving systemically via the vascular system of the plant [75]. Infected plants become less productive. The disease can result in up to 10% crop loss in strawberry yield. Plants may even die in severe infections [76]. Plants that are infected systemically produce the first infected leaves, and they served as the primary inoculum source in newly planted fields [77]. Disease symptoms may be confused with fungal diseases such as *Mycosphaerella fragariae* and a new pathovar of *X. arboricola* pv. *fragariae* [78]. *Xanthomonas fragariae* overwinter in plant debris by serving as the source of infection. The bacterium is resistant to desiccation and can easily survive on dry leaves in the soil but not independently in the soil. It creates a secondary infection in moist conditions. Infection of plants occurs both passively and actively. It is spread through rain, irrigation water. Daytime temperatures of around 20°C and cold nights, combined with high humidity or the presence of water, provide a favorable environment for infection and disease to develop [79]. Bacteria may survive for up to 2 weeks on metal and wood materials. That is why agricultural machines may carry the pathogen during an important period of time if not suitably disinfected. Machinery contamination with the bacterial ooze may cause the spread of *X. fragariae* infections [80]. *X. fragariae* is gram-negative bacterium. It is rod-shaped (0.4 × 1.3 μm size), non-spore-forming, and non-capsulated bacterium. Most cells of the bacterium are non-motile, but some of them have a single polar flagellum. Colonies are circular, entire, convex in shape, and glistening, translucent to pale-yellow on beef-extract-peptone agar [81]. Direct isolation of the bacterium on artificial nutrient media is difficult because of very slow growth. Wilbrink's medium with nitrate (Wilbrink-N) is recommended for the most suitable growth medium for isolation of the bacteria [82, 83]. Rapid screening tests based on serological (e.g., indirect immunofluorescence, (ELISA), and molecular methods are used in diagnosis. For confirmation of diagnosis, positive results in serological and molecular tests should be obtained. Several polymerase chain reaction (PCR) detection tests have been improved targeting diverse loci of the bacterial genome [84, 85]. To approve the incidence of *X. fragariae* in symptomatic plant material and latent *X. fragariae* infections and several of these tests have also been used [84–90]. Planting using certified disease-free propagation material is recommended for preventing disease occurrence [91]. Using immune strawberry cultivars such as *F. moschata* instead of susceptible ones (*Potentilla fruticose, P. Glandulosa, F. vesca,* and *F. virginiana*) is recommended [92, 93]. Eliminating infected leaves is important to reduce inoculum sources of bacteria. Copper compounds can be used for the chemical control of *X. fragariae*. Because of resistance established by the bacterium, these compounds must be applied at higher concentrations [84, 91]. Streptomycin and oxytetracycline antibiotics have shown efficacy, but these treatments are not largely registered because of high cost and resistance problems. Induction of systemic resistance using analogs of salicylic acid is also another meaning of control, but still, new developed efficient methods are needed to control angular leaf spot disease of strawberry [84, 91].

### 5.2 *X. arboricola pv fragariae*, Bacterial leaf blight

Bacterial leaf blight disease is caused by *X. arboricola pv fragariae* (*Xaf*). The disease was first observed on strawberry plants in northern Italy in 1993 [94]. The causative agent was reported as a new pathogen [95]. *Xaf* was determined as a quarantine organism in 2002 [96]. In 2007, it was removed quarantine list by EPPO (The European and Mediterranean Plant Protection Organization) [97]. *Xaf* reported on strawberry plants in Turkey [98]. Both *Xaf* and *X. fragariae* were reported to cause angular leaf spot or bacterial leaf blight symptoms infections on strawberry tissue [99]. Early leaf lesions of *X. arboricola* pv *fragariae* were not water-soaked on the contrary of *X. fragariae*. Disease caused dry, brown leaf spots. These lesions are large brown colored and V-shaped along the leaf margin and veins. İnfected leaves completely become wilted and turn completely yellow colored. The disease did not affect flowers, peduncles, or fruits of strawberry plants [100]. *X. arboricola* pv *fragariae* (*Xaf*) is gram-negative bacterium. The bacteria are obligate aerobes. On NA growth medium, colonies are yellow, glistening, circular, convex, 1 mm diameter. On YPGA-medium, colonies are yellow, glistening, mucoid, convex, or pulvinate, 1–3 mm diameter. Bacterial cells are 1.7–1.9 μm length, 0.5–0.65 μm width in size. Bacteria causes soft rot of potato slices. It is negative in the arginine dihydrolase test and positive in inducing HR on tobacco plant. The maximum growth temperature was determined as 39°C. Assimilation of glycerol, D-trehalose, L-glutamic acid, maltose, L-fucose, succinic acid, cellobiose, Tween-80, and D-galactose are determined as positive. *Xanthomonas arboricola* pv. *fragariae* is positive in the hydrolyzation of gelatin, esculin, and starch. Molecular (Real-time PCR assay) and Serological (indirect immunofluorescence, ELISA) tests are developed for the detection of strawberry bacterial blight pathogen (*Xaf*). PCR test (Xaf pep) was designed by replicating the pep-prolyl endopeptidase gene region (unique to Xaf) [101]. There is no effective control method of X. *arboricola* pv *fragariae* (*Xaf*). Because of the latent infections, routine testing of strawberry plant propagation material is recommended for preventing possible disease occurrence [98].

## 6. Pests on strawberry plants

Many pests damage strawberry plants and important pests are included in this section.

*Frankliniella occidentalis* Perg., *F. intonsa* Trby. (Thysanopthera: Thripidae): *Frankliniella occidentalis* and *F. intonsa* are thought as the dominant species in strawberry fields. Depending on the temperature, it can give 22 offspring per year. After the thrips spend the winter on the soil and various plants as adults, they pass to the strawberry with the formation of flower buds. In particular, flower-time populations are increasing [102]. They begin to feed with the opening of flower buds in strawberries. They damage the strawberry flower and fruit by absorbing the plant sap. As a result of suction, flower drop, low yield, small and seed fruit formation, and tanning are observed. Fruits become deformed and lose their market value [103].

On the other hand, it causes secondary damage by infecting viruses such as Thrips stylets and Tomato Spotted Wilt Virus [104]. Since it reaches a high population in a short time, there are difficulties in chemical control [103]. Weed cleaning is important in strawberry fields as a cultural precaution in the control of thrips. *Orius sp.* (Both: Anthocoridae), *Coccinella septempunctata* (Col: Coccinellidae), *Syrphus sp.* (Dip: Syrphidae), *Chrysoperla carnea* (Neur.: Chrysopidae), *Adalia bipunctata* (Col: Coccinellidae) are known. The most effective predator is *Orius sp.* has been reported. Chemical control should be done when 10 thrips/flower is determined in the flower counts in strawberry fields [102].

*Tetranychus urticae* and *T. cinnabarinus* are some of the most common pests on strawberries (**Figure 5**).

Common names of them are used as two-spotted spider mites or greenhouse red spider mites [105]. With the warming of the air in the spring, the spider females, which pass from the surrounding weeds to the strawberries, lay their eggs in the web they weave on the lower surface of the leaves. Therefore, the density of the networks gives information about the population. Small yellow spots and tanning are seen on the damaged leaf because of feeding by the red spider. A female can lay 100–150

**Figure 5.**
*Adults and nymphs of* Tetranychus urticae *(left) and* T. cinnabarinus *on strawberry plant and* Tetranychus urticae *adult and egg on strawberry plant (right).*

eggs in her lifetime. It completes one offspring in 10–20 days. The strawberry plant is controlled, and the damaged leaves are removed from the environment as a cultural precaution. It can be counted as controlling weeds on the edge of the garden. Biological control of *P. persimilis, Neoseiulus fallacis* (Garman) (Acari: Phytoseiidae), *Neoseiulus californicus* (McGregor), and *Galendromus occidentalis* (Nesbitt); are crucially important [106–108].

*Anthonomus rubi* Herbst (Coleoptera: Curculionidae): Adults of *Anthonomus rubi,* which started to appear from the flower bud period of strawberries. It especially feeds on the young leaves and flowers of strawberries. Females lay 1 egg in an unopened bud. During egg-laying, it pierces the flower stem and prevents the circulation of the sap [105]. The flower bud it lays eggs does not develop, dries up, remains on the branch, and finally falls. This form of damage is unique to *A. rubi*. The main damage is caused by females cutting the flower bud stalks while laying eggs and feeding the larvae inside the bud. The egg hatches after 5–10 days. The larval period is 14–20 days. The pupal period is about 8 days [105]. Adults emerge by piercing the flower buds at the end of June and mid-July. After feeding for a few days, it enters the summer-winter diapause; the pest gives 1 offspring per year. In high populations, bud damage can be 5–90%, and yield loss can be 60% or more. The damage rate is higher in early varieties. The main host is strawberry. Raspberry, blackberry, rose, and wild rose from the Rosaceae family are other important hosts [105].

*Phytonemus pallidus* (Banks): *Phytonemus pallidus* (Banks) is a pest that causes serious yield losses in strawberries [109]. It causes damage by feeding on the newly emerging young leaves, especially in the crown of the strawberry plant (**Figure 6**). Hardening, wrinkling, discoloration, and a brittle structure are observed in the sucked leaves. An increasing population is stunted and decreases both in size and number of fruits [109]. However, it can cause the death of the strawberry plant [109].

Each female individual lays approximately 90 eggs under suitable conditions and becomes an adult from an egg in 2 weeks [109]. For this reason, it can reach dense populations in a short time. As a cultural precaution in control, using healthy plant material and alternating. It is important to pay attention to the cleanliness of the garden. Strawberry plants should be examined in the spring, especially young and mature leaves. The control method should be decided according to the population situation. In its biological control, *Amblyseius cucumeris* and *A. reticulatus* Oudemans (Acarina: Phytoseiidae) species are effective in reducing the population as predators [109]. If an average of 10 individuals per leaf is detected, recommended plant protection products should be used [110].

**Figure 6.**
Phytonemus pallidus *(banks) on strawberry plants.*

*Lygus spp.* (Hem.: Lygaeidae): *Lygus elisus* (Van Duzee), *L. hesperus* (Knight), and *L. lineolaris* (Palisot de Beauvois) have been reported as species that cause extensive damage to strawberries [109]. These species are polyphagous species that damage flowers and young fruits in strawberries. Deformations and formation of seeded strawberry fruits are seen in damaged fruits. Strawberries lose their market value in dense populations and cause great economic losses. Lygus population is more concentrated in strawberry fields with weeds. The pest, which spends the winter as an adult, gives 3–4 offspring depending on the conditions during the year. It takes 30–40 days from egg to adult. In the cultural control, foreign vote control should be carried out around the strawberry production area, and plant protection products should be used, which are recommended by paying attention to the pest population in the controls made before flowering in the chemical control [110].

*Spodoptera littoralis* (Boisduval) (Lepidoptera: Noctuidae): It is an important polyphagous pest species commonly found in the world [111–114]. Although the main host of the pest is cotton, it can cause economically significant losses in many industrial and field crops, vegetables, and fruits [112–116]. They damage the leaves and fruits of the strawberry. Especially in the case of products such as cotton and corn around the strawberry field, it causes an increase in the pest population [115].

*Chaetosiphon fragaefolli* (Hom.: Aphididae): In general, adults and nymphs of Aphids live in colonies near the veins in the crown, fresh shoot, leaf, and underside of the strawberry [109]. Adults and nymphs of aphids feed by sucking plant sap on strawberry leaves [109]. As a result of feeding, damage in the form of curls, deformities, and yellow spots occurs on leaves and fresh shoots. These pests spend the winter in the egg period they lay on the branches and shoots of fruit trees.

*Chaetosiphon fragaefolii* spends all stages of its life on strawberries, including over-wintered eggs, nymphs, wingless adults, and winged adults [105]. In addition, they cause fumagine by the development of saprophytic fungi on the sweet matter they secrete during feeding. By covering the plant surface with fumagine, the respiration of the plant is prevented. As a result of the inhibition of plant respiration, the development of the strawberry plant is weakened. As a result of adversely affecting plant development, yield and quality loss occur. For this reason, strawberry loses their market value. Cytorhabdoviruses such as strawberry aphid, Strawberry crinkle virus (CV), and Strawberry mild yellow edge virus (MYEV) cause significant damage by infecting healthy plants [117–119]. The main hosts of strawberry aphids are wild and cultivated strawberry plants (*Fragaria spp.*, *Potentilla anserina*, *F. virginianana, F. vesca*). In its control, clean seedlings should be used in the greenhouse and in the open field as cultural measures, Aphid-infested plants and weeds should be cleaned, Plant stems and weeds remaining on the ground after harvest should be destroyed. In terms of biological control, species of predators, especially Coccinellidae, Chrysopidae, and Syrphidae families, parasitoids *Aphidius spp.* are important natural enemies. Chemical control should be decided according to the population density [105].

Leafhopper (Cicadellidae): These species, which are polyphagous pests, cause damage by absorbing the plant sap from the vascular system of the plant. The toxic saliva they secrete causes injury to the plant. Short petiole and small leaf formation are seen in the damaged strawberry plant [109]. In addition, deformations in the mid-vein angle of the leaves are observed. These symptoms can be confused with Strawberry vein banding virus symptoms, but they are not the harmful Strawberry vein banding virus vector. Control: Leaf surface can be checked by visual inspection method, adults can be detected with methods such as sweep net, yellow sticky trap. In case of increased density, recommended insecticides can be used [109].

Cercopidae families of some species cause serious damage in areas with high humidity. Nymphs surround themselves with this substance by secreting a white foamy substance 1–2 cm wide. The pest that takes its name from this substance causes serious production losses [109]. If the pest is not controlled, the damage will continue until the harvest. Nymphs pierce plant roots and feed on plant sap. After feeding on the roots, it moves towards the green part of the strawberry. As a result of the feeding of nymphs, plant growth stops, small, irregularly shaped fruit formation is observed in strawberries. It causes a loss in yield. Insecticides should be used for chemical control [109].

*Otiorhynchus spp*., (Col.:Curculionidae): *Otiorrhynchus spp.* They lay their eggs on the soil surface, and the hatched larvae feed on strawberry roots and cause damage. *Otiorrhynchus spp*. Larval damage rate increases in sandy soils and plastic mulching. Adults also feed on strawberry leaves. Strawberry damaged because of feeding weakens, bushes, fades, and eventually, the plant dies. On the other hand, Black root rot pathogen infection can be seen in the feeding area [109]. As a cultural precaution in its control, it should avoid growing strawberries in contaminated areas. Plastic mulching should be avoided as it increases larval damage. Chemical control should be carried out in line with the recommended practices [110].

Various soil pests feed on the roots of strawberries and can cause plants to wilt and dry out. Among these pests (*Agriotes spp.), Tipula spp*. (leatherjackets), (Melolonthidae), centipedes (eg *Scutigerella spp.*), (*Agrotis spp.)* and (Noctuidae spp.). In order to decide on the control, it is necessary to have information about the populations of these pests in the soil. Strawberry plants can be recommended to be cultivated and ventilated before planting. Recommended insecticides should be used in chemical control [110].

In general, snails feed on plants at night and hide among plant wastes in the garden during the day. As a result of feeding, small and medium-sized deep holes are formed in ripe strawberry fruits. This sign of damage can be confused with other insect pests [109]. Generally, these holes are seen in silver color in snail damage. Snails generally become active depending on the soil temperature. Suitable for snail development in humid weather conditions. Elimination of favorable conditions for its survival as a cultural precaution in its control helps in the control of this pest. For example, rocks, wood, leaves, dry leaves, and excessive mulching provide shading. Fermented food traps and commercial food traps are used as biotechnical control [109].

*Drosophila suzukii* Matsumara 1931 (Dip: Drosophilidae): *Drosophila suzukii* is a polyphagous pest that was first identified in Japan in 1916. It damages all fruits with a soft structure, such as strawberries, during ripening and harvest [120]. Female individuals can easily lay eggs on the strawberry fruit, especially with their saw-like ovipositor (**Figure** 7). The larvae that emerge from the eggs feed on the fruit flesh of the strawberry and cause the main damage to the fruit (**Figure** 7). The larvae complete their development and become pupae on or outside the strawberry fruit. Later, adults emerge from the pupae (**Figure** 7). Damaged strawberry fruit.

softens and collapses (especially when you touch it, it feels empty) and loses its market value (**Figure** 7). The pest can multiply rapidly in a short time and reach a high population. *Drosophila suzukii* emerges from egg to adult in 8 days at 25°C. An average female lays 400–600 eggs [121]. Irrigation should be done with a drip irrigation system in gardens. Equipment that may cause the spread of the pest to other places should be kept clean [105]. Biological control: *Anthocoris nemoralis* (Hemiptera: Anthocoridae) and *Orius spp.* (Hemiptera: Anthocoridae) and as pupal parasitoids, *Pachycrepoideus vindemmiae* (Pteromalidae) and *Trichopria drosophilae* (Diapriidae) are known to be effective on *D. suzukii* population [105]. Biotechnical Control: Traps should be hung from the time of Strawberry Fruit coloration. As traps, 8–10 holes with a maximum size of 3 mm are drilled on 0.5 lt transparent plastic containers. 100–300 ml of apple cider vinegar is put into the plastic container. Strawberry is hung on its habitus with the help of support. Insecticides recommended for chemical control should be used [105].

**Figure 7.**
Drosophila suzukii *ovipositor structure and egg (left),* Drosophila suzukii *female and male adult individuals (middle), Drosophila suzukii damage and larvae in strawberry (right).*

## 7. Weeds in strawberry fields

As with other plants, strawberry like soil rich in water, air, nutrients, and organic matter. Therefore, the well-cultivated soil is an advantage for the development of weeds. Weeds are a serious problem, and weed control is one of the biggest challenges for strawberry growers. Because strawberry plants grow relatively slowly and are weak competitors, weeds can quickly invade strawberry fields [122]. Numerous annual and perennial weeds species cause damage to strawberry plants (**Tables 1** and **2**) [122].

| Scientific name | Family | References |
|---|---|---|
| **a. Annual broadleaf weeds** | | |
| *Amaranthus retroflexus* L. | Amaranthaceae | [122] |
| *Amaranthus albus* L. | Amaranthaceae | [122] |
| *Anagallis arvensis* L. | Primulaceae | [122] |
| *Capsella bursa- pastries* (L.) | Cruciferae | [122] |
| *Chenopodium* spp. | Chenopodiaceae | [122] |
| *Euphorbia* sp. | Euphorbiaceae | [122] |
| *Fumaria* spp. | Papaveraceae | [122] |
| *Galium aparine* L. | Rubiaceae | [122] |
| *Geranium* spp. | Geraniceae | [122] |
| *Heliotropium* spp. | Boraginaceae | [122] |
| *Lactuca serriola* L. | Composetea | [122] |
| *Lamium amplexicaule* L. | Labiatae | [122] |
| *Malva sylvestris* L. | Malveceae | [122] |
| *Matricaria chamomillia* L. | Composetea | [122] |
| *Melilotus officinalis* L. Ders. | Leguminosae | [122] |
| *Oxalis* sp. | Oxalidaceae | [122] |
| *Polygonum aviculare* L. | Polygonaceae | [122] |
| *Polygonum persicaria* L. | Polygonaceae | [122] |
| *Portulaca oleraceae* L. | Portulacaceae | [122] |
| *Raphanus raphanistrum* L. | Cruciferae | [122] |
| *Rumex* sp. | Polygonaceae | [122] |
| *Silybum marianum* (L) Gaertn | Composetea | [122] |
| *Solanum nigrum* L. | Solanaceae | [122] |
| *Sonchus* sp. | Composetea | [122] |
| *Stellaria media* (L.) Vill. | Caryophyllaceae | [122] |
| *Tribulus terrestris* L. | Zygophyllaceae | [122] |
| *Trifolium* spp. | Leguminosae | [122] |
| *Urtica urens* L. | Urticaceae | [122] |
| *Vicia* spp. | Leguminosae | [122] |

| Scientific name | Family | References |
|---|---|---|
| **b. Annual narrow-leaf weeds** | | |
| *Allopecurus myosuroides* Huds. | Graminaceae | [122] |
| *Echinochloa crus gali* L. | Graminaceae | [122] |
| *Echinochloa colonum* L. | Graminaceae | [122] |
| *Poa annua* L. | Graminaceae | [122] |
| *Poa trivalis* L. | Graminaceae | [122] |

**Table 1.**
*Annual weeds that are problems in strawberry fields.*

| Scientific name name | Family | References |
|---|---|---|
| **a. Perennial broadleaf weeds** | | |
| *Convolvulus arvensis* L. | Convolvulaceae | [122] |
| **b. Perennial narrow-leaf weeds** | | |
| *Cynodon dactylon* L. | Graminaceae | [122] |
| *Cyperus rotundus* L. | Cyperaceae | [122] |
| *Digitaria sanguinalis* L. | Graminaceae | [122] |
| *Sorghum halopense* L. | Graminaceae | [122] |

**Table 2.**
*Perennial weeds that are a problem in strawberry fields.*

Since it is a high-value product, a well-integrated control, that is, cultural, mechanical, and physical control, should be carried out together for strawberry growers for weed control [123]. Crop rotation is an important part of the weed control program in many crops. Since we use different soil treatments and different herbicides when we plant crop plants alternately, it has an important role in the control of annual and perennial weeds [124]. Before using the agricultural tools and machines we use in another area, they should be cleaned, and the transportation of weed seeds should be prevented. A carried weed seed will increase in number and spread exponentially in the following years [124]. Transportation of weed seeds with irrigation water may occur [125]. In order not to carry weed seeds on the sides of the canal to the area where we cultivate, weeds should be controlled, and their transportation should be prevented [124]. Mechanical methods of weed control include manual weeding, hoeing, tillage between rows, and mowing [124]. Black plastic mulch controls most weeds; however, black mulch usually does not warm the soil as much as clear mulch. Clear mulch provides earliness with soil warming, but clear mulch does not control weeds [126].

## 8. Virus diseases on strawberry plants

Over 30 viruses and virus-like diseases distressing the genus, Fragaria has been reported. Some of these viruses have different races within themselves. Strawberry varieties can cause symptoms of different severity for each race, or these disease agents can be found asymptomatically in plants. While many viruses do not show obvious symptoms in commercial strawberry cultivars, frequently observed symptoms can be observed as plant stunting, crop and yield loss, and dieback. The viruses seen in strawberries can be found in mixed infections, causing reductions in yield and fruit quality, thereby reducing the market value of the product. The most reliable method used to detect the presence of strawberry viruses is the classical molecular and biological method. It is the use of classical clone grafting, aphid transport, and PCR methods on indicator plants of *F. vesca* and *F. virginiana* clones. However, since symptom outputs take 14–21 days in this method, the use of indexing after less time-consuming and reliable methods such as Enzyme-Linked Immunosorbent Assay (ELISA) and Polymerase Chain Reaction-Polymerase Chain Reaction (PCR) helps to obtain more reliable results [127, 128]. Strawberry viruses can be transmitted by aphids, nematodes, and some other vectors, while aphids are the most important vectors. Strawberry aphids known to infect the plants by being carried by *Chaetosiphon fragaefolii* viruses are [129] *Strawberry crinkle cytorhabdovirus* (ScRV), *Strawberry mottle virus* (SMoV), *Tomato ringspot virus* (ToRSV), *Strawberry vein banding virus* (SVBV), *Strawberry pseudo mild yellow edge virus* (SPMYEV), *Raspberry ringspot virus* (RpRSV), *Arabis mosaic virus* (ArMV), *Strawberry latent ringspot virus* (SLRSV), *Strawberry latent C virus* (STLCV), *Tomato black ring virus* (TBRV), and *Strawberry mild yellow edge virus* (SMYEV), which are known to infect the plant by being transmitted by some nematode species, and whiteflies [127, 130–132]. Among the aphid-borne viruses, SCrV, SMOV, SVBV, and SMYEV are the utmost significant viral diseases observed in strawberry production areas [133–135].

*Strawberry crinkle cytorhabdovirus* (ScRV); The family Rhabdoviridae is included in the genus Cytorhabdovirus. They are positive-sense ssRNA viruses [136]. *Strawberry latent virus*, strains A (mild form), and B and *Strawberry vein chlorosis virus* are synonyms. It is one of the most harmful strawberry viruses worldwide. Severe strains in susceptible varieties cause leaflets to be uneven in size, distorted and wrinkled, resulting in the formation of small irregularly shaped chlorotic spots on the veins. It is on the EPPO quarantine list. The presence of the agent has been reported in Asia, Africa, America, Oceania, and many European Union member countries [132, 137].

SCrV has a limited host variety within the Fragaria species. In addition to cultivated strawberries such as *F. x ananassa*, its presence has also been determined in wild species such as *F. vesca*, *F. virginiana,* and *F. chiloensis* [138]. Nicotiana glutinosa has been reported as experimentally transduced hosts in *Physalis floridana* [139]. SCrV is locally transmitted by the strawberry aphid *Chaetosiphon fragaefolii*. It has been reported that the shoot tip meristem culture method, following the application of temperature (38°C) in obtaining SCrV-free plants, increases the percentage of success in obtaining a virus-free plant [140].

*Strawberry mottle virus* (SMoV), *Strawberry mottle virus* (SMoV) from the Secoviridae family, is also called *Strawberry mottle sadwavirus*. The presence of the agent has been reported in Asia, America, Oceania, and many European Union member countries. It is one of the most common viruses in many areas of strawberry cultivation in the world. There are many breeds of SMoV; weak breeds are observed as asymptomatic, while strong breeds can cause yield losses of up to 30% Strawberry aphid, *Chaetosiphon fragaefolii*, is the main vector for SMoV, while *Chaetosiphon jacobi* and *C. minor* can also transmit the virus; *C. gossypii* can also carry SMoV. The transport of SMoV occurs semi-persistently within a few minutes during the feeding period [141]. *Fragaria vesca* and *F. virginiana* clones show symptoms at different rates after inoculation with inoculation due to their sensitivity to the agent. While symptom development is observed in *F. vesca* 7–10 days after infection, this period might be lengthier in mild and moderate breeds. In indicator plants, the symptoms are barely noticeable on the leaves, or the leaves may be slightly mottled, severely stunted, and deformations leading to plant death can be observed. In the control of the agent, the use of certified virus-free plants, the fight against aphids, the isolation of infected production areas is important. Since SMoV is one of the most sensitive to temperature among strawberry viruses, it is possible to obtain virus-free plants with the combination of thermotherapy and meristem culture. It has been reported that 2–3 weeks of thermotherapy is sufficient to obtain SMoV-free plants [127].

*Strawberry vein banding caulimovirus* (SVBV); the virus exists merely in Fragaria spp. Its chief host *is Fragaria vesca*. It can also infect commercial strawberry cultivars, but symptoms are commonly merely seeming when strawberry exists at the same time as the latent *C 'rhabdovirus'* [137]. The agent exists in some countries in Asia, Europe, America and Oceania, and *M. persicae, Macrosiphum rosae, Amphorophora rubi*, *Chaetosiphon fragaefolii, Aulacorthum solani, C. tetrarhodum, C. thomasi, C. jacobi, A. rubifolii, M. ornatus, Aphis idaei*, *Acyrthosiphon pelargonii* have been reported to be vectors of the agent. The most effective vector of these species is *Chaetosiphon* spp. While the virus can be transmitted to indicator plants by inoculation and *Cuscuta subinclusa*, it cannot be transmitted mechanically. Depending on the indicator plant, contamination occurs within 2–5 weeks. Symptoms appear as epinastie on the midrib and petiole on the youngest leaf. Some or all the affected leaves show varying lengths of yellowish vein banding along the main vein. The second and third leaves formed after the commencement of symptoms show more severe symptoms. In the diagnosis of the agent, the UC-12 clone of *F. vesca* and the UC-12 clone of *F. virginiana* are used as the most effective indicators for detecting SVBV. In routine serological testing, the agent is diagnosed using an antiserum specific for SVBV [129].

*Strawberry latent C 'rhabdovirus'* (STLCV); The agent reported to be in the Rhabdoviridae family has not been defined morphologically. Its presence has been reported in America and Canada [137]. While the agent is usually asymptomatic in cultivated strawberries, in case of mixed infection with other viruses, it can cause moderate to severe leaf deformation such as excessive stunting, curling of leaves, or severe symptoms observed in other viruses with weakening of the plant. In some indicator clones of STLCV *F. vesca*, it can cause severe epinasty and shrinkage in newly formed leaves and petioles, while mild and transient symptoms are observed in other clones. Determination of the presence of the agent can be accomplished by inoculating the agent into clones of *F. vesca* or *F. virginiana* [129]. It has been reported that in the USA, no other virus component in a complex form cause severe stunting in a short time as this factor in cultured strawberries [142].

Strawberry mild yellow edge disease: It has been determined to be caused by a virus complex called *Strawberry mild yellow edge luteovirus* (race or syn; *Soybean dwarf luteovirus*) and *Strawberry mild yellow edge-associated potexvirus* [143]. By itself, it is not the principal pest for most cultivars, but solitary infection rarely occurs. The agent is only Fragaria spp. There are types. While some clones of wild species *F. chiloensis, F. vesca,* and *F. virginiana* show symptoms in nature, *F. ovalis* can carry the agent without symptoms. Many strawberry cultivars are also asymptomatic carriers of the agent. It has been reported that the so-*called Strawberry mild yellow edge-associated 'potexvirus'* virus was experimentally transferred to *Chenopodium murale* and *C. quinoa* but did not remain in the plants for a long time [144].

This disease can be seen in many countries where the strawberry plant is grown. These are Australia, Europe, Israel, Japan, South Africa, and Northwest America. There are different views on economic losses due to the existence of different strains of the virus. However, because of many studies, it has been reported that product loss is between 0 and 30%. It is asymptomatic in cultivated strawberry cultivars [145]. In natural conditions, these two viruses in strawberries have been reported to be spread by the strawberry aphid *Chaetosiphon fragaefolii*. The spread of the disease can also occur with vectors or material spread from tissue culture. There is no information about propagation by seed. It has been reported that the complex of the disease with other pathogens such as *Strawberry crinkle rhabdovirus, Strawberry veinbanding caulimovirus*, *Strawberry mottle agent* [137] can cause serious losses in plant growth, fruit quality, and yield [145]. The control of the virus can be achieved with the use of thermotherapy or meristem culture and certified virus-free production material. Control of vectors is also an important factor in the prevention of disease agents [127, 145].

Viruses Carried by Nematodes; *Arabis mosaic nepovirus* (ArMV); the agent in the Nepovirus genus of the Comoviridae family is also called *Raspberry yellow dwarf virus.* ArMV is an RNA virus with a wide host range. The presence of the agent has been reported in Asia, Africa, America, Oceania, and many European Union member countries [146, 147]. ArMV is an RNA virus with a wide host range. It was observed that 93 species from 28 different families were infected by mechanical inoculation [148]. The main hosts are strawberry, hops, raspberry (*Rubus idaeus*), *Sambucus nigra, Rheum* spp., and *Vitis* spp. The virus has also been reported in sugar beet, celery, gladiolus, horseradish, and lettuce. Several further wild and cultivated species are reported as hosts. ArMV is transmitted by seed by nematodes over short distances [149]. *Xiphinema diversicaudatum* was suspected to carry hop strains of ArMV [150].

In the diagnosis of the agent, some indicator plants are used because they produce typical symptoms. *Chenopodium quinoa* and *C. amaranticolor* produce systemic blotches that follow chlorotic local lesions [151]. *Cucumis sativus* may produce chlorotic local lesions and systemic vascular banding or yellow spots in infected cotyledons. *Phaseolus vulgaris* shows symptoms as chlorotic local lesions, systemic necrosis, and deteriorations, *Petunia hybrida* as local chlorotic lesions, tiny necrotic rings, streaks, or vascular opening. However, these symptoms are more pronounced, especially in the hops variant (ArMV-H). ArMV is mostly seen in mixed infection with *Strawberry latent ringspot nepovirus* (SLRSV), which is also a nematode-transmitted agent [137]. Diagnosis of the agent is mostly made by ELISA. The presence of ArMV in a single nematode can also be detected by electron microscopy. Another way to detect ArMV is to use cDNA clones in dot hybridization tests [152, 153]. Distribution of virus-free production material for the control of ArMV with a strict certification program. In areas with infected nematode-vector populations, fallow and/or soil fumigation for at least a year is required, as replanting virus-free material without additional precautions will be ineffective [149].

*Raspberry ringspot nepovirus*; *Raspberry ringspot virus* (RPRSV), which is in the genus Nepovirus of the family Secoviridae, is a single-stranded RNA virus. Asia (Kazakhstan, Uzbekistan) and Europe (France, Serbia, Romania, Germany, Norway, Bulgaria, Italy, Greece, Denmark, Hungary, Ireland, Estonia, Latvia, Czech Republic, Luxembourg, Poland, Austria, Finland, Portugal, Belgium, Russia, Spain, Albania, Switzerland, Turkey, Ukraine, England) have also been reported [147, 154]. However, the main host is *Rubus idaeus, Fragaria* spp., Fragaria x ananassa, some Rubus spp., and Prunus spp. species. *Vitis vinifera* is another important host. Experimentally, It has also been reported to be transmitted by *Chenopodium giganteum,* Cucurbita spp., Nicotiana, Petunia spp., *Solanum lycopersicum*, *Spinacia oleracea, Vigna unguiculata* species [155]. The agent can be transmitted to some herbaceous plants mechanically and by infected seeds. RpRSV can be transmitted by both species of the nematode genus Longidorus. Scottish and Dutch variants of RpRSV are most effectively transmitted by *L. elongates* [156], while the English variant is transmitted by *L. macrosoma* [157]. Other nematode species (*Xiphinema diversicaudatum* and other Longidorus species) were suspected to carry variants of RpRSV, but transmissions were not considered acceptable [158]. Although the symptoms of RpRSV in strawberries vary according to the season and the breed of the agent, it can usually result in severe stunting and death. In *Fragaria vesca,* seedlings show yellow spots in the first year of infection but no spots after that. While symptoms vary in susceptible varieties, symptoms are less common in summer at high temperatures. RpRSV causes a serious disease that reduces both growth and fruiting and even kills plants. The use of fumigants such as dazomet or dichloropropane-dichloropropene for vector nematodes in raspberry and strawberry production areas provides prevention and control of virus transmission. Rubus spp., Ribes spp., and Fragaria species, the use of healthy certified production material free from RpRSV are one of the best control methods [127, 159, 160].

*Strawberry latent ringspot 'nepovirus'* (SLRSV); SLRSV has a very large host range. The agent, which is mostly found latent in berry fruits such as strawberries and raspberries, can sometimes cause infections resulting in mottling and back-drying [161]. It is found naturally in many berry species as well as in cherries, grapes, plums, peaches, *Sambucus nigra*, asparagus, celery, gladiolus, daffodils, and roses, as well as in many wild species, usually asymptomatic [162]. The presence of the agent has been reported in some countries in Europe, Oceania, North America, and in Israel and Turkey from Asian countries. SLRSV can be mechanically transported to herbaceous plants. It is reported that naturally, both larvae and adults of *Xiphinema diversicaudatum* can carry SLRSV for up to 84 days and up to 70% of seed transmission in several plant species [163]. While the causative agent is usually asymptomatic, varying degrees of mottling and retrograde death can be observed in some strawberry cultivars. Reliable diagnosis is possible with SLRSV specific antisera. *Chenopodium murale*, *C. quinoa*, and *C. amaranticolor* show symptoms as systemic chlorosis, necrotic or chlorotic local lesions, and sometimes pale chlorotic mottling or necrosis. Cucumber shows local lesions in the form of systemic intervascular chlorosis or necrosis or shows no symptoms at all. During the summer, the later leaves are asymptomatic but contain the virus, while in the winter, the symptoms may persist on the newly arrived leaves. *Nicotiana rustica*, *Nicotiana tabacum*, and *Petunia hybrida* are infected without symptoms. The virus can sometimes be found in soil together with *Arabis mosaic nepovirus*, which is also carried by *X. diversicaudatum* [137].

## 9. Conclusion

Several diseases, nematodes, weeds, and pests cause damage to strawberry foliar, roots, and fruits, and are responsible for crop losses. Numerous plant-parasitic nematodes cause damage on strawberry plants that some feed on roots, others in foliar parts. Parasitic nematodes cause yield losses, crop size, and quality. Fungal pathogens such as *Colletotrichum acutatum* and *Botrytis cinerea* are the most common pathogens in the strawberry field and may cause damage to more than 80% of strawberry flowers and fruits. *Xanthomonas fragariae* and *X. arboricola* pv *fragariae* are important bacterial diseases of strawberries. Many pests such as *Frankliniella occidentalis* damage strawberry plants and other important pests are included in this section. Weeds are also a serious problem, and weed control is one of the biggest challenges for strawberry growers. Because strawberry plants grow relatively slowly and are weak competitors, weeds can quickly invade strawberry fields. Other important diseases are viruses and virus-like diseases that some of these viruses can cause symptoms of different severity for each race, or these disease agents can be found asymptomatically in plants. In this chapter, information about diseases, pests, weeds, and nematodes that cause yield loss in strawberry plants are brought together. Thus, it is a significant chapter for the benefit of producers, researchers, and students. It is also important to apply integrated pest management strategies to control diseases and pests in strawberry plant cultivation.

## Conflict of interest

The authors declare no conflict of interest.

## Author details

Refik Bozbuga[1*], Selman Uluisik[2†], Pınar Aridici Kara[3†], Semiha Yuceer[3†], Hale Gunacti[3†], Pakize Gok Guler[3†], Elen Ince[3†], Hatice Nilufer Yildiz[3†] and Ozcan Tetik[3†]

1 Faculty of Agriculture, Plant Protection Department, Eskisehir Osmangazi University, Eskisehir, Turkey

2 Burdur Food Agriculture and Livestock Vocational School, Burdur Mehmet Akif Ersoy University, Burdur, Turkey

3 Biological Control Research Institute, Adana, Turkey

*Address all correspondence to: refik.bozbuga@ogu.edu.tr

† Authors contributed equally to this work. Fruit ripening (by S Uluisik), plant parasitic nematodes (by R Bozbuga), fungal diseases (by H Gunacti and S Yuceer), weeds (by O Tetik), pests (by PA Kara), virus diseases (by PG Guler and E Ince), bacterial diseases (by HN Yildiz) in strawberry plants are written in this book chapter.

## References

[1] Giovannoni J. Molecular biology of fruit maturation and ripening. Annual Review of Plant Physiology and Plant Molecular Biology. 2001;**52**:725-749

[2] Alexander L, Grierson D. Ethylene biosynthesis and action in tomato: A model for climacteric fruit ripening. Journal of Experimental Botany. 2002;**53**:2039-2055

[3] Merchante C, Vallarino JG, Osorio S, Aragüez I, Villarreal N, Ariza MT, et al. Ethylene is involved in strawberry fruit ripening in an organ-specific manner. Journal of Experimental Botany. 2013;**64**:4421-4439

[4] Li C, Jia H, Chai Y, Shen Y. Abscisic acid perception and signaling transduction in strawberry: A model for non-climacteric fruit ripening. Plant Signaling & Behavior. 2011;**6**:1950-1953

[5] Given NK, Venis MA, Gierson D. Hormonal regulation of ripening in the strawberry, a non-climacteric fruit. Planta. 1988;**174**:402-406

[6] Osorio CFS, Gutierrez JR, Kitamura J, Giavalisco P, Nakajima M, Fernie AR, et al. Medina-Escobar gibberellin biosynthesis and signaling during development of the strawberry receptacle. The New Phytologist. 2011;**191**:376-390

[7] Concha CM, Figueroa NE, Poblete LA, Oñate FA, Schwab W. Figueroa CR methyl jasmonate treatment induces changes in fruit ripening by modifying the expression of several ripening genes in *Fragaria chiloensis* fruit. Plant Physiology and Biochemistry. 2013;**70**:433-444

[8] Chai YM, Zhang Q, Tian L, Li CL, Xing Y, Qin L, et al. Brassinosteroid is involved in strawberry fruit ripening. Plant Growth Regulation. 2013;**69**:63-69

[9] Gu TT, Jia SF, Huang XR, Wang L, Fu WM, Huo GT, et al. Transcriptome and hormone analyses provide insights into hormonal regulation in strawberry ripening. Planta. 2019;**250**:145-162

[10] Symons GM et al. Hormonal changes during non-climacteric ripening in strawberry. Journal of Experimental Botany. 2012;**63**:4741-4750

[11] Estrada-Johnson E, Csukasi F, Pizarro CM, Vallarino JG, Kiryakova Y, Vioque A, et al. Transcriptomic analysis in strawberry fruits reveals active auxin biosynthesis and signaling in the ripe receptacle. Frontiers in Plant Science. 2017;**8**:889

[12] Aharoni A, O'Connell AP. Gene expression analysis of strawberry achene and receptacle maturation using DNA microarrays. Journal of Experimental Botany. 2002;**53**:2073-2087

[13] Nardi CF, Villarreal NM, Opazo MC, Martinez GA, Moya León MA, Civello PM. Expression of FaXTH1 and FaXTH2 genes in strawberry fruit. Cloning of promoter regions and effect of plant growth regulators. Scientia Horticulturae. 2014;**165**:111-122

[14] Liu DJ, Chen J, Lu W. Expression, and regulation of the early auxin-responsive Aux/IAA genes during strawberry fruit development. Molecular Biology Reports. 2011;**38**:1187-1193

[15] Kim J, Lee JG, Lee EJ. Analysis of eight phytohormone concentrations, expression levels of ABA biosynthesis genes, and ripening-related transcription

factors during fruit development in strawberry. Journal of Plant Physiology. 2019;**239**:52-60

[16] Shen YY, Rose JKC. ABA metabolism and signaling in fleshy fruits. In: Zhang DP, editor. Abscisic Acid: Metabolism, Transport and Signaling. Dordrecht: Springer; 2014. pp. 271-286

[17] Luo Y, Ge C, Ling Y, et al. ABA and sucrose co-regulate strawberry fruit ripening and show inhibition of glycolysis. Molecular Genetics and Genomics. 2020;**295**:421-438. DOI: 10.1007/s00438-019-01629-w

[18] Siebeneichler TJ, Crize RL, Camozatto GH, Paim BT, Messias RS, Rombaldi CV, et al. The postharvest ripening of strawberry fruits induced by abscisic acid and sucrose differs from their in vivo ripening. Food Chemistry. 2020;**317**:126407

[19] Molina-Hidalgo FJ, Medina-Puche L, Cañete-Gomez C, Franco-Zorrilla JM, Lopez-Vidriero I, Solano R, et al. The fruit specific transcription factor FaDOF2 regulates the production of eugenol in ripe fruit receptacles. Journal of Experimental Botany. 2017;**68**:4529-4543. DOI: 10.1093/jxb/erx257

[20] Jia HF, Chai YM, Li CL, Lu D, Luo JJ, Qin L, et al. Abscisic acid plays an important role in the regulation of strawberry fruit ripening. Plant Physiology. 2011;**157**:188-199. DOI: 10.1104/pp.111.177311

[21] Liao X, Li M, Liu B, Yan M, Yu X, Zi H, et al. Interlinked regulatory loops of ABA catabolism and biosynthesis coordinate fruit growth and ripening in woodland strawberry. PNAS. 2018;**115**:E11542-E11550

[22] Ozkan Y, Ucar M, Ozturk B. Pre-harvest gibberellic acid (GA3) treatments play an important role on bioactive compounds and fruit quality of sweet cherry cultivars. Scientia Horticulturae. 2016;**211**:358-362

[23] Gundogdu M, Berk S, Yıldız K, et al. Influence of foliar application with gibberellic acid on phenolic and bioactive compounds of strawberry fruits. Erwerbs-obstbau. 2021;**63**:15-23. DOI: 10.1007/s10341-021-00543-z

[24] Trainotti L, Pavanello A, Casadoro G. Different ethylene receptors show an increased expression during the ripening of strawberry: Does such an increment imply a role for ethylene in the ripening of these non-climacteric fruit? Journal of Experimental Botany. 2005;**56**:2037-2046

[25] Villarreal NM, Bustamante CA, Civello PM, Martínez GA. Effect of ethylene and 1-MCP treatments on strawberry fruit ripening. Journal of the Science of Food and Agriculture. 2010;**90**:683-689. DOI: 10.1002/jsfa.3868

[26] Villarreal NM, Marina M, Nardi CF, Civello P, Martinez GA. Novel insights of ethylene role in strawberry cell wall metabolism. Plant Science. 2016;**252**:1-11. DOI: 10.1016/j.plantsci.2016.06.018

[27] Lopes PZ, Fornazzari IM, Almeida AT, Galvão CW, Etto RM, Inaba J, et al. Effect of ethylene treatment on phytochemical and ethylene-related gene expression during ripening in strawberry fruit Fragaria x ananassa cv. Camino Real. Genetics and Molecular Research. 2015;**14**(4):16113-16125

[28] Elmi F, Pradas I, Tosetti R, Cools K, Terry LA. Effect of ethylene on postharvest strawberry fruit tissue biochemistry. Acta Horticulturae. 2017;**1156**:667-672. DOI: 10.17660/ActaHortic.2017.1156.97

[29] Hirch M, Langer SE, Marina M, Rosli HG, Civello PM, Martinez GA, et al. Expression profiling of endo-xylanases during ripening of strawberry cultivars with contrasting softening rates. Influence of postharvest and hormonal treatments. Journal of the Science of Food and Agriculture. 2020;**101**:3676-3684

[30] Reis L, Forney CF, Jordan M, Pennell KM, Filmore S, Schemberger MO, et al. Metabolic profile of strawberry fruit ripened on the plant following treatment with an ethylene elicitor or inhibitor. Frontiers in Plant Science. 2020;**11**:995. DOI: 10.3389/fpls.2020.00995

[31] Nambeesan S, Mattoo AK, Handa AK. Polyamines and regulation of ripening and senescence. In: Paliyath G, Murr DP, Handa AK, Lurie S, editors. Postharvest Biology and Technology of Fruits, Vegetables, and Flowers. Wiley-Blackwell, Ames, IA; 2008. pp. 319-340

[32] Jiaxuan G, Wang S, Xiaoyang Y, Dong R, Li Y, Mei X, et al. Polyamines regulate strawberry fruit ripening by abscisic acid, auxin, and ethylene. Plant Physiology. 2018;**177**(1):339-351. DOI: 10.1104/pp.18.00245

[33] Mo A, Xu T, Bai Q, Shen Y, Gao F, Guo J. FaPAO5 regulates Spm/Spd levels as a signaling during strawberry fruit ripening. Plant Direct. 2020;**4**(5):e00217

[34] Haghshenas M, Nazarideljou M, Shokoohian A. Phytochemical and quality attributes of strawberry fruit under osmotic stress of nutrient solution and foliar application of putrescine and salicylic acid. International Journal of Horticultural Science and Technology. 2020;7(3):263-278. DOI: 10.22059/ijhst.2020.298283.347

[35] Abd-Elgawad MMM. Plant-parasitic nematodes of strawberry in Egypt: A review. Bulletin of the National Research Centre. 2019;**43**:7. DOI: 10.1186/s42269-019-0049-2

[36] Decraemer W, Hunt DJ. Structure and classification. In: Perry RN, Moens M, editors. Plant Nematology. CABI Publishing: Ghent; 2013. pp. 3-39

[37] Samaliev HY, Mohamedova M. Plant-parasitic nematodes associated with strawberry (*Fragaria aiianassa* Duch.) in Bulgaria. Bulgarian Journal of Agricultural Science, Agricultural Academy. 2011;**17**(6):730-735

[38] LaMondia JA. Interaction of *Pratylenchus penetrans* and *Rhizoctonia fragariae* in strawberry black root rot. Journal of Nematology. 2003;**35**:17-22

[39] Nyoike TW, Mekete T, McSorley R, Weibelzahl-Karigi E, Liburd OE. Confirmation of *Meloidogyne hapla* on strawberry in Florida using molecular and morphological techniques. Nematropica. 2012;**42**:253-259

[40] Watson TT, Noling JW, Desaeger JA. Fluopyram as a rescue nematicide for managing sting nematode (*Belonolaimus longicaudatus*) on commercial strawberry in Florida. Crop Protect. 2020;**132**:105108. DOI: 10.1016/j.cropro.2020.105108

[41] Watson TT, Desaeger JA. Evaluation of non-fumigant chemical and biological nematicides for strawberry production in Florida. Crop Protect. 2019;**117**:100-107. DOI: 10.1016/j.cropro.2018.11.019

[42] Dias-Arieira CR, Molina R d O, Costa AT. Nematoides' causadores de doenças em frutíferas. Revista Agro@Mbiente On-line. 2008;**2**:46-56. DOI: 10.18227/1982-8470ragro.v2i1.230

[43] Gomes CB, Cofcewicz ET. Nematoides. In: Fortes JF, Osorio VA.

(Eds.), Morango: Fitossanidade, Frutas Do Brasil. Empresa Brasileira de Pesquisa Agropecuária, Brasília. 2003; p. 36

[44] Silveira SGP, da Curi SM, Elias R, Prates HS. Levantamento de nematoidé *Meloidogyne hapla* na cultura do morangueiro no estado de São Paulo. Pesquisa Agropecuária Brasileira. 1989;**24**:583-586

[45] Souza JT, de Campos VP, Maximiniano C. Ocorrência e distribuicao de nematoides associados a hortalicas e plantas medicinais. Summa Phytopathologica. 1998;**24**:283-291

[46] LaMondia JA, Cowles RS. Comparison of *Pratylenchus penetrans* infection and *Maladera castanea* feeding on strawberry root rot. Journal of Nematology. 2005;**37**:131-135

[47] Bozbuga R, Lilley CJ, Knox JP, Urwin PE. Host-specific signatures of the cell wall changes induced by the plant parasitic nematode, *Meloidogyne incognita*. Scientific Reports. 2018;**8**:17302. DOI: 10.1038/s41598-018-35529-7

[48] Caillaud MC, Dubreuil G, Quentin M, Perfus-Barbeoch L, Lecomte P, de Almeida EJ, et al. Root-knot nematodes manipulate plant cell functions during a compatible interaction. Journal of Plant Physiology. 2008;**165**:104-113. DOI: 10.1016/j.jplph.2007.05.007

[49] Jones JT, Haegeman A, Danchin EGJ, Gaur HS, Helder J, Jones MGK, et al. Top 10 plant-parasitic nematodes in molecular plant pathology. Molecular Plant Pathology. 2013;**14**:946-961. DOI: 10.1111/mpp.12057

[50] Chen P, Tsay TT. Effect of crop rotation on *Meloidogyne* spp. and *Pratylenchus* spp. populations in strawberry fields in Taiwan. Journal of Nematology. 2006;**38**:339-344

[51] LaMondia JA. Influence of rotation crops on the strawberry pathogens Pratylenchus penetrans, *Meloidogyne hapla*, and *Rhizoctonia fragariae*. Journal of Nematology. 1999;**31**:650-655

[52] Krezanoski CE, Hahn MH, Maceda A, Zawadneak MAC, Calegario RF, Duarte HSS. Plant-parasitic nematode survey of strawberry in Paraná, Brazil. Crop Protection. 2020;**138**:105338. DOI: 10.1016/j.cropro.2020.105338

[53] Talavera M, Miranda L, Gomez-Mora JA, Vela MD, Verdejo-Lucas S. Nematode management in the strawberry fields of southern Spain. Agronomy. 2019;**9**:17. DOI: 10.3390/agronomy9050252

[54] Vega JM, Paez J, Lopez-Aranda JM, Medina JJ, Miranda L, Montes F. Nematodes on strawberries in Southern Spain: Effects on yield and control of *Meloidogyne hapla*. Nematology. 2002;**4**:309

[55] Park SD, Khan Z, Yeon IK, Kim YH. A survey for plant-parasitic nematodes associated with strawberry (*Fragaria ananassa* Duch.) crop in Korea. Plant Pathology Journal. 2005;**21**:387-390

[56] Poiras L, Cernet A, Buvol A, Poiras N, Iurcu-Straistraru E. Preliminary analysis of plant parasitic nematodes associated with strawberry and raspberry crops in the Republic of Moldova. Muzeul Olteniei Craiova Oltenia Studii şi comunicări Ştiinţele Naturii. 2014;**30**:98-100

[57] de Brum D, Marchi PM, Gonçalves MA, Cruz FF, Antunes LEC, Gomes CB. Reaction of strawberry cultivars to root-knot and root-lesion

nematodes. Horticultura Brasileira. 2019;**37**:65-68. DOI: 10.1590/s0102-053620190110

[58] Ries SM. RPD No. 704—Gray Mold of Strawberry. 1995. Available from: http://ipm.illinois.edu/diseases/series700/rpd704/ [Accessed: October 4, 2021]

[59] Garrido C, Carbu M, Fernandez Acero FJ, Gonzalez-Rodriguez VE, Cantoral JM. New insights in the study of strawberry fungal pathogens. In: Husaini AM, Mercado JA, editors. Genomics, Transgenics, Molecular Breeding and Biotechnology of Strawberry. Vol. 5. Ikonobe, Japan: Global Science Books. 2011. pp. 24-39

[60] EPPO. 2015. Available from: https://www.eppo.int/plant_quarantine/A2_list

[61] Maas JL. Botrytis Fruit Rot and Blossom Blight: Compendium of Strawberry Disease. St Paul, Minnesota, USA: The American Phytopathological Society; 1998. pp. 28-31

[62] Harrison GL, Stewart DA. The effect of leaf age on susceptibility of strawberry to infection by *Mycosphaerella fragariae*. Journal of Phytopathology. 2010;**124**:112-118

[63] Moorman GW, Lease RJ. Benzimidazole- and dicarboximide-resistant *Botrytis cinerea* from Pennsylvania green houses. Plant Disease. 1992;**76**:477-480

[64] Freeman S, Katan T. Identification of Colletotrichum species responsible for anthracnose and root necrosis of strawberry in Israel. Phytopathology. 1997;**87**:516-521

[65] Madden LV, Wilson LL, Yang X, Ellis MA. Splash dispersal of *Colletotrichum acutatum* and *Phytophthora cactorum* by short duration simulated rains. Plant Pathology. 1992;**41**:427-436

[66] Gooding HJ, McNicol RJ, Maclntyre D. Methods of screening strawberries for resistance to *Sphaerotheca macularis* (wall ex Frier) and *Phytophthora cactorum* (Leb. and Cohn). Journal of Horticultural Science. 1981;**56**:239-245

[67] Nelson MD, Gubler WD, Shaw DV. Inheritance of powdery mildew resistance in greenhouse-grown versus fi eld-grown Californian strawberry progenies. Phytopathology. 1995;**85**:421-424

[68] Okayama K, Nakano T, Matsutani S, Sugimura T. A simple and reliable method for evaluating the effectiveness of fungicides for control of powdery mildew (*Sphaerotheca macularis*) on strawberry. Annals of the Phytopathological Society of Japan. 1995;**61**:536-540

[69] Maas JL. Leaf Spot: Compendium of Strawberry Diseases. St. Paul, Minnesota, USA: The American Phytopathological Society; 1998. pp. 21-24

[70] Legard DE, Xiao CL, Mertely JC, Chandler CK. Management of Botrytis Fruit rot in annual winter strawberry using Captan, Thiram and Iprodione. Plant Disease. 2001;**85**:31-39

[71] Maas JL. Compendium of Strawberry Diseases. 2nd ed. MN, USA: The American Phytopathological Society (APS); 1998. pp. 16-81

[72] Kennedy BW, King TH. Studies on epidemiology of bacterial angular leafspot on strawberry. Plant Disease Reporter. 1962;**46**:360-363

[73] EPPO/OEPP. Data sheet on Xanthomonas fragariae. In: Smith IM,

McNamara DG, Scott PR, Harris KM, editors. Quarantine Pests for Europe. Data Sheets on European Communities and for the European and Mediterranean Plant Protection Organisation. Wallingford, UK: CAB International; 1992. pp. 829-833

[74] Allan-Wojtas P, Hildebrand PD, Braun PG, Smith-King HL, Carbyn S, Renderos WE. Low temperature and anhydrous electron microscopy techniques to observe the infection process of the bacterial pathogen *Xanthomonas fragariae* on strawberry leaves. Journal of Microscopy. 2010;**239**:249-258

[75] Hildebrannd DC, Schroth MN, Wilhelm S. Systemic invasion of strawberry by *Xanthomonas fragariae* causing vascular collapse. Phytopathology. 1967;**57**:1260-1261

[76] Mass JL, Pooler MR, Galletta GJ. Bacterial angular leaf spot disease of strawberry: Present status and prospects for control. Advances in Strawberry Research. 1995;**14**:18-24

[77] Milholland RD, Ritchie DF, Daykin ME, Gutierrez WA. Multiplication and translocation of *Xanthomonas fragariae* in strawberry. Advances in Strawberry Research. 1996;**15**:13-17

[78] Janse JD, Rossi MP, Gorkink RFJ, Derks JHJ, Swings J, Scortichini M. Bacterial leaf blight of strawberry (*Fragaria ananassa*) caused by a pathovar of *Xanthomonas arboricola*, not similar to *Xanthomonas fragariae* Kennedy & King. Description of the causal organism as *Xanthomonas arboricola* pv. fragariae. Plant Pathology. 2001;**50**:653-665

[79] Vermunt A, Van Beuningen A. Monitoring van *Xanthomonas fragariae* in de aardbeiteelt en de ontwikkeling van een hygiëneprotocol. Productschap Tuinbouw en Plantum NL. 2008. Available from: https://books.google.nl/books/about/Monitoring_van_Xanthomonas_fragariae_in.html?id=em3yoAEACAAJ&redir_esc=y

[80] Maas JL. Strawberry disease management. In: Naqvi SAMH, editor. Diseases of Fruits and Vegetables, Diagnosis and Management. Vol. II. Dordrecht (NL): Kluwer Academic Publishers; 2004. pp. 441-483

[81] Bradbury JF. *Xanthomonas fragariae*. CMI Descriptions of Pathogenic Fungi and Bacteria. No. 558. Wallingford, UK: CAB International; 1997

[82] Koike H. The aluminium-cap method for testing sugarcane varieties against leaf scald disease. Phytopathology. 1965;**55**:317-319

[83] Dye DW. The inadequacy of the usual determinative test for the identification of *Xanthomonas* spp. New Zealand Journal of Science. 1962;**5**:393-416

[84] Roberts PD, Berger RD, Jones JB, Chandler CK, Stall RE. Disease progress, yield loss, and control of *Xanthomonas fragariae* on strawberry plants. Plant Disease. 1997;**81**:917-921. DOI: 10.1094/PDIS.1997.81.8.917

[85] Vermunt A, Van Beuningen A. Monitoring van *Xanthomonas fragariae* in de aardbeiteelt en de ontwikkeling van een hygiëneprotocol. Productschap Tuinbouw en PlantumNL. 2008. Available from: https://books.google.nl/books/about/Monitoring_van_Xanthomonas_fragariae_in.html?id=em3yoAEACAAJ&redir_esc=y

[86] Zimmerman C, Hinrichs-Berger J, Moltman E, Buchenauer H. Nested

PCR (polymerase chain reaction) for detection of *Xanthomonas fragariae* in symptomless strawberry plants. Journal of Plant Diseases and Protection. 2004;**111**:39-51

[87] Weller SA, Beresford-Jones NJ, Hall J, Thwaites R, Parkinson N, Elphinstone JG. Detection of *Xanthomonas fragariae* and presumptive detection of *Xanthomonas arboricola* pv. fragariae, from strawberry leaves, by real-time PCR. Journal of Microbiological Methods. 2007;**7**(2):379-383

[88] Vandroemme J, Baeyen S, Van Vaerenbergh J, De Vos P, Maes M. Sensitive real-time PCR detection of Xanthomonas fragariae in strawberry plants. Plant Pathology. 2008;**57**:438-444. DOI: 10.1111/j.1365-3059.2007.01813

[89] Turechek WW, Hartung S, McCalister J. Development and optimization of a real-time detection assay for *Xanthomonas fragariae* in strawberry crown tissue with receiver operating characteristic curve analysis, Phytopathology. 2008;**98**(3):359-368. DOI: 10.1094/ PHYTO-98-3-0359

[90] Mahuku S. Goodwin PH, Presence of *Xanthomonas fragariae* in symptomless strawberry crowns in Ontario detected using a nested polymerase chain reaction (PCR). Canadian Journal of Plant Pathology. 1997;**19**:366-370. DOI: 10.1080/ 07060669709501061

[91] Braun PG, Hildebrand PD. Effect of sugar alcohols, antioxidants and activators of systemically acquired resistance on severity of bacterial angular leaf spot (*Xanthomonas fragariae*) of strawberry in controlled environment conditions. Canadian Journal of Plant Pathology. 2013;**35**:20-26

[92] Kennedy BW. Infection of Potentilla by *Xanthomonas fragariae*. Plant Disease Reporter. 1965;**49**:491-492

[93] Maas JL, Compendium of Strawbeny Diseases. 41-42. St Paul, Minnesota, USA: The American Phytopathological Society; 1984

[94] Scortichini M. Una nuova batteriosi della fragola causata da *Xanthomonas campestris*. Frutticoltura. 1996;**58**:51-53

[95] Janse JD, Rossi MD, Gorkink RFJ, et al. Bacterial leaf blight of strawberry (Fragariae × ananassa) caused by a pathovar of *Xanthomonas arboricola*, not similar to *Xanthomonas fragariae* Kennedy & King. Description of the causal organism as Xanthomonas arboricola pv. fragariae (pv. nov., comb. nov.). Plant Pathology. 2001;**50**: 653-665

[96] Scortichini M, Rossi MP. Genetic diversity of *Xanthomonas arboricola* pv. Fragariae strains and comparison with some other *X. arboricola* pathovars using repetitive PCR genomic fingerprinting. Journal of Phytopathology. 2003; **151**:113-119

[97] Anonymous. A new pathogen, *Xanthomonas arboricola* pv. Fragariae, causing bacterial leaf blight of strawberry: Addition to the EPPO alert list. OEPP/EPPO Reporting Service. 2002;**4**:10-11

[98] EPPO. 39th Meeting of the Panel on Phytosanitary Measures, Paris, 2007. Available from: http://www.eppo.org/MEETINGS/2007_meetings/phytomeasures.htm

[99] Ustun N, Tjou-Tam-Sin NNA, Janse JD. First report on bacterial leaf blight of strawberry caused by *Xanthomonas arboricola* pv. fragariae Janse et al. in Turkey. Journal of Plant Pathology. 2007;**89**:109-112

[100] Van Den Mooter M, Swings J. Numerical analysis of 295 phenotypic

features of 266 *Xanthomonas* strains and related strains and an improved taxonomy of the genus. International Journal of Systematic Bacteriology. 1990;**40**:348-369

[101] Weller SA, Beresford-Jones NJ, Hall J, Thwaites R, Parkinson N, Elphinstone JG. Detection of *Xanthomonas fragariae* and presumptive detection of *Xanthomonas arboricola* pv. fragariae, from strawberry leaves, by real-time PCR. Journal of Microbiological Methods. 2007;**70**:379-383

[102] Anonymus. Zirai Mücadele Teknik Talimatları (in Turkish) Cilt 4, T.C. Ankara: Tarım ve Köyişleri Bakanlığı Tarımsal Araştırmalar Genel Müdürlüğü; 2008

[103] Coll M, Sinakya S, Schouster I, Nenner Y, Steinberg S. Decision-making tools for Frankliniella occidentalis management in strawberry: Consideration of target markets. Entomologia Experimentalis et Applicata. 2007;**122**(1):59-67

[104] Sakurai T, Inoue T, Murai T. Intraspecific variation in transmission of TSWV by *Frankliniella occidentalis* results from distinct virus accumulation. In: Thrips and Tospoviruses: Proceedings of the 7th International Symposium on Thysanoptera. 2001. pp. 51-57

[105] Anonymus, [Internet]. Bitki Zararlıları Zirai Mücadele Teknik Talimatları, Tarım ve Orman Bakanlığı, Ankara. 2021. Available from: https://www.tarimorman.gov.tr/TAGEM/Belgeler/yayin [erişim: October 03, 2021]

[106] Kazak C, Karut K, Kasap I, Kibritci C, Sekeroglu EI. The potential of the Hatay population of *Phytoseiulus persimilis* to control the carmine spider mite Tetranychus cinnabarinus in strawberry in Silifke-Içel. Turkey. Phytoparasitica. 2002;**30**:451-458

[107] Easterbrook MA et al. Biological control of strawberry tarsonemid mite *Phytonemus pallidus* and two-spotted spider mite *Tetranychus urticae* on strawberry in the UK using species of Neoseiulus (Amblyseius) (Acari: Phytoseiidae). Experimental and Applied Acarology. 2001;**25**:25-36

[108] Çakmak I, Çobanoğlu S. Amblyseius californicus (McGregor, 1954) (Acari: Phytoseiidae), a new record for the Turkish fauna. Turkish Journal of Zoology. 2006;**30**:55-58

[109] Maas JL. Compendium of Strawberry Diseases. 2nd ed. USA: APS Press; 1998. pp. 82-95

[110] EPPO. [Internet]. Available from: https://gd.eppo.int/.2021 [Accessed: October 04, 2021]

[111] Aydın H, Gürkan MO. The efficacy of spinosad on different strains of *Spodoptera littoralis* (Boisduval) (Lepidoptera: Noctuidae). Turkish Journal of Biology. 2006;**30**(2):5-9

[112] Aydın H. *Spodoptera littoralis* (Boisduval) (Lepidoptera: Noctuidae)'e Karşı Bazı Tanacetum spp. (Compositae) Ekstraktlarının Biyolojik Aktivitelerinin Belirlenmesi. Ankara: Ankara Üniversitesi, Fen Bilimleri Enstitüsü, (Basılmamış) Yüksek Lisans Tezi; 2002. 100 s

[113] Ünlü L, Kornoşor S. Şanlıurfa ilinde saptanan Noctuidae (Lepidoptera) familyası türleri ve morfolojik özellikleri. Harran Üniversitesi Ziraat Fakültesi Dergisi. 2003;**7**(3-4):19-28

[114] Hadim N, Gürkan MO. "Pamuk yaprak kurdu (*Spodoptera littoralis* Boisd.) (Lepidoptera:Noctuidae)'nda sentetik pyretroitlere karşı ortaya çıkan direncin biyokimyasal mekanizmaları, 60". Türkiye II. Bitki Koruma Kongresi

(27-29 Ağustos 2007) Bildirileri, Isparta. 2007. p. 342

[115] Yıldırım EM, Başpınar H. Aydın ili çilek alanlarında saptanan Noctuidae (Lepidoptera) familyası türleri, yayılışı, zararı ve popülasyon dalgalanmaları üzerinde çalışmalar. ADÜ Ziraat Fakültesi Dergisi. 2008;**5**(2):115-121

[116] Karakoç ÖC, Gökçe A. Bitki ekstraktlarının *Spodoptera littoralis* (Lepidoptera: Noctuidae)'e olan kontak toksisiteleri. Türk. entomol. derg. 2012;**36**(3):423-431

[117] Nickel O, Silva SW, Fajardo TVM. Strawberry crinkle virus em morangueiro. Tropical Plant Pathology. 2009;**34**:267

[118] Bernardi D, Botton M, Nava DE, Zawadneak MAC. Guia Para a identificação e Monitoramento de Pragas e Seus Inimigos Naturais Em Morangueiro. Embrapa: Brasíli; 2015. p. 49

[119] Benatto A, Mogor AF, Penteado SC, Pereıra LS, Salas FJS, Zawadneak MAC. Influence of Trichomes in strawberry cultivars on the feeding. Behavior of *Chaetosiphon fragaefolii* (Cockerell) (Hemiptera: Aphididae). Neotropical Entomology. 2018;**47**:569-576

[120] Arıdıcı-Kara P. "Doğu Akdeniz Bölgesi Meyve Bahçelerinde Zararlı *Drosophıla suzukii* Matsumura (Diptera: Drosophilidae)'nin Biyoekolojisi ve Mücadelesine Esas Bazı Kriterlerin Belirlenmesi". Adana: Doktora Tezi, Çukurova Üniversitesi, Fen Bilimleri Enstitüsü; 2020

[121] Arıdıcı- Kara P, Ulusoy MR. *Drosophila suzukii* (Matsumura) (Diptera: Drosophilidae)' nin Doğu Akdeniz Bölgesinde Yayılış Alanları ve Konukçuları. Ç.Ü Fen ve Mühendislik Bilimleri Dergisi. 2020;**39**(5):125-137

[122] Anonymous. Çilek (*Fragaria Vesca*) Alanlarındaki Yabancı Otlar. Bitki Hastalıkları ve Yabancı Ot Zirai Mücadele Teknik Talimatları. TAGEM. Bitki Sağlığı Araştırmaları Daire Başkanlığı (In Turkish). 2021:11-13

[123] Anonymous. Integrated Weed Management. 2021. Available from: https://www2.ipm.ucanr.edu/agriculture/strawberry/Integrated-Weed Management

[124] Strawberry IPM Weed Management Guide. 2021. Available from: https://www2.gnb.ca/content/dam/gnb/Departments/10/pdf/Agriculture/mallFruitsPetitsfruits/StrawberryIPM.pdf

[125] Tetik Ö. Çukurova Bölgesi, Aşağı Seyhan Ovası tarım alanlarında sulama suyu ile taşınan ve sulama kanalları etrafında bulunan yabancı ot türlerinin belirlenmesi (in Turkish) [master thesis]. Adana, Turkey: Çukurova Üniversitesi Fen Bilimleri Enstitüsü; 2010. 105 p

[126] Johnson MS, Fennimore SA. Weed and crop response to colored plastic mulches in strawberry production. HortScience. 2005;**40**(5):1371-1375

[127] Maas JL. Compendium of Strawberry Diseases. 2nd ed. St. Paul, MN, USA: APS Press; 1998

[128] Xiang Y, Bernardy M, Bhagwat B, Wiersma PA, DeYoung R, Bouthillier M. The complete genome sequence of a new polerovirus in strawberry plants from eastern Canada showing strawberry decline symptoms. Archives of Virology. 2015;**160**:553-556

[129] Converse RH. In: Converse RH, editor. Virus Disease of Small Fruits. Washington, DC, USA: United States Department of Agriculture; 1987. pp. 1-288

[130] Thompson JR, Jelkmann W. Strain diversity and conserved genome elements in strawberry mild yellow edge virus. Archives of Virology. 2003;**149**:1897-1909

[131] Eulense GE. Differential transmission of strawberry mottle virus by *Chaetosiphon thomas* Hille Ris Lambers and *Chaetosiphon fragaefolii* (Cockerell) [M.S. thesis]. Corvallis, OR: Oregon State University; 1981. p. 30

[132] Frazier NW. Differential transmission of strains of strawberry vein banding virus by four aphid vectors. Plant Disease Reporter. 1960;**44**:436-437

[133] Ioannis E, Tzanetakis RR, Martin. Incidence of major strawberry viruses in North America. Acra Horticulturae. 2014;**1049**:595-598

[134] Diekmann M, Frison EA, Putter T. FAO/IPGRI Technical Guidelines for the Safe Movement of Small Fruit Germplasm, Food and Agriculture Organization of the United Nations. Rome: Rome/International Plant Genetic Resources Institute; 1994

[135] Babini AR, Cieślińska M, Karešová R, Thompson JR, Cardoni M. Occurrence and identification of strawberry viruses in five European countries. In: ISHS Acta Horticulturae. 656: X International Symposium on Small Fruit Virus Diseases. 656, 39-43. DOI: 10.17660/ActaHortic.2004.656.4

[136] Leone G, Lindner JL, Schoen CD. Attempts to purify strawberry viruses by nonconventional separation methods. Acta Horticulturae. 1991;**308**:121-130

[137] EPPO/CABI. Strawberry crinkle cytorhabdovirus. Data Sheets on Quarantine Pests. Prepared by CABI and EPPO for the EU under Contract 90/399003

[138] Sylvester ES, Frazier NW, Richardson J. Strawberry Crinkle Virus. In: CMI/AAB Descriptions of Plant Viruses. No. 163. Wellesbourne, UK: Association of Applied Biologists, Wellesbourne, UK; 1976

[139] Hunter BG, Richardson J, Dietzgen RG, Karu A, Sylvester ES, Jackson AO, et al. Purification and characterization of strawberry crinkle virus. Phytopathology. 1990;**80**:282-287

[140] Frazier NW, Sylvester ES, Richardson J. Strawberry crinkle. In: Converse RH, editor. Virus diseases of small fruits, USDA Agriculture Handbook. Vol. 631. USA: United States Government Printing; 1988

[141] Lazic T, Dulic-Markovic I. Diagnosis of strawberry mottle virus. (Dijagnostika virusa sarenila lista jagode.). Jugoslovensko Vocarstvo. 2003;**37**(1/2):59-65

[142] Bolton AT. Effects of three virus diseases and their combinations on fruit yield of strawberries. Canadian Journal of Plant Sciences. 1974;**54**:271-275

[143] Randles JW, Rattigen JP. Luteovirus genus. Archives of Virology, Supplement. 1995;**10**:379-383

[144] Jelkmann W, Martin RR, Lesemann DE, Vetten HJ, Skelton F. A new potexvirus associated with strawberry mild yellow edge disease. Journal of General Virology. 1990;**71**:1251-1258

[145] Converse RH, Martin RR, Spiegel S. Strawberry mild yellow-edge. In: Converse RH, editor. Virus diseases of small fruits, Agriculture Handbook. No. 631. Washington DC, USA: US Department of Agriculture; 1987. pp. 25-29

[146] CABI/EPPO. Arabis mosaic virus. [Distribution map]. Distribution Maps of

Plant Diseases. Wallingford, UK: CABI, Map 731 (Edition 3); 2015

[147] EPPO. EPPO Global database. In: EPPO Global Database. Paris, France: EPPO; 2021. Available from: https://gd.eppo.int/

[148] Schmelzer K. Investigations on viruses of ornamental and wild woody plants. 2nd part. Viroses of forsythia, Lonicera, Ligustrum and laburnum. Phytopathologische Zeitschrift. 1963;**46**:105-138

[149] McNamara DG. The spread of arabis mosaic virus through non-cultivated vegetation. Plant Pathology. 1980;**29**:173-176

[150] Valdez RB, McNamara DG, Ormerod PJ, Pitcher RS, Thresh JM. Transmission of the hop strain of arabis mosaic virus by *Xiphinema diversicaudatum*. Annals of Applied Biology. 1974;**76**(1):113-122

[151] Murant AF. Arabis Mosaic virus. CMI/AAB Descriptions of Plant Viruses. No. 16. Wellesbourne, UK: Association of Applied Biologists; 1970

[152] Jelkmann W, Maiss E, Breyel ME, Casper R. Production and use of cDNA clones from arabis mosaic virus. Annals of Applied Biology. 1988;**113**:483-491

[153] Steinkellner H, Himmler G, Laimer M, Mattanovich D, Bisztray G, Katinger H. Construction of cDNA of arabis mosaic virus and its use for diagnosis. Mitteilungen Klosterneuburg, Rebe und Wein, Obstbau und Früchteverwertung. 1989;**39**:242-246

[154] CABI, EPPO. Strawberry mild yellow edge virus. [distribution map]. In: Distribution Maps of Plant Diseases, Wallingford, UK: CAB International. 2004; Map 937. DOI:10.1079/DMPD/20066500937

[155] OEPP/EPPO. EPPO data sheets on quarantine organisms No. 98, raspberry ringspot virus. Bulletin OEPP/EPPO. Bulletin. 1983;**13**(1)

[156] Taylor CE. Transmission of raspberry ringspot virus by *Longidorus elongatus* (de Man) (Nematoda: Dorylaimidae). Virology. 1962;**17**:493-494

[157] Debrot EA. Studies on a strain of raspberry ringspot virus occurring in England. Annals of Applied Biology. 1964;**54**:183-191

[158] Trudgill DL, Brown DJF, McNamara DG. Methods and criteria for assessing transmission of plant viruses by longidorid nematodes. Revue de Nématologie. 1983;**6**:133-141

[159] Pinkerton JN, Kraus J, Martin RR, Schreiner RP. Epidemiology *of Xiphinema americanum* and tomato ringspot virus on red raspberry, *Rubus idaeus*. Plant Disease. 2008;**92**:364-371

[160] OEPP/EPPO. Quarantine procedures No. 31, Rubus viruses: Inspection and test methods. Bulletin OEPP/EPPO Bulletin. 1991;**21**:241-244

[161] Pelet F. Small fruit viruses. Revue Suisse de Viticulture, Arboriculture, Horticulture. 1989;**21**:113-116

[162] Schmelzer K. SLRV from Euonymus, Robinia and Aesculus. Phytopathologische Zeitschrift. 1969;**66**:1-24

[163] Murant AF. Strawberry latent ringspot virus. In: CMI/AAB Descriptions of Plant Viruses. No. 126. Wellesbourne, UK: Association of Applied Biologists; 1976

Chapter 12

# Avoid Sharing of Strawberries with Birds, Rodents and Other Vertebrate Pests

*Muhammad Sarwar*


## Abstract

Strawberries *Fragaria ananassa* Duch., are delicious and packed with nutrition, so not just only humans love to eat them, but are tasty and nutritious to many other creatures. Unfortunately, birds, rodents and a range of other pests may harshly injury to strawberry plants. Thus, purpose of this chapter is enhancing protections against these vertebrate culprits to mark the dissimilarity among an awkward and normal plants. Numerous species of birds consume strawberries that are among the most common pests and as well challenging to manage for the reason that they are very moveable. Frightening maneuvers, such as noisemakers and fake owls or hawks, to some degree are operative, however they might be regularly changed since birds become accustomed to them. Physical hurdles, such as netting and wire mesh are further operative and plants must be completely covered with them, though it is labor-intensive to install. Various rodents might eat strawberries including rats, mice and squirrels, however rabbits may likewise create harms. Physical obstacles will typically retain these pests outside of orchards, even though to this tactic, there are few challenges. Several rodents particularly mice are enough smaller to slipup under or through fences. Barriers used to retain smaller animals out should be prepared from good netting and the bottommost might be tight completely alongside the earth, or be submerged a few centimeters underneath, and likewise netting row shelters above plants might be effective. Deer will also feast on strawberries, but a few more control options are available for these animals than for birds. Physical barriers, like wire mesh row covers can also be effective and electric fences around an entire garden might keep deer out. A combination of methods used in this work are best to protect strawberry plants, but nothing is 100% guaranteed when a foraging pest is truly motivated.


**Keywords:** strawberry, vertebrate, damage, rodent, bird, pest

## 1. Introduction

Strawberries *Fragaria ananassa* Duch. (Rosales: Rosaceae), are a unique fruit, mostly made up of 91% water when is fresh and ripe. Their primary nutrient is carbohydrates, but contain a very small amount of fat and protein, however these make up less than 1% of their nutrient composition. In general, depending on the size



IntechOpen

of the strawberry, they usually contain about 10 calories per berry for large or medium-sized berries. They are also fairly high in natural sugar content, which makes up the carbohydrate content. The strawberries, along with other berries such as blueberries and blackberries, are functional foods that have an actively beneficial effect on the body. Strawberries are packed full of all sorts of nutrients that can provide a healthy mental state and create the requirements for fighting off diseases like osteoporosis [1, 2].

In addition to these nutrients, strawberries contain some omega-3 fats and also fiber, which is an important nutrient for good digestive health. They also contain fair amounts of manganese, folate and potassium. Strawberries are full of vitamin C (which supports a healthy immune system), vitamin K, vitamins B1 and B6, potassium and magnesium. They also contain compounds called antioxidants that help to prevent chronic health issues. And more, strawberries are with 91% water content, so, a great source of hydration. For peoples, strawberry tea brewed from the leafy green tops (the green leafy part right on top of berry) can actually help to ease the pain of arthritis [3].

Strawberries (**Figure 1**) are delicious, packed with nutrition, so not just only humans love to eat them, but are also tasty and nutritious to many other creatures. In addition to being tasty, juicy treat, berries are also packed with antioxidants compounds that can help to pets attain a healthy and strong physique [4]. Their sweet texture makes them a popular food for birds, rodents and other small animals to eat. Strawberries have a good quantity of potassium, which actually support with helping to birds stay happy and healthy [5].

Several rodents will feed strawberries if gardeners let them to do so, comprising squirrels, chipmunks and mice. Rabbits might likewise create complications, wherein physical obstacles will generally retain these pests outside to orchards, though there are challenges to this approach. Several rodents particularly mice are enough smaller to slipup under or through fences. Barriers used to retain smaller animals out should be prepared from good netting and the bottommost might be tight completely alongside the earth, or be submerged a few centimeters underneath, and netting row shelters above plants might be effective likewise [6].

Globally, these vertebrate pests are of momentous health and financial significance, therefore, managing these vermin will not merely advantageous to diet security, however as well human and animal health [7]. Whether strawberry fields are large or small, birds and rodents pests will compete for crop. The sweet, sugary berries are irresistible to hungry pests, so some proactive measures are needed to keep unwanted visitors at bay. A combination of methods used in tandem works best to protect strawberry plants, but nothing is 100% guaranteed when a foraging pest is truly motivated. In this article, some simple suggestions will be given for letting our animal friends to find other sources of sustenance and keeping strawberries for humans.

**Figure 1.**
*Strawberries.*

## 2. Faunae attracted to strawberry plants

Peoples are not only the ones who delight in strawberries taste, but foliage, fruit, root and stem of strawberry plants appeal to numerous birds and animals species. These birds and animals can create variable amounts of harm fluctuating from a mere annoyance to eradicate whole strawberry plant. There are a limited choices existing to support in protecting the plants without spoiling the birds and animals.

Different species of berry consuming birds comprise robins, orioles and finches that can harshly damage or ingest strawberry plants by seriously dropping of produce. Several deterrents are available to keep birds from eating the strawberries without harming to feathered friends. Netting covering all sides of the plant protects the fruit from birds and is readily available at garden centers and home improvement stores. Other deterrent options include fake snakes that frighten to the birds away from the plant. These deterrents must be put into place before the plant produces fruit [8].

Rats, voles and mice are some other animals, which nourish on strawberry plants. Mice and rats commonly gnaw on roots and bark of plant, whereas voles devour fruits, however may as well consume to the roots. Rodents chemical repellent containing Thiram (sulfur fungicide) helps to retain these pests far from plants of strawberry. One more choice is to usage domestic mousetraps lured with peanut butter. Chives, garlic, leek, onion and shallots protect against moles, and lavender is effective as a mouse repellent.

### 2.1 Birds invader of strawberries

Dozens species of birds feed on strawberries, although some birds are pickier eaters than others. Of the hundreds of bird species, the two dozen or so are the most likely to eat berries. Crows are black birds of the Corvidae family known for their intelligence, adaptability and loud harsh vocalizations. The Corvus is a widely distributed genus comprising crows (generally smaller than ravens), ravens (much larger) and rooks (smaller than crows and have light-colored bills and distinct wedge-shaped tails). The coloration of the livery is dominated by shades of black, with some species having plumage with metallic iridescences and others that have white or gray areas on the neck or torso: a robust and slender appearance, equipped with a small rounded head with a strong conical beak, elongated and pointed, with a slightly curved end towards the bottom: the legs are strong and the tail is short and wedge-shaped. Crows usually feed on the ground and eat almost anything, they frequently cause damage to crops, their common sights in treetops, fields and roadsides, and in habitats ranging from open woods and empty beaches to town centers. The house crow (*Corvus splendens*), has a gray neck collar (**Figure 2**) and is a common bird of the crow family found in many parts of the world [9].

**Figure 2.**
*Crow.*

Crows prefer larger fruit, such as apples, but they will also take a bite out of strawberries. Crows seldom eat a whole berry, they take a single peck at a strawberry and leave the rest for insects and rodents. Methods for control include hunting, chemical immobilization, harassment and scare tactics, and trapping [10].

The common grackle (*Quiscalus quiscula*), have a long, dark bill, pale yellowish eyes and a long tail, their feathers appear black with purple, green, or blue iridescence on the head, and primarily bronze sheen in the body plumage (**Figure 3**). Common grackles nest in places other than their usual treetops, including birdhouses, old woodpecker holes and barns. The common grackle forages on the ground, in shallow water or in shrubs, eating insects, minnows, frogs, eggs, berries, seeds, grain, and even small birds and mice [11]. Grackles eat small strawberries whole, but they may just slash larger berries.

Starlings are small to medium-sized passerine birds in the family Sturnidae, having plumages of many species are typically dark with a metallic sheen (**Figure 4**). Their preferred habitat is fairly open country, and they eat insects and fruits. Asian species are most common in evergreen forests and many species are important dispersers of seeds. The common starling or European starling (*Sturnus vulgaris*), has glossy black plumage with a metallic sheen, which is speckled with white at some times of year. The legs are pink, and the bill is black in winter and yellow in summer [12]. Starlings can do massive damage to strawberry beds in home and field gardens. They descend in flocks and eat everything they can find. They eat smaller fruit whole and peck out the interiors of larger strawberries.

Finches and sparrows are some of the very small species of birds, but sparrows are a bit larger than finches. House finches bear thick, large beaks having grayish shade, while house sparrows hold a greatly more tapering bill, which is lesser than finches and yellow or black in color, subjected to the bird's breeding stage and gender. House sparrows pattern of color is usually shadier than that of house finches, along with profounder brown, and extra black on the wings and back. Male and female house finches both have noteworthy brown flashing on their belly and flanks in comparison to house sparrows [13]. House finches and house sparrows peck holes in strawberries that leave the fruit susceptible to decay.

Sparrows belong to the family Emberizidae, and arise in brown feathers, dull gray heads and black beaks. Males (**Figure 5a**) have brighter black, white and brown markings around its bill and on throat; and females (**Figure 5b**) and young birds are

**Figure 3.**
*Grackle.*

**Figure 4.**
*Starling.*

colored pale brown and gray all over their bodies. The house sparrow is strongly associated with human habitation and can live in rural or urban settings. They largely feed on the seeds of plants and weeds, grains, but commonly eat insects and many other foods, and even household foods such as bread. The most common type of sparrow that one witnesses is the house sparrow (*Passer domesticus*) [14].

Finches are described one of many (hundreds) of small passerine bird species of family Fringillidae. They have tout conical bills adapted for eating of seeds and nuts, and often have colorful plumage. Their small and sharp beaks, help them to eat the food or drink nectar of the flower. They occupy a great range of habitats where they are usually resident and do not migrate. Finches are typically inhabitants of well-wooded areas, but some can be found on mountains or even in deserts. The house finch (*Haemorhous mexicanus*), in most cases, adult males heads, necks and shoulders are reddish (**Figure 6a**), while female adults have brown upperparts and streaked underparts (**Figure 6b**). House finches forage on the ground or in vegetation normally. They primarily eat grains, seeds and berries, being voracious consumers of weed seeds [15].

Robins are any of several small similar bird species of thrush group having a red or reddish breast. The European robin (*Erithacus rubecula*), simply known as robin redbreast of family Muscicapidae, are with an orange breast and face lined with gray, brown upper-parts and a whitish belly, and bill and eyes are black (**Figure 7**). Its diet

(a) (b)

**Figure 5.**
*(a) Sparrow (male) and (b) sparrow (female).*

(a) (b)

**Figure 6.**
*(a) Finch (male) and (b). Finch (female).*

**Figure 7.**
*European robin.*

generally consists of around 40% small invertebrates mainly insects, and 60% wild and cultivated fruits and berries [16].

The American robin (*Turdus migratorius*), is a migratory songbird of the true thrush Turdidae family, and has a brown back, reddish-orange breast, white throat with black streaks, white belly and under tail coverts, and the bill is mainly yellow (**Figure 8**). Its diet consists of invertebrates, fruits and berries [17].

Robins eat a wide variety of food, including worms, seeds, nuts, suet, invertebrates and fruits. Robins eat strawberries whole and they can do substantial damage to a small bed of strawberries.

Oriole is any bird of several brightly colored passerine of the family Icteridae, having black and yellow or orange feathers. They are often found in orchards, prairies, farmland, urban parks, suburban landscapes, forest edge, open woodland, wooded wetlands, leafy deciduous trees and stands of trees along rivers. Among orioles (Passeriformes: Icteridae), bullock's oriole (*Icterus bullockii*) adult males (**Figure 9a**) are characterized by strongly contrasting orange and black plumage, a black throat patch and a white wing bar. The underparts, breast and face are orange or yellow. Adult females (**Figure 9b**), have gray-brown upperparts, duller yellow on the breast and underparts, and an olive crown. These birds forage in trees and shrubs, also making short flights to catch insects, and they mainly eat insects, berries and nectars [18].

Adults Baltimore orioles (*Icterus galbula*), constantly have white blocks on the wings. The adult males are yellow-orange on the rump, underparts and patches on shoulder, while entire rest of male's feathers are black (**Figure 10a**). The adult females are yellow-brown on the upper parts, have darker wings, and on the belly and breast dull orange-yellow (**Figure 10b**). It forages in trees and shrubs, also making short flights to catch insects. Baltimore orioles seem to prefer only ripe, dark-colored fruits, seek out the darkest mulberries, the reddest cherries and berries, and will ignore green cherries even if they are ripe. They crack the closed bill into soft fruits, then open their mouths to cut a juicy band from which they drink with their tongues [19].

Black-capped chickadee (*Poecile atricapillus*) in the tit family Paridae, is a passerine bird. It has a black bib and cap with white sides to the face. The underparts of

**Figure 8.**
*American robin.*

a. b.

**Figure 9.**
*(a) Bullock's oriole (male) and (b) Bullock's oriole (female).*

a. b.

**Figure 10.**
*(a) Baltimore oriole (male) and (b) Baltimore oriole (female).*

black-capped chickadee are white with rusty brown on the flanks, tail is normally slate gray and back gray (**Figure 11**). It lives in deciduous and mixed forests, wherein seeds and berries become more important in birds hop along tree branches searching for food, and sometimes hanging upside down or hovering [20].

Blue jay (*Cyanocitta cristata*) is a passerine bird of the family Corvidae. Predominantly, its coloration is blue, with a white chest and underparts, and a blue crest. It has a black, U-shaped collar around its neck and a black border behind the crest (**Figure 12**). The blue jay feeds mainly on seeds and nuts, fruits and other berries. In both deciduous and coniferous forests, it normally picks up food from trees, shrubs and the ground [21].

Black-headed grosbeak (*Pheucticus melanocephalus*) in family Cardinalidae, is similar in size to a common starling. The males have black wings and head, breast is dark to tawny orange, tail with prominent white patches and yellow belly (**Figure 13a**). The females have brown neck and head; breast is white; white streaks down the middle of head, eyes and on her cheeks; wings and tail are grayish-brown with two white wing bars, yellowish wing edges; and back with sparrow-like black streaks

**Figure 11.**
*Black-capped chickadee.*

**Figure 12.**
*Blue jay.*

a. b.

**Figure 13.**
*(a) Black-headed grosbeak (male) and (b) black-headed grosbeak (female).*

**Figure 14.**
*Bohemian waxwing.*

**Figure 15.**
*Brown thrasher.*

(**Figure 13b**). It prefers to live in deciduous and mixed wooded areas with large trees and thick bushes, wetlands and suburban areas. Forages mostly in shrubs and trees, searching for food among foliage, and also may forage on ground. Feeds on seeds of various weeds and eats berries of many plants as well as some cultivated fruits [22].

Bohemian waxwing (*Bombycilla garrulus*), is also a starling-sized passerine bird of family Bombycillidae. It has mainly buff-gray plumage, black face markings, under tail feathers are rusty, white rectangles on wings and a pointed crest. Its wings are patterned with white and bright yellow, and some feather tips have the red waxy appearance (**Figure 14**). Mainly feeds on insects, especially flying insects, and mostly berries and fruits as they become available, wherein takes berries while perched or hovering. Also eats seeds of birch and other trees, and will drink oozing sap [23].

Brown thrasher (*Toxostoma rufum*), is a bird of the Mimidae family usually nesting on small trees and shrubs. It has brown upper parts and white under part with dark streaks, slightly down curved bill, and staring yellow eyes (**Figure 15**). The habitat of brown thrasher is thickets brush shrubbery and thorn scrub. It forages a diet that includes insects, berries, nuts and seeds. Sometimes they visit feeders or the ground below to pick up fallen seeds. There is a better chance they will visit if dense cover is close by. They can also be attracted by planting of shrubs that produce berries [24].

Although birds will eat strawberries anywhere they are planted, damage is greater in urban gardens than on farms. Bird damage is greater in a home gardener's bed of strawberries than in a farmer's field of strawberries.

### *2.1.1 Preventing of birds eating strawberries*

Protect fruit from birds with netting and put egg shells around strawberries as the shells are excessively harsh for the birds to get nearby to them. Place pea straw around fruit lying on the dirt to keep them going into rotten. Paint stones to look like strawberries, one peck of those and the birds will learn to leave the real ones alone. Start to grow them in the greenhouse away from birds and eat the fruits as soon as these ripe.

The mere presence of birds in the area does not mean those birds are actively eating berries. It is not compulsory to just panic for the reason that growers notice birds in the surrounding area of berries. Several birds are mostly insect eaters and not at all involved in berries. However, those species that are fruit and seed eaters can well

*DOI: http://dx.doi.org/10.5772/intechopen.104682*

be eating on berries. Of the hundreds of bird species, the two dozen or so are the maximum expected to consume berries. Before taking counter measures, make sure there are actually witnessing that the birds are eating fruit. And prior to take any procedures for chasing away of birds, growers must consider the information that those birds can be existing to consume other pests, which are injurious to berries. The similar actions may be used to retain birds far away from the berries and also keeping them for consumption of insects that might be the more severe enemies.

For strawberries grown commercially, bird damage is usually insignificant. Several groups of scientists have done a skillful learning of birds harm to marketable strawberry territories. They observed that birds usually eat approximately 3–4% of a strawberry crop and their fecal contamination affect only around 0.01% of the crop. Within the internal parts of a commercial field, birds consume insects more than strawberries. But, in the margin of the strawberry field, occurrence of birds is a clear encouraging. This is only round the periphery of a farmer's strawberry field where birds injury compensates benefits by birds. Homebased growers practice a dissimilar story with opinion that birds can destroy some entire crop of home gardeners. However, there are points that both commercial growers and home gardeners can organize to retain birds injury to a lowest level.

### *2.1.2 Protecting of berries from birds*

There are several humane ways to keep berry crop safe from neighborhood birds. Here are a few of the best approaches against the invasions of birds. The sweet, sugary berries are irresistible to hungry pests, so some proactive measures are needed to keep unwanted visitors at bay. A combination of methods used in tandem works best to protect strawberry plants, but nothing is 100% guaranteed when a foraging pest is truly motivated. For homebased strawberry beds, the way out is to shelter strawberries with bird-resistant material. Collect some 1 × 1 woody decking bits (25 mm × 25 mm) from neighborhood home development center or timber backyard. Practice the decking fragments to construct an arrangement similar a table wide and long sufficient to cover whole strawberry bed, and enough tall to lodge the highest plants. Create a table, but not a box, so that there is no lower edge to press them. Fasten netting cloth on five sides over the box, by leaving the base exposed. At that moment, just place the box over strawberries and lift it up only to reap. This configuration can as well be used to shelter other plants while strawberries are not bearing fruit.

If gardeners are commercially developing of strawberries, there are two options for caring field from birds; should have all of strawberries in a continuous single field, because larger is the field, the more secure plants are in its middle. Eliminate birds habitation nearby to strawberry field and do not leave shrubbery, grasslands or bird houses round strawberry plants.

Consider two times prior to stating fighting on birds that consume to strawberries. Several cities and towns are bird reservations, and it is each time unlawful to destroy birds in these localities. Toxic substance does not discriminate among bird species, if gardeners place out toxin for a species, it is legal to destroy, for instance a grackle, however if an endangered species consumes the poisoned bait and expires, growers might be guilty of a federation or state criminality. Even though birds are swarming and dropping down on plant, gardeners could require to apply for a migratory bird depredation permit that might be attained from wildlife service to take care of large numbers of birds with lethal methods. Non-toxic and non-lethal methods are better for controlling of birds on commercial fields [25].

#### 2.1.2.1 Visual scare and noisemaker devices

Visual scaring devices comprise streamers, spinners, plastic owl and plastic snake models, scare-eyes (balloons having eyes dyed on them), and aluminum pie pans. Birds may come to be habituated to visual deterrents, hence it is compulsory to use variable colors and types of devices positioned in changeable sites to retain birds from becoming used to these.

Noisemaker devices for birds deterrence contain exploders, cannons and sirens. Noisemakers work the finest while they are applied at irregular interims. When noisemakers are used every day at the same time, birds will turn into habituated to them. Tape recorders of bird distress calls are extra operational than generic sources of noise, however, it is essential to acquire tapes of the precise types for birds initiating losses. Noise does frighten to birds off and having a radio near or in berry patch will create enough noise to frighten hungry birds away. Compatibly, once the birds get used to the noise, they would not be shy about investigating of garden.

For protecting of strawberry beds from birds, gather several small pebbles, paint them red with craft paint and sprinkle them in and around strawberry plants. Of course the birds will be attracted, after a while get tired of pecking the rocks and leave the strawberries alone. Growers can likewise effort cheating to the birds through finely image of strawberry-shaped rocks to appear alike strawberries and then allocate them evenly all over strawberry plants. The cheerful red color will appeal to resident avian raiders looking for some free food to these locations. Fortuitously, few pecks will show the trick and halt the plans of birds. Not desiring to be tricked constantly, they will learn soon that strawberries, which their mind states should be charming are really quite uneatable. As soon as the experience has been learned, they will vacate to strawberry bed alone.

#### 2.1.2.2 Flash tape and CDs or aluminum pie plates

Movement and shine of any objects will frighten to most birds off. Flash tape basically is foil tape or strip of Mylar, which flutters in the wind, so frightening off to the birds. The birds do not like to the luster of tape and something that moves is worthy for retaining birds outside the area. Foil tape is humane, inexpensive and comparatively self-effacing. However, when the birds are enough hungry, they will take risk for moving adjacent the tape to acquire a strike of fresh berries.

The idea behind CDs or aluminum pie plates on a thread system is the similar to the Mylar flash tape, because glittery, stirring stuffs will scare to any voracious birds from the region. For using of this technique, only tie some cord through the hole in a CD, or create a hole in a pie plate and string or thread through it and suspend it from a pole or fence nearby to berries. This method also provides a great opportunity to recycle, since growers can use old CDs and used aluminum pie plates. And, as with the flash tape method, if birds are very hungry, this method would not deter them for long.

#### 2.1.2.3 Netting

Birds love to ripe berries and for keeping of birds from eating of berries, it is a good idea to cover the plants with bird's netting as soon as the berries begin to ripen. Small birds may get inside the netting and get caught. Also, the netting is easy to lift off for picking the berries, fairly inexpensive and easy to replace. Be sure to weight down the netting with stones, boards or rocks, so it does not fly away. And do not leave any gaps where the birds can hop in underneath the net. This is possibly the greatest fool-proof technique for protecting more of the berry yield. As a result of netting draping above small fruit trees and berry bushes, growers can check to birds from getting at the great bulk of the berries. For plants such as strawberries, floating row cover frames can support netting to protect the berries from birds. Loftier berry bushes can be secured through pop-up screens, traded by garden supplying retailers.

#### *2.1.2.4 Fake predators and pinwheels*

A realistic-looking fake predators on a post will help to scare pesky birds away from berries. The clue behind it is forthright that birds realize an imaginary predator (scarecrow, bird of prey, human dummy or snake) nearby the berries and they would not move close to them. Orchard supply stores sale a number of diverse cardboard or inflatable owls, scarecrows, snakes or hawks to practice in the orchard. For making of this effort, frames are needed to exchange the scarecrow after every few days to a fresh location in the area. Straightly, the birds will eventually figure out that the scarecrow is not stirring. Birds have snakes as natural enemies, so placing a rubber snake around the yard is a great way to keep the birds at a distance. This is especially true of birds looking for a place to nest as they know that snakes will eat the eggs. If some part of the scarecrow moves, such as a tie that flutters in the breeze, it will work all the better.

Outdoor pinwheels are more than simple decorations, but when the fins have a metallic or reflective surface like the metal, solar powered ornament, they can naturally scare birds away. The sudden flashes of light startle the birds and will help to keep them away from strawberries. Pinwheels are also quite durable and only need occasional inspection to ensure they are still upright and working.

#### *2.1.2.5 Birds bath and feeders*

Birdbaths will bring a wide variety of backyard birds to garden or yard and many of these birds feast on insect pests. A general concept is that while birds attack a berry area, they are not so voracious as greatly as they are thirsty. The higher content of water in berries makes them a perfect object for thirsty birds. By means of having a birdbath in adjacent, growers provide to them what they want really and birds vacate berries lone. This will perform even well if growers can add the voice of water, by the usage of a fountain or dripper. If the birds really are just hungry, growers have provided them with a full meal instead of just a drink.

As with the bird bath idea, inviting birds to the garden can help growers to control insect pests. The belief behind this clue is that when growers provide to the birds their very particular diet, they will vacate crop lonely. Place one or two feeders in nearby vicinity to berry area and the birds will move to the feeders as a replacement of looting to plant. But, if growers are not careful about keeping of the feeders full once hungry birds have been attracted, they could very well notice the luscious berries nearby and feast on them.

#### *2.1.2.6 Smells to scare away birds*

There are many smells that can repel to birds away from strawberries, for example, garlic, cayenne pepper and peppermint oil are all known to be offensive to birds and all happen to be organic products. Growers can create a mixture of cayenne pepper combined with water to spray on strawberries. The scent will cause the birds to react and pull away quickly, however, growers need to reapply the cayenne pepper spray frequently to keep the scent strong and remember to wash strawberries before eating them.

A chemical so-called methyl anthranilate is there, which flavors bad to birds, however it is unpleasant for persons. But, this might not be lawful wherever growers are living and it may not be used in production of organic strawberry. On occasion, this injuries to foliage of plants on which it is dripping. And occasionally, the top-secret to keep birds beyond strawberry area, is exchanging of varieties. Implanting of alpine gold strawberries that when ripe are yellow, could fool to birds searching for red strawberries.

## 2.2 Rodents invader of strawberries

There are numerous animal species that also love to these mouthwatering red berries. And that is why growers may often find their berries drilled or half-eaten in a specific style. Mice and some other rodents might be a key problematic for any horticulturist. They can remove seeds out of the ground, and eat gnaw on the foliage and eliminate any berries and fruit the plants produce. Soft fruits, such as strawberries, are mostly vulnerable to rodents attack. Strawberry plants deliver delicious and fresh fruits from a yard. They likewise act as an open offer for several pests, such as rodents that will feast on entire portions of the plant, as well as its roots. These vermin can result variable grades of injury to strawberry and in risky situations lurk to the lifespan of the plant.

Various rodents will eat strawberries, including mice, squirrels, chipmunks and rats that love to eat strawberries, Fortuitously, maximum of rodents live inside a smaller expanse and certainly not wander further than nearly 100 feet from anywhere they are born. If farmers can remove rodents habitation round strawberry plants, they would not have a problematic with mice and rats that steal to fruit [26].

### *2.2.1 Mice*

Characteristically, mouse is a small slender body mammal known to have a pointed snout, small rounded ears and a thinly furred scaly tail (**Figure 16**) in family Muridae. Mice are a disastrous problematic equally outside and inside, and they may invade garden and home, nourishing on plants and foods. They are good jumpers and climbers as well as master swimmers, living in temperatures as low as 14°F and squeezing into openings greatly lesser than their body mass. Mice gnaw on the barks and roots of trees along with vegetable and fruit plants. They will likewise passage to other soft parts of the plants, for instance, fruits and foliage. Three utmost communal mice species expected to produce pest burdens for farm holders are house mouse (*Mus musculus* Linnaeus), white-footed mouse (*Peromyscus leucopus* Rafinesque) and deer mouse (*Peromyscus maniculatus* Wagner) [27, 28].

**Figure 16.**
*Mouse.*

These furry wander often run across wild strawberries during their travels. They often eat every part of the strawberry, including the stalk, the berry and the roots of the plant. Grow mice repell plants near to the strawberries. Mint, lavender and catnip are all disliked by mice and other rodents, possibly because all these plants attract to cats.

### *2.2.2 Rats*

Rats are various medium-sized, long-tailed rodents of the family Muridae and usually a large muroid rodent include the rat (**Figure 17**), while a smaller muroid is mouse. The best-known rat species are black rat also known as the ship rat (*Rattus rattus*) and the brown rat or wharf rat (*Rattus norvegicus*). Indeed, rats have long been considered deadly pests, they devour a wide range of plant and animal foods, and infest any area that provides shelter and easy sources of food and water. Rats are unwanted in gardens because of the damage they can cause to fruits, vegetables, seeds, bulbs, plants and containers. They can also expose people and pets to various diseases and parasites. Rodents can cause damage in garden, which can range anywhere from feasting on fruits and a wide range of garden vegetables, and damage to earth banks. The only rat species likely to find in garden is the brown (Norway) rat (*R. norvegicus*) [29].

Out of all foods the rodents consume, their top two loves are generally for fruits and berries. The wild rats and mice consume these foods at every opportunity, even before they are ripe. As a result, strawberries trees as well as blackberry and raspberry shrubberies can work as magnetic form for the animals. After unconsumed berries and fruits are left on their trees or bushes to rot, or left in exposed trash cans, these rodents are drawn by the sweetness and smell. Strawberries can be a good source of dietary fiber, water and nutrients as part of rat's weekly fresh foods rotation. Rats and mice are known to nibble on everything from grass and weeds to small twigs and bits of bark. Plant seeds are another favorite food among these animals [30].

The best way to keep rodents out of garden is to eliminate any harborage points around garden, and remove any potential nesting places by keeping of gardens clean and tidy. Remove piles of wood, clippings etc., and cut back overgrown areas. Rodenticides will be helpful to destroy a rodent populace. An appetizing bait formulation appeals to rats and mice having necessary odors, texture and taste with constituents such as oats. These features retain them eating until they have swallowed a fatal dosage by engaging to their natural wish for chewing [31, 32].

**Figure 17.**
*Rat.*

### *2.2.3 Voles*

Voles (Rodentia: Cricetidae) look similar to house mice, but have a rounded snout and head, smaller ears and shorter tail than the average house mouse. Their coats are dark brown with gray fur on their bellies (**Figure 18**). As herbivores, voles will eat plants, fruits, grasses, stems, leaves and berries. Voles feast on plants of all species including strawberries and other crops. There are a number of vole species, the

**Figure 18.**
*Vole.*

maximum frequent being meadow vole (*Microtus pennsylvanicus* Rhoads), prairie vole (*Microtus ochrogastor* Wagner) and woodland or pine vole (*Microtus pinetorum* Le Conte) [33]. They usually feed on the above ground parts of strawberries including the foliage, seeds and fruits, and also gnaw bark from plants. They gladly chew on fruits throughout the plant's developing period, however will passage onto strawberry roots in the fall, spring and winter until their abdomens are filled. Clusters of grassland surrounded by grass pieces and brown or green rice-shaped feces is a symbol that voles are invading to orchard [34].

### 2.2.4 Moles

Moles are small mammals of the family Talpidae in the order Eulipotyphla, and have cylindrical bodies, velvety fur, very small inconspicuous eyes and ears, reduced hind limbs, and short powerful forelimbs with large paws adapted for digging (**Figure 19**). Moles have polydactyl forepaws each has an extra thumb (also known as a prepollex) next to the regular thumb. Moles are known pests to human activities such as agriculture, lawn and gardening through damage to young plants by disturbance of the soil. They can undermine plant roots, indirectly causing damage or death, but do not eat plant roots. Moles on occasion are accused of inflicting damage to strawberry roots, although they are insectivores, do not classified as rodents and not eat plants. However, they can disturb the roots by burrowing tunnels below the plant. Additionally, plant-feeding rodents, for instance, voles usage old mole burrows to acquire plant roots without digging of new tunnels. Moles have short tails and pointy small snouts, and their back and front limbs are short with a spade like outline. Their anterior feet are bigger than the posterior and have lengthy claws prepared for burrowing through soil [35].

Moles are managed through traps such as mole-catchers, smoke bombs and poisons such as calcium carbide, which produces acetylene gas to drive moles away. The most common method now is Phostoxin or Talunex tablets that contain aluminum phosphide and are inserted in the mole tunnels, where they turn into phosphine gas [36].

**Figure 19.**
*Mole.*

### 2.2.5 Squirrels and chipmunks

Squirrels of family Sciuridae, are members in order Rodentia of the family Sciuridae that include small or medium-size rodents (**Figure 20**). The squirrel family

**Figure 20.**
*Squirrel.*

mainly includes tree squirrels, ground squirrels, chipmunks and flying squirrels plus others. Squirrels typically have slender bodies with long very bushy tails and large eyes, and generally their fur is soft and silky. In most squirrel species, the hind limbs are longer than the fore limbs, while all species have either four or five toes on each paw. Squirrels are occasionally considered pests because of their propensity to chew on various edible and inedible objects, and cause economic losses to homeowners, nut growers and forest managers. One factor all of those squirrels have in common is their desire to devour fresh berries from strawberry plants. Tree squirrels can live on trees; however, they frequently select orchard beds as their favorite dwelling for exploration of diet. Strawberries appeal to these pests as the fruits start to develop. The squirrels attack to orchard, robbing the berries earlier to growers have a chance to reap them [37, 38].

Squirrels are excellent climbers and can reach the plants via telephone or electrical wires, or jumping from tree branches. Both ground (*Otospermophilus beecheyi*) and tree squirrels (Western gray squirrel *Sciurus griseus* and Eastern fox squirrel *Sciurus niger*) enjoy foraging on the ground for vegetables and fruits. Ground squirrels primarily eat vegetable plants in the seedling stage, but they can consume entire young plants. The burrows of the ground squirrel can leave large mounds in garden or lawn and chew through any small tree roots or plant roots they encounter. Tree squirrels eat most soft vegetables and fruits grown in gardens, and tomatoes, strawberries and corn often see the worst damage. Tree squirrels of all species love feeding on fruit and nut tree, and they eat both mature and immature fruits [39].

Squirrels are also a nuisance when growing of strawberries, as they like to feed on ripe berries. For keeping them off from eating berries, wrap netting around the plants during fruiting. Once harvested the strawberries, remove the netting, so the birds can eat any unwanted insect pests on the plants. Active measures are needed to be taken in order to ensure that squirrels do not damage strawberry crop. Modify the habitat surrounding to strawberry patch to make it less desirable to squirrels. Remove bird feeders, clean up brush piles and keep tight lids on outdoor trashcans. Trapping is often necessary to remove squirrels from residential structures. Effective baits include fruit, peanut butter, nuts, seeds and vanilla extract.

Eastern chipmunk (*Tamias striatus*) is a small, brownish, ground-dwelling squirrel, have short, pointy head marked with two white stripes, one above and one below the eye, five black lines with white striping down to the back (**Figure 21**), and holds food

**Figure 21.**
*Chipmunk.*

with the front feet. Chipmunks have an omnivorous diet primarily consisting of seeds, nuts and other fruits, and buds. They also commonly eat grass, shoots and many other forms of plant matter as well. Around humans, chipmunks can eat cultivated grains and vegetables, and other plants from farms and gardens, so they are sometimes considered pests. Chipmunks mostly forage on the ground, but they climb trees to obtain nuts. Fruits and vegetables also make up the chipmunk's diet, and they often seek out strawberries, blackberries and other berries that grow close to the ground. Chipmunks can be pretty cute, but the cuteness fades quickly when they begin to destroy the strawberry harvest. They can be voracious eaters and will reduce to strawberry patch to a wasteland in record time, if allowed to get away with it. Their digging and acrobatic abilities make it hard to fence them out [40].

Both snap traps and live traps are an option for getting rid of chipmunks. Snap traps will kill them, while live traps make its catching, so can be transported to a more suitable location. Chipmunks are fond of nuts and seeds, so peanut butter and sunflower seeds are good bait for traps. A chicken wire hedge as a minimum of 30 inches tall submerged about 6 inches deep will benefit to guard the plants. The bottommost of the hedge that will be concealed ought to be curved at a 90-degree angle out away from the plants. This will aid to stop the animals from get into the plants by digging below the wire. Refrain from using poisons since they can also harm to pets and other wildlife.

### *2.2.6 Raccoons*

The raccoon is a medium-sized mammal and the largest member of the family Procyonidae in order Carnivora. Raccoons (*Procyon lotor*) have masks on their faces that frequently illustrates them as robbers. They have soft, ringed tails and their bodies are variable in shades of gray, with hints of light brown (**Figure 22**). The forepaws of raccoons look like to small human hands with five toes that mark raccoons very agile [41].

The original habitats of the raccoon are deciduous and mixed forests, but due to their adaptability, they have extended their range to mountainous areas, coastal marshes and urban areas, where some homeowners consider them to be pests. It is usually nocturnal and omnivorous, its diet consists a variety of different foods, eating about 27% vertebrates, 33% plants and 40% invertebrates, but it prefers fruits and nuts. Wild populations prefer areas with trees and water nearby in addition to garbage cans. Raccoons are notorious for their love of sweet corn, and will also eat strawberries, melons, potatoes, peas, tree fruits and grubs. They will dig holes in maturing melons and munch on ripening tree fruits. Like squirrels, they also have been known to clean out bird feeders. They eat both mature and immature fruits of strawberries [42].

Raccoons are great climbers, swimmers, jumpers and runners. Their five-toed paws make them very dexterous and their agility can allow them to outsmart every human's concoction used to deter them. On fruit trees, place baffles on tree trunks to

**Figure 22.**
*Raccoon.*

prevent raccoons from climbing them. Prune overhanging limbs that raccoons can use to launch themselves into the fruit tree. Some homemade repellents that have been used successfully are a dusting of baby powder or blood meal on fruit trees. Raccoons can be scared temporarily by loud, intermittent noises or lights, or try using of motion sensors that activate jets of water, but the best deterrent is a good fence throughout the strawberry growing area.

### *2.2.7 Hamsters*

Hamsters are rodents of order Rodentia belonging to the family Cricetidae and the best-known species of hamster is golden or Syrian hamster (*Mesocricetus auratus*). Physically, they are stout-bodied with distinguishing features that comprise elongated cheek pouches extending to their shoulders, used to carry food back to their burrows, as well as a short tail and fur-covered feet (**Figure 23**). Depending on the species, they have silky fur long or short, colored gray, black, white, honey, brown, red, yellow or a mix [43].

Hamsters are more crepuscular (primarily active during the twilight period), than nocturnal or diurnal and remain concealed during the day to escape being trapped by predators. Primarily, they feed on seeds, nuts, fruits and occasionally eat burrowing insects. They carry food in their spacious cheek pouches to their underground storage chambers [44].

Wild hamsters are omnivores and exist on a mixed diet of vegetation, grains and grasses as well as some insects. Strawberries are a perfectly healthy, safe snack for hamsters and can be a good part of their balanced diet. Hamsters can eat strawberries quite comfortably without hurting to themselves. The fruits have a great nutritional value for hamsters since it also provides them with many health benefits.

**Figure 23.**
*Hamsters.*

### *2.2.8 Guinea pigs*

Guinea pig (*Cavia porcellus*), also known as the cavy in family Caviidae have a short and smooth coat that may be cream, white, reddish or chocolate brown, tan, black, or a combined pattern. There is a crest of longer hairs at the neck, but tail is not visible externally. It resembles to other cavies in having a robust body with large head and eyes, and short ears and limbs. There are four toes on the forefeet and three on the hind feet having hairless soles and short sharp claws (**Figure 24**). The guinea pig's natural diet is grass and their molars are particularly suited for grinding of plant matter [45, 46].

Herds of animals move together, eating grass or other vegetation, they do not build nest or burrow, but frequently seek shelter in the burrows of other animals. They are herbivores, feed on mainly grasses and greens, and most likely love to chew on some strawberries, and can eat strawberry's stems, tops and leaves. Similar to strawberries, strawberry stems might get even more excited to pest than the fruit itself. Leaves of strawberries are another great treat for cavy, as they contribute to facilitate a proper

**Figure 24.**
*Guinea pig.*

digestion of the fruit. And it is needless to say that guinea pigs adore chewing on various green leaves just for fun. Additionally, guinea pig's health will utterly benefit from the high levels of compounds that strawberries contain [47, 48].

For protecting of strawberry plants from guinea pig, gardeners can build a wooden cage and install a wire mesh over the cage to avoid approaching animals from entering into this enclosure. Unfortunately, fencing would not always help to keep guinea pig and other rodents away from strawberry plants. It is because these culprits have the ability to dig underground holes and come up to garden for feeding on these mouthwatering berries.

### *2.2.9 Marmots*

Marmots also known as groundhogs are squarish stocky animals; they can climb and swim; have short strong legs; coarse, grizzled, gray-brown fur; small ears; a short, bushy tail; and curved claws. They have large, competent, chisel-like teeth that make them efficient, but not quite endearing. Marmots are occasionally destructive to gardens and pasturelands, and not only eat garden crops, but sometimes also eat worms and insects. Greens food like lettuce, dandelions, daisies, alfalfa, clovers, red mulberry and hackberry leaves they eat from the garden. The food in the trees category includes bark and twigs, and they love eating vegetables like celery, carrots, broccoli, peas, corn and beans. The fruits that they feed on include berries, cherries and apples, and these food varieties can draw in groundhogs. Accordingly, the more groundhog food has in garden, the more probable these critters will be to make tunnels. This is particularly obvious since groundhogs travel no farther than 150 feet from their homes for food [49, 50].

Apart from other rodents, marmots also love to the taste of strawberries and some of them can even damage the crop just to nibble on a few. Gardeners might even face mysterious disappearances of berries or may have already noticed every morning. Their habit of burrowing makes them serious nuisance animals around farms and gardens, and are a thorn on the side of many gardeners.

Groundhog (*Marmota monax*), also known as woodchucks, thickwood badger, red monk, land beaver or whistling pigs, are members of the squirrel family Sciuridae known as marmots. Groundhogs have four incisor teeth, are well-adapted for digging, with short, powerful limbs and curved thick claws, and tail is comparably shorter (**Figure 25**). It is typically found in forests, small woodlots, fields, pastures and hedgerows, and constructs dens in well-drained soil. It cleans its face similar to the manner of the squirrels and licks its fur like the manner of a cat. They are solitary hibernating mammals and eat almost every type of green plants [51].

Mostly groundhogs are herbivorous, eat primarily wild grasses and other vegetation, including berries and agricultural crops. Groundhogs live in tunnels, which can have more than one entry and burrows can be simply recognized by way of dunes of

**Figure 25.**
*Marmot.*

dug soil along with entries. Their dirt mounds and tunnel holes obstruct to farmhouse tools and can pose a risk to livestock and horses. An adult groundhog can consume about a pound to a pound-and-a-half of vegetation daily. They eat everything from flowers to vegetables, and consume grasses, clovers, leaves, twigs, berries, beans, plantain, soybeans, blossoms, twigs, bark and bugs. Damage to fruit and ornamental trees is caused by gnawing, so, look for paw imprints with four toes on the front paws and five toes on the back with one set lower than the rest [51].

Eliminate woodpiles and other places where groundhogs nest, and keep undergrowth and grass cover low to deter groundhogs. Groundhogs are always looking for vacant burrows, so, close down their tunnel systems. The most important feature to help keep groundhogs out of garden is a good fence and if the steel mesh is embedded around six inches into the ground, this can help reduce the chances of the groundhog being able to dig under the fence. Excluding of woodchucks from the garden will also keep out deer, rabbits and raccoons. Once is selected a groundhog trap, lure the animal inside with a carefully designated and positioned bait such as strawberries in order to lead a successful catch.

### *2.2.10 Rodents control for strawberries*

Eradicate rodents habitation through clearing of tall bushes and grasses surrounded by 50 feet from plantings. Make certain that grasses nearby to field or garden are clipped during the developing period and earlier to frost. At no time let firewood trash, or boards to hoard nearby field or garden and at no time left strawberries filled by weeds. Rodents can be controlled with traps, repellants and baits options.

#### *2.2.10.1 Trapping*

Trapping of moles, voles and mice can be done by usage everyday mousetraps for home based routine. Peanut butter positioned under the pressure trigger is helpful to stop the rodents from acquiring the diet without trap setting. Just the once baited, place the trap alongside the base of rodent's burrow or a noticeable runway and align the trap at an exact angle. Once a day, check the trap and eliminate any deceased rodents, and re-bait, if needed until no anymore is catching. Retaining pieces of apples in the rodent's burrow during the spring will aid to conclude if orchard is once again infested with these pests. If the apple pieces display symbols of gnawing, then once again, place newly baited traps near to burrow [52].

#### *2.2.10.2 Repellents*

Marketable rodent repellents comprising of thiram are helpful to retain moles, voles, mice and other pests far from strawberry plants. These repellants have their individual particular guidelines that must be trailed to take full advantage of efficiency and avoid harm to the plants. One more choice is to remove the rodent's hiding areas and habitat through observance of areas with high grasses clipped and yard neat and free of wildflowers. Caring of strawberry plants with fences, such as barrier is another choice to retain rodents from ingestion of plants. But, these fences normally shield to stems and leaves of strawberry plant, and ensure nil to shelter the roots [53].

#### *2.2.10.3 Baits*

There are certain rodenticides existing for field usage and can only be used by certified pesticide applicators. However, aluminum phosphide tablets and gas cartridges could be used to fumigate pest burrows. Agrarians may use bait stations to evaluate rodent populations decline, afterward use of rodenticide across the plantation flooring and below the canopies of trees. Globally, rodent pests are maximally frequently controlled chemically and there is an augmented interest in biological control by avian predators. Several avian predators (*Tyto alba*, *Elanus axillaris*, *Falco tinnunculus*, *Falco cenchroides*, *Bubo bengalensis*, *Buteo rufinus*) are commonly quoted in the biological control of rodents; however, barn owl (*T. alba*) is the most cited species [54].

## 2.3 Other vertebrate pests invader of strawberries

There are a few other herbivorous and mammals that become sometimes sporadic pests on crops. These include some species of rabbits and deer that depredate on strawberries.

### *2.3.1 Rabbits*

Also known as bunnies, rabbits (Animalia: Chordata: Mammalia), are small mammals of the family Leporidae in the order Lagomorpha, along with hare (larger in size than rabbits and have proportionately longer ears) and pika (an even coat of fur, short rounded ears, short limbs, round body and no external tail), having similar herbivorous diets. Rabbits have large hind leg bones and front foot has four toes plus a dewclaw, while each hind foot has four toes (**Figure 26**). Rabbit habitats include forests, woods, meadows, grasslands, wetlands and deserts. They live in burrows, as a result of their appetites and the breeding rate their depredation can be problematic for agriculture. Fumigation of warrens, barriers (fences), shooting, snaring and ferreting have been used to control rabbit populations [55].

The rabbits executing the maximum of injury in gardens and yards are black-tailed jackrabbit (*Lepus californicus*), desert cottontail rabbit (*Sylvilagus audubonii*) and desert brush rabbit (*Sylvilagus bachmani*). These brown to gray rabbits a lot look

**Figure 26.**
*Rabbit.*

similar, however they differ in size significantly. Jackrabbits are the biggest, measuring 17–21 inches long and weighing 3–7 pounds. Desert cottontails develop to 1–1/2 to 2–3/4 pounds weight and 12–15 inches long. The brush rabbit is the tiniest among them, at 1–1/4 to 1–4/5 pounds weight and 11–13 inches in length. Rabbits also leave behind dark round fecal pellets in abundance. Rabbits normally feed at night, but a desperate animal may feed any time of day. Rabbits are cute, but they will make strawberry plants their food. Rabbits munch on foliage and other tender parts of strawberry plants [56].

Noisemakers, ultrasonic devices and lights are only effective for short-term protection. Installing of barrier may not retain these pests far away from plants, as they can excavate below the fence. Chicken wire may be used to make a barrier, if its bottom is curved at an angle far from the orchard and then concealed no less than 6 inches, to keep rabbits away from strawberry plants. The chicken wire barrier must be as a minimum of 24 inches tall and borders the whole plant. Garlic may act as a rabbit repellent, especially if it is mixed with oil and fish emulsion, and used as a spray for plants. Probably, a limited gnawed foliage would not destroy to vine, however rabbits quickly can destroy a fresh vine in initial spring. Protect to fresh vines with a plastic cover or encircle these in chicken wire. Chicken wire having holes lesser than 1 inch can efficiently prevent to rabbits, which can squeeze through bigger places. As soon as plants are mature, it is generally safe to get rid of protecting coverings. Rabbit deterring sprays that contain rotten egg, cayenne pepper or castor oil are in use, however they need to be reapplied repeatedly, particularly after irrigation or rain. Blood meal mixed into a gallon or water or an Epsom salt spray will keep the rabbits from eating young berry plants [57, 58].

### 2.3.2 Hungry herd of deer

Deer or true deer are hoofed ruminant mammals from the family Cervidae (diverse family after bovids) forming the Artiodactyla order. Deer are widely distributed in a variety of biomes, ranging from tundra (treeless mountain tract) to the tropical rainforest, prairie and savanna (open space). Deer are browsers, feed primarily on foliage of trees, grasses, shrubs, forbs and sedges, and secondarily on lichens during winter in northern latitudes. Spotted deer also known as chital deer or axis deer (*Axis axis*), males are bigger than females, antlers are only present on males, upper portions are golden to rufous and covered completely with white spots (**Figure 27**). The throat, ears, rump, insides of legs, tail and abdomen are whole white, whereas a noticeable black stripe goes alongside to back bone [59].

Deer are stylish creatures that can cause extensive damage to strawberry plants. Deer forage on the foliage of strawberries and can consume the entire plant leaving only a stub. Deer can cause serious damage to strawberry plants unless precautions are taken. Deer feed on the leaves and fruit of strawberry plants, and can also trample

**Figure 27.**
*Deer.*

plants. There are various methods that can be employed to have deer management and they can be fragmented down into two main types: nonlethal (chemical repellents, fencing to form a physical barrier and wildlife birth control using vaccines administered by dart or shot) and lethal (hunting for meat known as venison or a trophy using a bow and arrow, rifle, muzzle loader or other approved weapons) to keep them away or minimize the damage. Both these methods are used to regulate the population of deer in a specific area if their populations remain high for long periods [60].

Deer damage to orchards by nourishing on roots, leaves and stems of plant. Fortuitously, onion, garlic, catnip, dill, chives, oregano, lavender, sage, rosemary, thyme and spearmint retain deer away from lawn, garden and shrubbery. Even though persons find the odor of these herbs amusing, deer distaste to their fragrance and give a wide berth to areas planted with them. Growers can plant a boundary of more or one of these herbs round whole garden, lawn or landscape expanse, so they may enclose bushes or plants that deer find pretty. Blood meal, feather meal and marketable deer repellents positioned nearby the strawberry plants will be helpful to retain deer not there. It is superlative to usage these repellents earlier to deer start feeding on the plants. One more choice is to shelter the strawberry plants through set up a barrier round the whole orchard. But, this could be a pricey decision dependent on the bulk of the orchard and kind of material selected for the barrier. Odor repellents may keep deer out of gardens and options include soap, dog hair and human hair [61, 62].

## 3. Conclusion

Strawberries are packed with full of vitamins C, B9 and K, plus healthy dietary fiber, manganese, and potassium. Planting of strawberries is a serious work, not to mention expensive. But, strawberry fields are to be protected from birds and other rodents like rats, mice and chipmunks. Just like any other animals, rodents afraid of humans, so to deal with them is by using of a battery-operated radio in filed at each night when these pests usually strike. Tune into a 24-hour radio station that broadcasts news rather than music. It will sound as if humans are near and talking to each other. Any other method of growers choice is expected to provide more or less achievement in guarding of berry yield, though the top durable key, definitely is to practice net above berry crop. Generally, the reap period for birds is a little one nearly the similar period it is for humans reaping berries. Hence, it needs not to have much worry in the initial days of berry development and maturing. However, likewise remember that wildlife of backyard is one of the delights for gardening. Birds are helpful by consuming of pest insects and they increase a component of living attractiveness to the orchard. Through trying these simple humane ways of guarding berries, growers can peacefully coexist with these feathered groups. The simplest method of protecting strawberries, works against all animal pests, but not just birds or rodents. An exclusion method of protection also prevents birds and rodents from wreaking havoc in strawberry garden. Over the top of the stakes, drape bird's net, holding the netting over the plants. Mound rocks or soil onto the edges of netting to anchor it to the earth for excluding rodents, birds and other animals from strawberry area.

Birds love to eat strawberries because they are sweet and juicy and the perfect way for birds is to get lots of nutrients. Insects such as tiny worms or flies can be found on strawberries, which can boost a bird's protein intake. All these vitamins and nutrients are essential to keep birds healthy and aid in preventing of diseases. Strawberries are safe for wild birds and although it is important to remember that store bought fruit may have been treated with pesticides. When a pet bird eats pesticide chemicals, it can cause internal damage or even death. For this reason, it is always recommend offering home-grown or organic strawberries to the pet bird if it can be done. If that is not possible, then strawberries should be cleaned before offering them at birds feeder by soaking in a diluted vinegar solution (4 parts water, 1 part vinegar) for 15 min and then thoroughly rinse with fresh water. Also, there can be used white vinegar or apple cider vinegar for washing of strawberries.

It is not necessary to kill all probable pests to protect whole strawberry crop. Diligent pests control can preserve strawberry crop and let Nature to take care of birds, rodents and other pests. Killing of endangered birds or rodents is a crime that can result in civil and criminal penalties. Determine, which species of animals are occurring as pests in orchards, fields or yards of any region, but avoid using control methods that could harm to endangered species before to begin a control program. Such consequences will enable managers to manage or modify landscapes at appropriate levels to increase avian and rodent fauna to real densities.

## Author details

Muhammad Sarwar
National Institute for Biotechnology and Genetic Engineering (NIBGE), Faisalabad, Pakistan

*Address all correspondence to: drmsarwar64@gmail.com

## References

[1] Hira S, Imran S, Mariha A, Sarwar MF, Sarwar MH, Sarwar M. Fundamental aspects of vitamin B complex in human nourishment and fitness. American Journal of Food Science and Health. 2020;**6**(4): 109-118

[2] Sarwar MF, Sarwar MH, Sarwar M. Deficiency of vitamin B-complex and its relation with body disorders. In: LeBlanc JG, editor. B-Complex Vitamins—Sources Intakes and Novel Applications. London, UK: IntechOpen Ltd.; 2021. p. 21

[3] Ibrahim DS, Abdel-Maksoud MAE. Effect of strawberry (*Fragaria* × *ananassa*) leaf extract on diabetic nephropathy in rats. International Journal of Experiential Pathology. 2015; **96**(2):87-93

[4] Sarwar MH, Sarwar MF, Khalid MT, Sarwar M. Effects of eating the balance food and diet to protect human health and prevent diseases. American Journal of Circuits, Systems and Signal Processing. 2015;**1**(3):99-104

[5] Sarwar M. Crops depredation by birds (Chordata: Aves) and deterrence of the forfeiture. Agricultural and Biological Sciences Journal. 2022;**8**(1):9

[6] Sarwar M. Some observations on species composition and deterioration of crop plantations and forest flora by porcupines in consort with control techniques. International Multidisciplinary Research Journal. 2018;**8**:08-14

[7] Sarwar M. Raiding of agricultural crops and forests by wild boar (*Sus scrofa* L.) and its mitigation tricks. Journal of Scientific Agriculture. 2019;**3**:1-5

[8] Roberts TJ. The Birds of Pakistan. Passeriformes: Pittas to Buntings. Vol. 2. Oxford: Oxford University Press; 1992. p. 682

[9] Nyari A, Ryall C, Peterson AT. Global invasive potential of the house crow *Corvus splendens* based on ecological niche modeling. Journal of Avian Biology. 2006;**37**:306-311

[10] Brook BW, Sodhi NS, Soh MCK, Lim HC. Abundance and projected control of invasive house crows in Singapore. The Journal of Wildlife Management. 2003;**67**(4):808-817

[11] Capainolo P, Perktaş U, Fellowes MDE. Late quaternary range dynamics in the common grackle *Quiscalus quiscula*. Ardea. 2020;**108**(1):95-104

[12] Peris S, Soave G, Camperi A, Darrieu C, Aramburu R. Range expansion of the European starling *Sturnus vulgaris* in Argentina. Ardeola. 2005;**52**(2):359-364

[13] Clement P, Harris A, Davis J. Finches and Sparrows: An Identification Guide. London: Christopher Helm; 1993. p. 480

[14] Anderson TR. Biology of the Ubiquitous House Sparrow: From Genes to Populations. Oxford: Oxford University Press; 2006. p. 560

[15] Ericson PGP. The phylogenetic relationships and generic limits of finches (Fringillidae). Molecular Phylogenetics and Evolution. 2012;**62**(2):581-596

[16] La HeraI D, Fandos G, Fernandez-Lopez J, Onrubia A, Perez-Rodríguez A, Perez-Tris J, et al. Habitat segregation by breeding origin in the declining populations of European robins wintering in southern Iberia. Ibis. 2018; **160**(2):355-364

[17] Morneau F, Lepine C, Decarie R, Villard MA, DesGranges JL. Reproduction of American robin (*Turdus migratorius*) in a suburban environment. Landscape and Urban Planning. 1995;**32**(1):55-62

[18] Rising J, Williams P. In: Poole A, Gill F, editors. Bullock's Oriole. Vol. 416. Philadelphia, PA: The Birds of North America, Inc.; 1999. pp. 1-20

[19] Rising J, Flood N. In: Poole A, Gill F, editors. Baltimore Oriole. Vol. 384. Philadelphia, PA: The Birds of North America, Inc.; 1998. pp. 1-32

[20] Otter KA. Ecology and Behaviour of Chickadees and Titmice: An Integrated Approach. Oxford: Oxford University Press; 2007. pp. 153-230

[21] Jewell SD. Weights and wing lengths in Connecticut Blue Jays. Connecticut Warbler. 1986;**6**(4):47-49

[22] Ortega C, Hill GE. Black-headed Grosbeak (*Pheucticus melanocephalus*). In: Poole AF, editor. The Birds of North America. Ithaca, NY, USA: Cornell Lab of Ornithology; 2010

[23] Topfer T. Suspected road salt poisoning in Bohemian Waxwings *Bombycilla garrulous* (Aves: Passeriformes: Bombycillidae). Vertebrate Zoology. 2010;**60**(2):171-174

[24] Zink RM, Dittmann DL. Evolutionary patterns of morphometrics, allozymes, and mitochondrial DNA in Thrashers (genus *Toxostoma*). The Auk. 1999;**116**(4):1021-1038

[25] Sarwar M. Keeping ecosphere safe from pestilences of insects by incorporating birds bio-control agents. Agricultural and Biological Sciences Journal. 2022;**8**(1):7

[26] Wilson DE, Reeder DM. Class Mammalia Linnaeus, 1758. In: Zhang ZQ, editor. Animal Biodiversity: An Outline of Higher-Level Classification and Survey of Taxonomic Richness. Vol. 3148. Zootaxa; 2011. pp. 56-60

[27] Sarwar M. Menacing populations determinant of mice (Rodentia: Muridae) and guideline for the interference. International Journal of Plant Science and Ecology. 2021;**7**(4): 91-98

[28] Sarwar M, Ashfaq M, Baig MY. The species complex, damage pattern and control of rodents (Mammalia: Rodentia) in sugarcane (*Saccharum officinarum* L.) fields. International Journal of Agronomy and Plant Production. 2011;**2**:145-150

[29] Sarwar M. Pattern of damage by Rodent (Rodentia: Muridae) pests in wheat in conjunction with their comparative densities throughout growth phase of crop. International Journal of Scientific Research in Environmental Sciences. 2015;**3**(4): 159-166

[30] Millin M, Rickert GT. Effect of a strawberry and spinach dietary supplement on spatial learning in early and late middle-aged female rats. Antioxidants (Basel). 2019;**8**(1):1

[31] Sarwar M. The Rodents (Mammalia: Rodentia)-gnawing away on crops and options for the integrated pest management at field. American Journal of Marketing Research. 2015;**1**(3): 136-141

[32] Sarwar M. The rodents (Mammalia: Rodentia)-gnawing away on stored grains and options for the integrated pest management in stores. American Journal of Food Science and Health. 2016;**2**(6): 161-168

[33] Sarwar M. Voles (Rodentia: Cricetidae) as pesky animals for cultivated areas and approaches to damage administration. Journal of Agricultural Science and Engineering. 2022;**8**(1):15

[34] Fritzell EK. Mammals in prairie wetlands. In: Vander VA, editor. Northern Prairie Wetlands. Ames: Iowa State University Press; 1989. pp. 268-301

[35] Mouchaty SK, Gullberg A, Janke A, Arnason U. The phylogenetic position of the Talpidae within Eutheria based on analysis of complete mitochondrial sequences. Molecular Biology and Evolution. 2000;**17**(1):60-67

[36] Kott O, Sumbera R, Nemec P. Light perception in two strictly subterranean rodents: Life in the dark or blue. PLoS One. 2010;**5**(7):e11810

[37] Thorington RW, Hoffmann RS. Family Sciuridae. In: Mammal Species of the World: A Taxonomic and Geographic Reference. Baltimore: Johns Hopkins University Press; 2005. pp. 754-818

[38] Nowak RM. Walker's Mammals of the World. 16th ed. Baltimore: Johns Hopkins University Press; 1999. p. 1936

[39] Patterson BD, Norris RW. Towards a uniform nomenclature for ground squirrels: The status of the Holarctic chipmunks. Mammalia. 2016;**80**(3): 241-251

[40] Baack JK, Switzer PV. Alarm calls affect foraging behavior in eastern chipmunks (*Tamias striatus*, Rodentia: Sciuridae). Ethology. 2003;**106**: 1057-1066

[41] Okuyama MW, Michito S, Go ABE, Mariko N, Mariko S, Ken-Ichiro S, et al. Timing of puberty and its relationship with body growth and season in male raccoons (*Procyon lotor*) in Hokkaido. Journal of Reproduction and Development. 2013;**59**(4):361-367

[42] Ikeda T, Asano M, Abe G. Present status of invasive alien raccoon and its impact in Japan. Global Environmental Research. 2004;**8**(2):125-131

[43] Gattermann R, Fritzsche P, Neumann K, Al-Hussein I, Kayser A, Abiad M, et al. Notes on the current distribution and the ecology of wild golden hamsters (*Mesocricetus auratus*). Journal of Zoology. 2001; **254**(3):359-365

[44] Logsdail C, Logsdail P, Hovers K. Hamsterlopaedia: A Complete Guide to Hamster Care. Lydney: Ringpress; 2002. p. 161

[45] Vanderlip S. The Guinea Pig Handbook (Barron's Pet Handbooks). Knoxville: B.E.S. Publishing; 2003. p. 168

[46] Graur D, Hide W, Li W. Is the Guinea-pig a rodent? Nature. 1991; **351**(6328):649-652

[47] Zacharisen MC, Levy MB, Shaw JL, Kurup VP. Severe allergic reactions to Guinea pig. Clinical and Molecular Allergy. 2005;**3**:14

[48] Manjeli Y, Tchoumboue J, Njwe RM, Teguia A. Guinea-pig productivity under traditional management. Tropical Animal Health and Production. 1998; **30**(2):115-122

[49] Kwiecinski GG. *Marmota monax*. Mammalian Species. 1998;**591**:1-8

[50] Stam MZ, Christine RM, Jerry AW, Gregory LF. Latitudinal differences in the hibernation characteristics of woodchucks (*Marmota monax*). Physiological and Biochemical Zoology:

Ecological and Evolutionary Approaches. 2010;**83**(1):135-141

[51] Aschemeier LM, Maher CR. Eavesdropping of woodchucks (*Marmota monax*) and eastern chipmunks (*Tamias striatus*) on heterospecific alarm calls. Journal of Mammalogy. 2011;**92**(3):493-499

[52] Sarwar M. The species diversity, seasonal variation and abundance of rodents (Mammalia: Rodentia) along with their burrows distribution pattern in wheat farmland. Journal of Basic and Applied Research International. 2015; **5**(1):48-55

[53] Sarwar M. Species complex, damage pattern and efficiency of rodenticides in controlling rodents attacking rice (*Oryza sativa* L.) fields. International Journal of Animal Biology. 2015;**1**(5):202-208

[54] Labuschagne L, Swanepoel L, Taylor P, Belmain S, Keith M. Are avian predators effective biological control agents for rodent pest management in agricultural systems? Biological Control. 2016;**101**:94-102

[55] Lieber RL, Blevins FT. Skeletal muscle architecture of the rabbit hindlimb: Functional implications of muscle design. Journal of Morphology. 1989;**199**(1):93-101

[56] Autifi MAH, El-Banna K, Ebaid AES. Morphological study of rabbit lung, bronchial tree, and pulmonary vessels using corrosion cast technique. Al-Azhar Assiut Medical Journal. 2015; **13**(3):41-51

[57] Memmott J. Food webs: A ladder for picking strawberries or a practical tool for practical problems? Philosophical Transactions of the Royal Society B: Biological Sciences. 2009;**364**: 1693-1699

[58] D'Ovidio D, Pierantoni L, Noviello E, Pirrone F. Sex differences in human-directed social behavior in pet rabbits. Journal of Veterinary Behavior. 2016;**15**:37-42

[59] Francis CM. A Field Guide to the Mammals of South-East Asia. London, UK: New Holland; 2008. p. 130

[60] Heffelfinger J. Deer of the Southwest: A Complete Guide to the Natural History, Biology, and Management of Southwestern Mule Deer and White-Tailed Deer. 1st ed. Texas, USA: Texas A & M University Press; 2006. pp. 1-57

[61] Feldhamer GA, McShea WJ. Deer: The Animal Answer Guide. Baltimore, USA: Johns Hopkins University Press; 2012. pp. 1-142

[62] Nugent G, Fraser KW. Pests or valued resources? Conflicts in management of deer. New Zealand Journal of Zoology. 1993;**20**(4):361-366

Section 6

# Postharvest Techniques of Strawberries

Chapter 13

# Post-Harvest Problems of Strawberry and Their Solutions

*Huma Qureshi Quarshi, Waseem Ahmed, Rafia Azmant, Nabila Chendouh-Brahmi, Abdul Quyyum and Asad Abbas*

## Abstract

Strawberry is a fruit with a short season of harvest. Strawberry is well-known among people all over the world for its distinct flavour, nutritional value, and delicacy. While on the other hand, preserving strawberry and shelf life extension has been a huge difficulty due to their perishable nature. Making effective and sustainable use of already available food processing and preservation technology needs time. Researchers must use advanced techniques like a cool store, modified atmospheric packaging (MAP), cool store, controlled atmospheric storage (CA), various packaging methods, and a variety of chemical and physical treatments to retain commodities for a longer period due to strategic market sales following harvest. Except for the preserving techniques, there is some polysaccharide-based edible coating which has a crucial role in delaying fruit softening, fruit decay, maintaining the increased levels of ascorbic acid and phenols, enhancing the activities of antioxidant enzymes, and reducing membrane damage. During the postharvest stages, there are numerous threats to keep in view regarding the safety and quality of strawberries. In this chapter, we will discuss the benefits and drawbacks of some of the various preservation technologies, as well as how they might be utilised to preserve and a prolonged period of freshly harvested strawberries.

**Keywords:** non-climacteric, strawberry, perishable, fruit softening, preservation

## 1. Introduction

The strawberry (*Fragaria ananassa)* belongs to the Rosaceae (flowering plant). It is among the most well-known non-climacteric fruits because of its nutritional and organoleptic characteristics [1]. The fleshy component is formed not from the plant's ovary but from the receptacle that stores the ovaries, making it an aggregate accessory fruit. Each "seed" (achene) outside the fruit is one of the flower's ovaries, which contains seeds. The genetic makeup of the currently grown strawberry fruit is an octoploid hybrid (8n), with 56 chromosomes. It's eaten in huge volumes, both freshly and in ready food like juice, pies, jam, ice cream, chocolates, milkshakes etc. Strawberries are commercially grown for sustenance as well as can be eaten fresh or processed into frozen, canned, or preserved fruit or juice.

Strawberry fruit slices are high in flavonoids, fibre, vitamins, potassium, and a wide range of phenolic acids, including hydroxycinnamic and hydroxybenzoic acids [2]. They are quite beneficial to one's health. They're high in vitamin C and manganese, and they also have a considerable quantity of folate (vitamin $B_9$) and potassium. Fruit antioxidants such as kaempferol, quercetin, and anthocyanins help to prevent the creation of deadly blood clots that are linked to strokes. The antioxidants work by neutralising free radicals in the body, preventing tumour formation, and reducing inflammation. Because of their nutritious content, strawberries had been recommended to those with high blood pressure and sugar. Strawberries have a low glycemic index (40) as compared to other fruits, making them a good choice for diabetic patients.

**Table 1** shows the nutritional values per 100 g of strawberry:

Strawberry production is getting importance by the consumers due to its nutritional facts. In 2019, global strawberry output totalled 8.9 million tons, with China accounting for 40% of the entire and the United States and Mexico rounding out the top three producers. China is the largest producer with 3.9 million tons of production annually in 2019. **Table 2** shows the top producers of strawberries in 2019.

Due to the ever-increasing demand for strawberries, several issues arise during the cultivation, production and harvesting phases. Strawberry is regarded as among the most problematic fresh food to preserve due to the difficulties in maintaining fruit freshness [3]. After production, the berries are harvested by hands and placed in trays for further operations. Moreover, it is highly perishable and has a limited postharvest life, owing to its fast metabolism and sensitivity to mechanical damage as well as infection by plant pathogens bacteria, fungi, and viruses [4].

Various operations like cooling at low temperature, edible films coating, UV radiations, fruit sanitization and many more are carried out for lowering the respiration rate and loss of water, maintaining fruit firmness, and restricting microbial spread are all objectives of postharvest operations to extend its life span. Cooling at low temperatures is one of the most efficient procedures for increasing fruit longevity [5]. In recent years,

| Nutritional components | Quantity | Nutritional components | Quantity |
|---|---|---|---|
| Energy | 136 kJ | Choline | 5.7 mg |
| Water | 90.95 g | Vitamin C | 58.8 mg |
| Carbohydrates | 7.68 g | Vitamin E | 0.29 mg |
| Sugars | 4.89 g | Vitamin K | 2.2 μg |
| Dietary fibre | 2 g | Calcium | 16 mg |
| Fat | 0.3 g | Iron | 0.41 mg |
| Protein | 0.67 g | Magnesium | 13 mg |
| Thiamine ($B_1$) | 0.024 mg | Manganese | 0.386 mg |
| Riboflavin ($B_2$) | 0.022 mg | Phosphorous | 24 mg |
| Niacin ($B_3$) | 0.386 mg | Potassium | 154 mg |
| Pantothenic acid ($B_5$) | 0.125 mg | Sodium | 1 mg |
| Vitamin $B_6$ | 0.047 mg | Zinc | 0.14 mg |
| Folate $B_9$ | 24 μg | — | — |

**Table 1.**
*The nutritional values per 100 g of strawberry.*

| Countries | Production (million tons) |
|---|---|
| China | 3.2 |
| United States | 1.0 |
| Mexico | 0.9 |
| Egypt | 0.5 |
| Turkey | 0.5 |
| Spain | 0.4 |
| World | 8.9 |

**Table 2.**
*Countries wise the production (million tons).*

there has been a surge in interest in edible films coating. Edible films are effective as they provide a physical border around the fruit, shielding it from moisture, fumes, and microorganisms that could compromise its quality [5]. Aside from that, starch-based films are edible, translucent, odourless, tasteless, and colourless, all of which are desirable qualities for food packaging [6]. The use of C ultraviolet (UV-C) entails eradicating microorganisms from vegetative tissues through the antibacterial effect generated by radiation, which is one of the strategies that has been gaining traction to extend post-harvest fruit lifespan [7]. Chlorine, mostly in the form of sodium hypochlorite, is commonly employed in fruit sanitization. It works by removing harmful organisms, resulting in a longer fruit lifespan [8]. A need to produce healthier foodstuff while reducing the usage of fungicides and other elevated toxic treatments emphasise the significance of investigating and presenting sustainable fruit conservation approaches in detail. The goal of the study is to assemble and discuss in detail the procedures that must be followed during harvest and postharvest activities of strawberries to minimise losses.

## 2. Factors involve in post-harvest handling of strawberries and their issues

### 2.1 Harvesting of strawberry

Strawberry fruit is non-climacteric, meaning it ripens while still attached to the parent plant. The fruit should be harvested after it has reached the desired shape and size, as determined by varietal characteristics. The berries should be harvested with care because ripe one's bruise readily. Strawberry fruit is carefully collected with its calyx by manually pinching the stem (0.5–1.0 cm above calyx) instead of pressing the fruit's fleshy section.

Three main factors affect the harvesting and postharvest strategies of strawberries.

- Phases of fruit ripening: Strawberries should not ripen after they have been picked. They should be harvested when they are ripe, red, and firm. Harvesting time is determined by the fruit's ripeness and market demand. Plant nutrients can help the fruit to retain its firmness and freshness.

- Cleanliness: Keep the fruits clean as much as possible. Keep watering the roads to avoid the dust and vulnerability to soil-born pathogens. Instruct labourers on the causes of contamination, the consequences (health issues, reputational damage, financial loss, etc.), and a detailed study to prevent contamination.

- Harvesting time: The harvesting time has a significant impact on the fruit's shelf life. Try to keep the fruit at its optimal temperature. Before the temperature rises, pluck the fruits early in the morning before 10 am to avoid heat injury and sunburn. Before transferring the fruit to the cold storage, keep it in the shade. Usage of the forced air-cooling system to reduce the temperature of the fruit to an acceptable level is important. At 0°C–2°C, the life span is extended.

- Other factors: Place the berries in suitable containers, reduce vibration damage during transportation, reduce the exposure duration to unfavourable conditions, storage etc.

Though manual harvesting is the most common way of strawberry harvesting, various investigations on mechanical and intelligent robotic harvesting systems have been conducted. Han et al. [9] created a computer vision-assisted robot system to recognise and handle mature strawberries planted in a bench-type farming system. The fruit detection algorithms were programmed utilising real-time position tracking methods in natural light. Strawberry harvesting [10], cucumber harvesting [11], aubergine harvesting [12], and de-leafing [13] are only a few of the studies that have used robotic technology in greenhouses.

### 2.2 Pre-cooling

Pre-cooling fresh fruits before storage and marketing have a big impact on their quality [14]. The respiration rate and all metabolic responses in newly harvested produce are reduced when field heat is removed. Pre-cooling is the elimination of heat from the field from the freshly harvested commodity. Strawberry precooling (rapid reduction of field heat) is required within 1 hour following harvest. Strawberry has a high rate of metabolism, so its marketability is reduced by 20%, 37%, 50%, or 70% with cooling delays of 2, 4, 6, or 8 hours, respectively. There are different methods involved in precooling such as forced air cooling, hydro cooling contact icing, room cooling, vacuum cooling. All these techniques vary in their ability to remove field heat. Forced-air cooling is the most popular approach for precooling berries. The containers are forced to circulate cold air quickly, enabling the cold air to come into interact directly with the warm berries. Strawberry pallets are arranged in such a way that cooler air must penetrate through the package opening as well as around each berry to reach the low-pressure area. Heat removal is a very important process to the low temperature (0°C–1°C) and relative humidity (90%–95%) because it extends the shelf life of produce, lessens physiological decline and reduces the occurrence of pathological degradation [15, 16].

A few elements influence the system's cooling rate and performance:

1. the difference in temperature between the cool air and the fruit,

2. the rate at which air flows,

3. the fruit's ability to be exposed to cool air,

4. the width and length of the air channel.

Although hydro-cooling is a faster form of precooling, strawberries are not commercially hydro-cooled due to deterioration concern by the water left on strawberries after hydro cooling [17]. As a result, the most frequent strategy for preserving strawberry fruit quality after harvest is to cool the fruits immediately after harvesting and then store them at a low temperature (0°C–4°C) [18].

### 2.3 Modified atmosphere packaging

Mmodified atmospheric packaging (MAP) is a technique of packaging that involves altering the gaseous atmosphere surrounding a food product within a package by using packing materials and layouts that have enough level of gas barriers to keep the altered environment at a safe level for food preservation. MAP methods preserve fresh food by reducing oxygen exposure and raising carbon dioxide content. The use of $CO_2$ as part of a modified environment has been found to help preserve strawberry sensory attributes [19]. The oxygen content should be kept low because oxidation is caused by the presence of oxygen and can result in discolouration, rotting, and off-flavours and textures. The model incorporates respiration, transpiration, and diffusional movement of $O_2$, $CO_2$, $N_2$, and $H_2O$ in a microperforated pack.

For strawberries, the optimal MAP composition is $O_2$ (5%–10%), $CO_2$ (15%–20%), and $N_2$ (70%–80%) [20]. Maintaining a reasonably low temperature is required for MAP storage since the temperature has a greater influence on the rate of respiration than gaseous concentration, with a 72%–82% drop in respiration rate for various $O_2$ and $CO_2$ combinations when the temperature was reduced from 23°C to 10°C [21]. Short exposure to increased levels of carbon dioxide has been shown to reduce the chemical and physical concepts affiliated with fruit decay, lowering tissue ATP levels and producing a low ethanol metabolism, in contrast to when the fruit is stored in the presence of air, which causes an increase in ATP and an explosion of fermentation processes in the tissue, leading to putrefaction [22]. Similarly, when strawberries are processed for 3 hours at 3°C with % $CO_2$, then packaged in MAP film, stored at 1°C for a day, transported at 1°C for 10 days, and lastly distributed at 4°C for 3 days [23]. When paired with MAP, $CO_2$ treatment preserved the quality of fruit (strawberry) by lowering weight loss, tissue softening index, as well as duration.

Polypropylene with different perforation sizes is used as a packaging material in modified atmosphere packaging and develop a gas combination with proportions that are like those employed in MAP [24]. Strawberries were stored at 2°C in polypropylene packets with various holes [25]. This perforated packing was suggested because there was minimal loss of marketable fruits, no symptoms of Botrytis-related decay, and a little fall in sugar content. The use of carbon dioxide has an influence on the spread of bacteria like Botrytis cinerea, which causes strawberry quality to deteriorate. According to the studies, using $CO_2$ concentrations of 5%–10% helps to limit Botrytis multiplication while preserving a uniform and appealing colour for the consumer [26, 27].

Before using strawberries in modified atmosphere packaging, they were treated with various gases and coated, and the overall impact of MAP was found to be increased. In comparison to simply MAP, which increased shelf life by 4–6 days, MAP (2.5% $O_2$ and 15% $CO_2$) coupled with an edible film coating and ozone treatments prolonged life span up to 8–10 days. The handling of ethylene presence in MAP is the most important technique to for managing the ripening process in fruits and vegetables. As in case of the strawberry, some research has been done in this approach, with the goal of clarifying the molecular pathways accountable for the tissue's response to this gas [28]. In terms of changed atmospheres, packages have been devised that directly regulate the quantities of various gases throughout the shipping and storage of the product, extending the useful life of strawberries by around 10 days compared to ordinary packages [29, 30].

Many studies show that the application of 1-methylcyclopropene in strawberry during MAP lower the senescence rate, with beneficial benefits without impairing quality at doses ranging from 0.5 to 5 $\mu L^{-1}$, while the effects of deterioration intensified at larger doses.

### 2.4 Controlled atmosphere storage

A continuously controlled gas environment is referred to as controlled atmospheric storage (CA). A controlled situation is an agrarian store system, manages the percentages of $O_2$, $CO_2$, and N, as well as the humidity and temperature of the storage room. CA suppresses some taste development while slowing ripeness and maintaining firmness. Fruit respiration can be minimised by lowering the oxygen concentration and fruit viability can be maintained by reducing the oxygen in an enclosed storage area. The $CO_2$ content is usually permitted to increase as well, which helps to maintain quality. Fruit emit carbon dioxide during respiration, which piles up in an enclosed environment and slows ripening. Several research has concluded that $CO_2$ levels of 15%–20% and $O_2$ concentrations of 5%–10% are the acceptable atmospheric composition for effective strawberry storage [31].

Castellanos et al. [32] found that investigating the impact of respiratory gases on fruit physiology is critical for developing optimal packaging to improve post-harvest shelf life. Alamar et al. [33] investigated the CA effect at the early and middle stages of strawberries. The results showed that applying CA for 2.5 days midway through storage at 5°C (2.5 days; 15 kPa $CO_2$ + 5 kPa $O_2$ after 2 days in the air) extended life span by 3 days (depending on the prevalence of disease). CA also inhibited inner ethylene synthesis, resulting in fruit that was lighter, vibrant, as well as firmer, implying a lessening in ripening.

Almenar et al. [19] experimented by storing the wild strawberry for 3 weeks at 3°C in different atmospheric conditions. The study claimed that the combination of 10% $CO_2$ and 11% $O_2$ can efficiently enhance the fruit shelf life by keeping quality criteria within a reasonable range and preventing the spread of Botrytis cinerea, without affecting consumer approval greatly.

Strawberry fruits maintained in CA storage retain their freshness too long than those kept in the refrigerator, and decaying loss is minimised in storage having high $CO_2$ concentration. Strawberries held in a controlled atmosphere of 12% $CO_2$ and 2% $O_2$ had higher fruit texture, total soluble solids, titratable acidity, and ascorbic acid concentration than strawberries stored in the open air [34]. This research also finds that the CA retain the maximum level of volatile compounds like esters and furanone during storing strawberry. Total terpenes, total alcohols, total acids, and cold stress resistance were all higher in CA-stored strawberries in comparison to air-stored strawberries.

The strawberry when stored in CA and MAP with high $CO_2$ content increases the quality attributes and decreases the incidence of microbes. Nakata and Izumi [35] experimented by storing strawberries at high concentrations of $CO_2$ (20%, 30% and 40%) for 10 days at 5°C, using CA and active MAP. The CA of 20%–40% $CO_2$ was efficient in limiting the exterior production of mould mycelia and inhibiting the increase in the fungus population. However, because of $CO_2$ injury, a CA of >30% $CO_2$ caused black staining over the berry skin. Strawberry fruit when placed in a MAP had consistent fungal levels throughout the days of storage. External development of mould mycelia characterised as Botrytis cinerea and surface black staining were produced in strawberry fruit in MAP flushed with 30% and 40% $CO_2$ after transfer to ambient conditions for 6 days at 10 8C.

### 2.5 Losses due to mechanical injury

Mechanical damage is a sort of stress that happens throughout the harvesting and manipulation of fruits after they have been harvested. Physiological and morphological changes accompany this stress, affecting the fruit quality. Soft fruits, such as strawberries, are especially subject to severe harm during and after harvesting due to their thin epidermis. Strawberry fruits may be crushed, impacted, punctured, or damaged during postharvest handling, resulting in a shortened shelf life and deterioration. Strawberry fruits soften quickly after harvest at the ripening stage due to pectin solubilisation and hemicellulose and cellulose hydrolysis, which causes the central lamella to degrade [36].

The most prevalent sort of mechanical injury that occurs during harvesting, handling, and transportation is bruising [37, 38]. Bruise damage occurs when an excessive amount of external force is applied to the fruit surface during interaction with a solid body or fruit versus other fruit [39, 40]. The most unfavourable damage is bruising, which serves as a portal for infections, particularly fungus [41].

Due to the strawberry's great vulnerability to damage during the picking stage, trained people and adequate equipment are required. In addition, daily inspection of the harvested product quality and improvement of the situation can help to reduce harvest losses. Overfilling of boxes must be avoided, and the number of layers within boxes must be kept as low as feasible, to minimise package losses during handling. The use of paperboards or plastic boards between the fruit layers is useful since it prevents the fruits from moving around and thereby reduces fruit damage. It appears that good packing and regular handling methods are the most important factors in ensuring the safe delivery of a product to a market.

### 2.6 Edible coatings

Very fragile fruits, such as berries and tropical fruits, are good candidates for coating treatment since they are costly and have a short storability. A thin layer of edible material, often not exceeding 0.3 mm, placed to the surface of meals in addition to or as a substitute for natural coating materials is classified as an edible coating. The use of an active edible covering to enhance the longevity of fruits and vegetables is a unique and potential method [42]. Edible coverings with semipermeable film can improve post-harvest fruit life by minimising moisture, respiration, gas exchange, as well as oxidative reaction rates [43].

Li et al. [44], uses three edible coatings i.e., alginate, chitosan (CS), and pullulan (polysaccharide) during cold storage (4°C) to postharvest strawberry fruit. The experiment concluded that during 16 days of storage, polysaccharide coatings severely hampered fruit weakening and rot, as well as minimised modifications in TSS and titratable acidity content. These coatings also kept ascorbic acid and total phenolic content greater than controls from day 2 onwards, and dramatically reduced fruit deterioration and respiration after 12 days of storage. The enzymatic activities like peroxidase (POX), catalase (CAT), superoxide dismutase (SOD), and ascorbate peroxidase (APX) were increased by polysaccharide coatings, preventing lipid peroxidation and reducing membrane damage. Furthermore, among these coatings, CS-based coatings had the most beneficial impacts on the quality of fruit and had the highest relative antioxidant enzyme activity.

Aloe vera gel is more effective than CS. Nasrin et al. [45], performed an experiment by coating the strawberry with CS and aloe vera gel, keep in a polypropylene box and stored at 6°C ± 1°C with 50 ± 5% relative humidity. A. vera gel-coated strawberry retained their colour, wetness, firmness, fresh appearance, and general acceptability for longer than CS or uncoated ones. Furthermore, AV gel coating on strawberries slowed the emergence of microbe occurrence for approximately 15 days. While the moulds impacted control and CS (1.5% or 2%) coated strawberries on days 6 and 9 of storage, respectively.

In the sector of the food industry, the use of a CS-based antimicrobial coating to prolong the shelf life of strawberries throughout storage appears to be quite promising. Chitosan-monomethyl fumaric acid (CS-MFA) is an excellent antimicrobial coating derived from CS and its derivatives. Khan et al. [46] applied CS-MFA coating on strawberries and stored it at 10°C. The total weight loss and deterioration were considerably lower in the CS-MFA samples than in the CS and control samples. When compared to CS, CS-MFA coated fruit had a significantly lower yeast and mould load. Finally, the CS-MFA enhanced microbial properties and extended shelf life from 4 to 8 days.

The edible coating, which contains natural bioactive chemicals, could be used as a substitute for artificial additives and can help save money on cold storage. The orange peel essential peel (OPEP) and carnauba wax have been proven to decrease the damages in strawberries. Saeed et al. [47] designed an experiment to make an edible antifungal covering with different concentrations of OPEO and carnauba wax to protect strawberries against blue mould (fungus). Orange peel oil had a maximum growth constraint of 96% against fungus. Strawberry storage life was increased owing to an OPEP coating that preserved the fruit's quality.

When strawberry was treated with polysaccharide edible coating (alginate and pectin) containing citral and eugenol showed that the fruit have higher firmness, TSS, as well as enzymatic activity, while having lesser weight loss and microbiological deterioration [4].

Surface colour parameters were significantly affected by the cellulose coating. The impact of a methylcellulose edible covering on the qualitative, chemical, physical, and mechanical attributes of strawberries was studied. Nadim et al. [48] used methylcellulose on the surface of the berry and stored it for 11 days at 4°C. The edible coating inhibits weight loss and decay while also retaining a tiny quantity of fruit sugar, preserving the firmness of the strawberries and enhancing their quality and storage characteristics.

## 2.7 Treatment with chemicals

Strawberry is counted in the list of fruits that are perishable and is especially prone to postharvest losses (up to 50%) owing to fungal disease outbreaks. For many years' different chemicals have been used to extend the durability of strawberries. However due to health issues now their uses are restricted. Liu et al. [49] treated the strawberry with different concentrations of melatonin for 5 min and stored it at 4°C with a relative humidity of 90% for 12 days. The results concluded that 1 mmol $L^{-1}$ melatonin treatment delayed the ripening of berries, extend the shelf life, improve the fruit quality and minimised the concentration of hydrogen peroxide and malondialdehyde However, the total phenolics and flavonoid levels were enhanced, leading to increased antioxidant potential. These results suggest that melatonin administration could be a viable tactic for extending strawberry fruit postharvest life and improving quality.

### 2.8 Combined treatments

When two or more post-harvest treatments are used together, they show a synergistic impact on the product's standard and usable viability. Feliziani et al. [50] compares the effect of CS, laminarin, extracts of *Abies* spp., *Polygonum* spp., and *Saccharomyces* spp., organic acids and calcium combination, and benzothiadiazole with fungicides. These chemicals were applied to the strawberry canopy every 5 days. In comparison to water-treated controls, alternative chemical treatments reduced strawberry postharvest loss by 30%, primarily against grey mould and Rhizopus rot instead of altering the colour or firmness of the fruit.

### 2.9 Physical treatment

The physical treatment includes ultraviolet radiations with a focus on UV-C radiations. UV-C irradiation aids in the prevention of fruit degradation and the postponement of ripening. This sort of radiation (UV-C) is used for not only reducing fungal deterioration but also increasing fruit phytochemical content after harvesting. Several research on the application of UV-C radiation has shown that it can reduce the biological load of the fruit without compromising sensory qualities like colour, firmness, texture, and humidity, among others [51–54].

However, the health risks of using pesticides on food after harvest has required research into non-chemical postharvest treatments. Non-chemical treatments such as gaseous ozone and UV-C irradiation have looked promising in the processing and preservation of certain fruits and vegetables. Gumede et al. [55] investigated the effect of ozone (gaseous form) and UV-C radiations in extending the shelf life and preserving the quality of strawberries. When compared with control treatments, fruit exposure to UV-C and continuous ozone exhibited a considerably lower deterioration rate. Furthermore, fruit mass loss was substantially reduced in ozonated atmospheres as compared to UV-C and control treatment. Fruit treated with ozone and UV-C had significantly greater antioxidant capacity. Gaseous ozone and UV-C irradiation have been demonstrated to be effective non-chemical postharvest treatments for strawberries in this study.

In several species, notably strawberry, UV-C radiation is effective in preventing disease development. UV-C radiation appears to be helpful not just because of its disinfectant properties, but also because it may boost plant defence mechanisms. Forges et al. [56] studied the effect of UV-C radiation administrated during cultivation. The UV-C-treated plants flowered earlier than the non-treated ones. Despite a modest reduction in leaf area, treated plants produced a greater amount of fruit at harvest. In reaction to UV, spontaneous infection of leaves with powdery mildew and fruit with Rhizopus was significantly reduced.

## 3. Conclusion

Strawberry is a non-climacteric fruit whose fruit develops and matures through a succession of physiological and molecular resulting in significant changes in fruit size, colour, texture, flavour, and fragrance and there seem to be a variety of postharvest losses after harvesting. To minimise postharvest losses various techniques, must employ to preserve the fruit quality, firmness as well as extend the shelf life. Each approach has benefits and drawbacks, and the use of one or more of these post-harvest procedures will be determined by the growing region's economic, technological, and social considerations. Farmers could reduce the decline in fruit quality as well as keep the freshness for longer by using proper postharvest methods.

## Author details

Huma Qureshi Quarshi[1], Waseem Ahmed[1*], Rafia Azmant[2], Nabila Chendouh-Brahmi[3], Abdul Quyyum[4] and Asad Abbas[5]

1 Department of Horticulture, The University of Haripur, Pakistan

2 Department of Chemistry, University of Karachi, Pakistan

3 Laboratory of 3BS, Faculty of Life and Nature Sciences, University of Bejaia, Bejaia, Algeria

4 Department of Agronomy, The University of Haripur, Pakistan

5 School of Horticulture, Anhui Agricultural University, Hefei, China

*Address all correspondence to: waseemuaf12@gmail.com

## References

[1] Nasrin TAA, Rahman MA, Hossain MA, Islam MN, Arfin MS. Postharvest quality response of strawberries with aloe vera coating during refrigerated storage. The Journal of Horticultural Science and Biotechnology. 2017a;**92**(6):598-605

[2] Giampieri F, Tulipani S, Alvarez-Suarez JM, Quiles JL, Mezzetti B, Battino M. The strawberry: Composition, nutritional quality, and impact on human health. Nutrition. 2012;**28**:9-19

[3] Marelli B, Brenckle MA, Kaplan DL, Omenetto FG. Silk fibroin as edible coating for perishable food preservation. Scientific Reports. 2016;**6**:25263

[4] Guerreiro AC, Gago CML, Faleiro ML, Miguel MGC, Antune MDC. The use of polysaccharide-based edible coatings enriched with essential oils to improve shelf-life of strawberries. Postharvest Biology and Technology. 2015a;**110**:51-60

[5] Sogvar OB, Saba MK, Emamifar A. Aloe Vera and ascorbic acid coatings maintain postharvest quality and reduce microbial load of strawberry fruit. Postharvest Biology and Technology. 2016;**114**:29-35

[6] Bersaneti GT, Mantovan J, Magri A, Mali S, Celligoi MAPC. Edible films based on cassava starch and fructooligosaccharides produced by Bacillus subtilis natto CCT 7712. Carbohydrate Polymers. 2016;**151**:1132-1138

[7] Acosta TA, Marques LOD, Nardello I, Navroski R, Mello-Farias PC. Ultraviolet-C radiation on *Psidium cattleyanum* L. conservation and its influence on physicochemical fruit characteristics. Journal of Experimental Agriculture International. 2018;**6**:1-7

[8] Nascimento MS, Silva N. Tratamentos químicos na sanitização de morango (*Fragaria vesca* L.). Brazilian Journal of Food Technology. 2010;**13**(1):11-17

[9] Han K, Kim S, Lee Y, Kim S, Im D, Choi H, et al. Strawberry harvesting robot for bench-type cultivation. Journal of Biosystems Engineering. 2012;**37**:65-74

[10] Hayashi S, Shigematsu K, Yamamoto S, Kobayashi K, Kohno Y, Kamata J, et al. Evaluation of a strawberry-harvesting robot in a field test. Biosystems Engineering. 2010;**105**:160-171

[11] Van Henten EJ, Van Tuijl BAJ, Hemming J, Kornet JG, Bontsema J, Van Os EA. Field test of an autonomous cucumber picking robot. Biosystems Engineering. 2003;**86**(3):305-313

[12] Hayashi S, Ganno K, Kurosaki H, Arima S, Monta M. Robotic harvesting system for eggplants trained in V-shape (part 2). Journal of Society of High Technology in Agriculture. 2003;**15**(4):211-216

[13] Van Henten EJ, Van Tuijl BAJ, Hoogakker GJ, Van Der Weerd MJ, Hemming J, Kornet JG, et al. An autonomous robot for de-leafing cucumber plants grown in a highwire cultivation system. Biosystems Engineering. 2006;**94**(3):317-323

[14] Ahmad MS, Siddiqui MW. Postharvest quality assurance of fruits: Practical approaches for developing countries. Food Science & Nutrition. 2015:224

[15] Pelletier W, Brech JK, Nunes MCN, Mond JPE. Quality of strawberries shipped by truck from California to Florida as influenced by postharvest temperature management practices. Horticulture Technology. 2011;**21**:482-493

[16] Caner C, Aday MS, Demir M. Extending the quality of fresh strawberries by equilibrium modified atmosphere packaging. European Food Research and Technology. 2008;**227**:1575-1583

[17] Thompson JF, Mitchell FG, Kasmire RF. Cooling horticultural commodities. In: Kader AA, editor. Postharvest Technology of Horticultural Crops. Davis, California, USA: University of California; 1992. pp. 56-63

[18] Tahir HE, Xiaobo Z, Jiyong S, Mahunu GK, Zhai X, Mariod AA. Quality and postharvest-shelf life of cold-stored strawberry fruit as affected by gum arabic (Acacia senegal) edible coating. Journal of Food Biochemistry. 2018;**42**:1-10

[19] Almenar E et al. Controlled atmosphere storage of wild strawberry fruit (*Fragaria vesca* L.). Journal of Agricultural and Food Chemistry. 2006;**54**(1):86-91

[20] Sandhya S. Modified atmosphere packaging of fresh produce: Current status and future needs. LWT - Food Science and Technology. 2010;**43**:381-392

[21] Barrios S, Lema P, Lareo C. Modeling respiration rate of strawberry (cv. San Andreas) for modified atmosphere packaging design. International Journal of Food Properties. 2014;**17**:2039-2051

[22] Blanch M et al. $CO_2$-driven changes in energy and fermentative metabolism in harvested strawberries. Postharvest Biology and Technology. 2015;**110**:33-39

[23] Choi HJ, Bae YS, Lee JS, Park MH, Kim JG. Effects of carbon dioxide treatment and modified atmosphere packaging on the quality of long distance transporting 'Maehyang' strawberry. Agricultural Sciences. 2016;**7**:813-821

[24] Giuggioli NR, Girgenti V, Baudino C, Peano C. Influence of modified atmosphere packaging storage on postharvest quality and aroma compounds of strawberry fruits in a short distribution chain. Journal of Food Processing and Preservation. 2014;**39**:3154-3164

[25] Sanz C, Olías R, Pérez AG. Quality assessment of strawberries packed with perforated polypropylene punnets during cold storage. Food Science and Technology International. 2002;**8**:65-71

[26] Franco-Gaytan I et al. Quality and shelf life of three strawberry (*Fragaria ananassa*) cultivars treated with high concentrations of $CO_2$ for short period. Agrociencia. 2018;**52**(3):393-406

[27] Park DS, Jeong CS. Effect of $CO_2$ and ClO2 gas pre-treatment for maintain shelf-life of summer strawberries. Korean Journal of Horticultural Science and Technology. 2015;**33**(5):705-711

[28] Elmi F et al. Effect of ethylene on postharvest strawberry fruit tissue biochemistry. Acta Horticulturae. 2017;**1156**:667-672

[29] Almenar E et al. Equilibrium modified atmosphere packaging of wild strawberries. Journal of the Science of Food and Agriculture. 2007;**87**(10):1931-1939

[30] Giuggioli NR et al. Influence of modified atmosphere packaging storage on postharvest quality and aroma compounds of strawberry fruits in a short distribution chain. Journal

of Food Processing & Preservation. 2015;**39**(6):3154-3164

[31] Bishop D. Controlled atmosphere storage. In: Dellino CVJ, editor. Cold and Chilled Storage Technology. Boston, MA: Springer; 1997

[32] Castellanos DA, Mendoza R, Gavara R, Herrera AO. Respiration and ethylene generation modelling of 'Hass' avocado and Feijoa fruits and application in modified atmosphere. International Journal of Food Properties. 2017;**20**:333-349

[33] Alamar MC, Collings E, Cools K, Terry LA. Impact of controlled atmosphere scheduling on strawberry and imported avocado fruit. Postharvest Biology and Technology. 2017;**134**:76-86

[34] Li L, Luo Z, Huang X, Zhang L, Zhao P, Ma H, et al. Label-free quantitative proteomics to investigate strawberry fruit proteome changes under controlled atmosphere and low temperature storage. Journal of Proteomics. 2015;**120**:44-57

[35] Nakata Y, Izumi H. Microbiological and quality responses of strawberry fruit to high $CO_2$ controlled atmosphere and modified atmosphere storage. HortScience. 2020;**55**(3):386-391

[36] Terry LA. Soft fruit. In: Rees D, Farrell G, Orchard J, editors. Crop Post-Harvest: Science and Technology. 1st ed. Boston: Blackwell Publishing Ltd; 2012. pp. 226-246

[37] Ahmadi E, Ghasemzadeh HR, Sadeghi M, Moghadam M, Zarifneshat S. The effect of impact and fruit properties on the bruising peach. Journal of Food Engineering. 2010;**97**:110-117

[38] Tabatabaekoloor R. Engineering properties and bruise susceptibility of peach fruits (*Prunus persica*). Agricultural Engineering International: CIGR Journal. 2013;**15**:244-252

[39] Li Z, Thomas C. Quantitative evaluation of mechanical damage to fresh fruits. Trends in Food Science and Technology. 2014;**35**:138-150

[40] Stropek Z, Gołacki K. The effect of drop height on bruising of selected apple varieties. Postharvest Biology and Technology. 2013;**85**:167-172

[41] Aliasgarian S, Ghassemzadeh HR, Moghaddam M, Ghaffari H. Mechanical damage of strawberry during harvest and postharvest operations. Acta Technologica Agriculturae. 2015;**18**:1-5

[42] Zhang C, Garrison TF, Madbouly SA, Kessler MR. Recent advances in vegetable oil-based polymers and their composites. Progress in Polymer Science. 2016;**71**:91-143

[43] Petriccione M, Mastrobuoni F, Pasquariello MS, et al. Effect of chitosan coating on the postharvest quality and antioxidant enzyme system response of strawberry fruit during cold storage. Food. 2015;**4**(4):501-523

[44] Li L, Sun J, Gao H, Shen Y, Li C, Yi P, et al. Effects of polysaccharide-based edible coatings on quality and antioxidant enzyme system of strawberry during cold storage. International Journal of Polymer Science. 2017;**2017**:1-8

[45] Nasrin TA, Rahman MA, Hossain MA, Islam MN, Arfin MS. Postharvest quality response of strawberries with aloe vera coating during refrigerated storage. The Journal of Horticultural Science and Biotechnology. 2017b:1-8

[46] Khan I, Tango CN, Chelliah R, Oh D-H. Development of antimicrobial

edible coating based on modified chitosan for the improvement of strawberries shelf life. Food Science and Biotechnology. 2019;**4**:1257-1264

[47] Saeed M, Azam M, Saeed F, Arshad U, Afzaal M, Bader ul Ain H, et al. Development of antifungal edible coating for strawberry using fruit waste. Journal of Food Processing & Preservation. 2021;**45**(11):e15956. DOI: 10.1111/jfpp.15956

[48] Nadim Z, Ahmadi E, Sarikhani H, Amiri Chayjan R. Effect of methylcellulose-based edible coating on strawberry Fruit's quality maintenance during storage. Journal of Food Processing and Preservation. 2014b;**39**(1):80-90

[49] Liu C, Zheng H, Sheng K, Liu W, Zheng L. Effects of melatonin treatment on the postharvest quality of strawberry fruit. Postharvest Biology and Technology. 2018;**139**:47-55

[50] Feliziani E, Landi L, Romanazzi G. Preharvest treatments with chitosan and other alternatives to conventional fungicides to control postharvest decay of strawberry. Carbohydrate Polymers. 2015;**132**:111-117

[51] Charles MT, Arul J. UV treatment of fresh fruits and vegetables for improved quality: A status report. Stewart Postharvest Review. 2007;**3**(3):1-8

[52] Nigro F, Ippolito A, Lattanzio V, Di Venere D, Salerno M. Effect of ultraviolet-C light on postharvest decay of strawberry. Journal of Plant Pathology. 2000;**82**(1):29-37

[53] Xie Z et al. Effects of preharvest ultraviolet-C irradiation on fruit phytochemical profiles and antioxidant capacity in three strawberry (Fragaria × ananassa Duch.) cultivars. Journal of the Science of Food and Agriculture. 2015;**95**(14):2996-3002

[54] Xie Z et al. Preharvest ultraviolet-C irradiation: Influence on physicochemical parameters associated with strawberry fruit quality. Plant Physiology and Biochemistry. 2016;**108**:337-343

[55] Gumede M, Mditshwa A, Tesfay SZ, Magwaza LS, Mbili NC. The effect of ozone and UV-C irradiation on strawberry postharvest quality. ISHS. 2018:15-22. DOI: 10.17660/ActaHortic.2020.1275.3

[56] Forges M, Bardin M, Urban L, Aarrouf J, Charles F. Impact of UV-C radiation applied during plant growth on pre- and post-harvest disease sensitivity and fruit quality of strawberry. Plant Disease. 2020;**104**(12):3239-3247. DOI: 10.1094/pdis-02-20-0306-re